Advances in Space Exploration
COSPAR Symposium Series

# VOLUME 7

# NON-SOLAR GAMMA-RAYS

# Advances in Space Exploration

## COSPAR Symposium Series

| | |
|---|---|
| Volume 1 | K. A. VAN DER HUCHT and G. S. VAIANA (Editors): New Instrumentation for Space Astronomy |
| Volume 2 | E. A. GODBY and J. OTTERMAN (Editors): The Contribution of Space Observations to Global Food Information Systems |
| Volume 3 | W. A. BAITY and L. E. PETERSON (Editors): X-Ray Astronomy |
| Volume 4 | H.-J. BOLLE (Editor): Remote Sounding of the Atmosphere from Space |
| Volume 5 | W. RIEDLER (Editor): Scientific Ballooning |
| Volume 6 | YASH PAL (Editor): Space and Development |
| Volume 7 | R. COWSIK and R. D. WILLS (Editors): Non-Solar Gamma-Rays |
| Volume 8 | A. P. MITRA (Editor): Low Latitude Aeronomical Processes |
| Volume 9 | V. V. SALOMONSON and P. D. BHAVSAR (Editors): The Contribution of Space Observations to Water Resources Management |
| Volume 10 | W. D. CARTER, L. C. ROWAN and J. F. HUNTINGTON (Editors): Remote Sensing and Mineral Exploration |

# NON-SOLAR GAMMA-RAYS

*Proceedings of a Symposium of the Twenty-second Plenary Meeting of COSPAR*
*Bangalore, India*
*29 May to 9 June 1979*

Edited by
R. COWSIK
*Tata Institute of Fundamental Research, Bombay, India*
and
R. D. WILLS
*Space Science Department of the European Space Agency,*
*European Space Research and Technology Centre, Noordwijk, The Netherlands*

Published for
THE COMMITTEE ON SPACE RESEARCH
by
PERGAMON PRESS

OXFORD · NEW YORK · TORONTO · SYDNEY · PARIS · FRANKFURT

| | |
|---|---|
| U.K. | Pergamon Press Ltd., Headington Hill Hall, Oxford OX3 0BW, England |
| U.S.A. | Pergamon Press Inc., Maxwell House, Fairview Park, Elmsford, New York 10523, U.S.A. |
| CANADA | Pergamon of Canada, Suite 104, 150 Consumers Road, Willowdale, Ontario M2J 1P9, Canada |
| AUSTRALIA | Pergamon Press (Aust.) Pty. Ltd., P.O. Box 544, Potts Point, N.S.W. 2011, Australia |
| FRANCE | Pergamon Press SARL, 24 rue des Ecoles, 75240 Paris, Cedex 05, France |
| FEDERAL REPUBLIC OF GERMANY | Pergamon Press GmbH, 6242 Kronberg-Taunus, Pferdstrasse 1, Federal Republic of Germany |



First edition 1980

**British Library Cataloguing in Publication Data**

Non-solar gamma-rays. - (Advances in space exploration; vol.7).
1. Gamma ray astronomy - Congresses
2. Gamma rays - Congresses
I. COSPAR. *Plenary Meeting, 22nd, Bangalore, 1979*
II. Indian National Committee for Space Research
III. Cowsik, R IV. Wills, R D V. Series
522'.6 QB471.A1 79-41335

ISBN 0-08-024440-8

*In order to make this volume available as economically and as rapidly as possible the authors' typescripts have been reproduced in their original forms. This method has its typographical limitations but it is hoped that they in no way distract the reader.*



*Printed in Great Britain by A. Wheaton & Co. Ltd., Exeter*

Symposium on Non-Solar Gamma-Rays

Organized by

THE COMMITTEE ON SPACE RESEARCH—COSPAR

and

THE INDIAN NATIONAL COMMITTEE FOR SPACE RESEARCH—INCOSPAR

THE INDIAN SPACE RESEARCH ORGANIZATION—ISRO

THE INDIAN NATIONAL SCIENCE ACADEMY—INSA

Sponsored by

THE COMMITTEE ON SPACE RESEARCH—COSPAR

THE INTERNATIONAL ASTRONOMICAL UNION—IAU

THE INTERNATIONAL UNION OF PURE AND APPLIED PHYSICS—IUPAP

# CONTENTS

## SESSION 5. Cosmic-Rays and Extragalactic Gamma-Rays

Chairman: *S. L. Mandelshtam*



## SESSION 6. Techniques and Prospects

Chairman: *M. Oda*

# OPENING REMARKS

B. V. Sreekantan
Chairman, Programme Organizing Committee

*Tata Institute of Fundamental Research, Homi Bhabha Road, Bombay 400 005, India*

We are indeed very grateful to the COSPAR for having agreed to hold its 22nd meeting in India and as part of it to have given us this unique opportunity of organizing this symposium on Non-solar Gamma Rays. The symposium has been sponsored also by IAU and IUPAP. It is, in my opinion, very appropriate that the meetings are being held in this beautiful garden city of Bangalore, which, apart from its natural beauty and agreeable weather, has been one of the most important centres of scientific research in the past sixty years and in the last decade has become the most important centre in India for research in various fields of astronomy and space sciences and also in the development of Space Technologies.

Thanks to the foresight of scientists like Dr. Homi Bhabha and Dr. Vikram Sarabai, India entered the field of experimental cosmic ray research in a very big way as early as in the late forties. This activity led to the development of indigeneous stratospheric balloon technology in the early fifties and of sounding rockets in the early sixties and enabled the Indian cosmic ray scientists to enter the area of space astronomies rather early in the development of these fields. Also the equatorial locations of the balloon and rocket launching stations at Hyderabad and Thumba respectively proved to be a definite advantage for experiments in space astronomies due to the considerably lower cosmic ray induced background - particularly for experiments on the measurement of the diffuse gamma ray background. There have been several international collaboration experiments carried out from India - Indo-Japan and Indo-USSR experiments in the field of gamma ray astronomy.

I am sure that among other benefits, this symposium would serve a very useful purpose in inspiring some of our young scientists who are participating in this meeting to undertake research in these new fast developing areas. Thanks to the very keen and personal interest taken by the member of the Programme Committee, especially Professor Larry Peterson, the symposium has a very attractive and stimulating programme. I would like to take this opportunity to express my most sincere thanks to them. I would also like to thank the various invited speakers who readily responded to our invitations. We are eagerly awaiting to hear their expert coverages of the various fields.

We were hoping that there would be a much larger number of participants from the different countries. However, the dates of the meetings have clashed with some other similar meetings - some important conferences in the related areas have already been held or will take place in a few weeks.

I do sincerely hope that this Bangalore symposium on gamma ray astronomy will be remembered as a significant landmark in the future development of this field.

SESSION 1

# Discrete Gamma-ray Sources—I

*Chairman:* **L. E. PETERSON**

# DISCRETE GAMMA-RAY SOURCES
# (E < 30 MeV)

V. Schönfelder

*Max-Plank-Institut für Physik und Astrophysik, Institut für extraterrestrische Physik, D-8046 Garching bei München, Federal Republic of Germany*

ABSTRACT

Low and medium energy gamma ray measurements - say between 0.5 and 30 MeV - are motivated by the knowledge that unique astrophysical phenomena may become observable in this range between X-ray energies and high-energy gamma rays. At present there exist only a few results on discrete gamma-ray sources in this difficult energy range, and even these results are sometimes conflicting. In this review observations of galactic and extragalactic sources are summarised and future aspects are outlined. It seems that the development of extragalactic gamma-ray astronomy may be especially promising at these low gamma-ray energies.

## INTRODUCTION

The exciting results obtained during the last few years in high-energy gamma-ray astronomy at energies around 100 MeV have once again demonstrated that entirely new information is obtained as soon as a new field is explored. The COS-B catalogue of high-energy gamma-ray sources now consists of more than 30 sources including at least one extragalactic object. So far only 4 of the galactic sources could be identified. All 4 sources are radio pulsars and gamma-ray data seem to indicate that gamma-ray emission is a prime process in the pulsar mechanism. All the other sources have no known counterparts at other wavelengths, and it seems that gamma-ray astronomy will now guide observations at other wavelengths.

In the low-energy gamma-ray range - say between 0.5 and 30 MeV - the situation is at present still entirely different. This range between X-ray and high-energy gamma-ray photons is one of the last nearly unexplored fields in astronomy. Gamma-ray emission at these low energies has been reported from less than a handful of objects and even these few results are sometimes conflicting. Nevertheless, there is no doubt that unique astrophysical phenomena will become observable as soon as this field is explored. There are at least two important differences between low- and high-energy gamma-ray astronomy:

1. The dominant gamma-ray production processes for continuum emission at low and medium gamma-ray energies may be different from those at high gamma-ray energies. At least $\pi^0$-decay gamma rays, which are supposed to be important at high gamma-ray energies, no longer contribute at these low energies. Therefore, continuum spectra from low-energy gamma-ray observations will give information on high-energy electrons rather than on high-energy protons. High-energy electrons produce

gamma-radiation via bremsstrahlung, inverse Compton collisions or as synchrotron radiation. Even thermal bremsstrahlung production seems to be possible at low gamma-ray energies.

2. In the low-energy gamma-ray range gamma-ray lines are expected. This is an extremely interesting aspect. Gamma-ray lines are on the one hand expected from the decay of radioactive nuclei, which were produced during explosive nucleosynthesis processes in supernova remnants. Observation of these lines would provide a unique means to observe nucleosynthesis processes directly. On the other hand gamma-ray lines are produced in nuclear interactions of energetic nucleons (typically 10 MeV per nucleon) with matter, e.g. with matter near the source or with interstellar matter. By observing gamma-ray lines from nuclear interactions, information can be obtained on the flux and spectral shape of these energetic particles which cannot be obtained by other means.

The reason for the slow progress in low-energy gamma-ray astronomy is mainly due to experimental difficulties. These difficulties are essentially threefold:

1. The sensitivity of experiments flown so far is too low. On the one hand this is due to insufficient detection areas or detection efficiencies of the instruments, and on the other hand the observation times have been too short because most experiments were flown on balloons where sources could be observed only for periods of some hours.

2. The directional resolution of existing experiments is insufficient. For most instruments it is worse than 10$^{\circ}$ and in some cases the detectors were even only omnidirectionally sensitive.

3. Most experiments suffer from high background counting rates.

As a consequence of these difficulties all results obtained in low- and medium-energy gamma-ray astronomy are still incomplete and sometimes conflicting.

In the following I shall review the observations, discussing first the galactic and then the extragalactic sources. At the end I shall try to illustrate how the development in the field of low-energy gamma-ray sources may proceed.

## GALACTIC GAMMA-RAY SOURCES

The only galactic source which has been investigated with much effort at low gamma-ray energies by various experimental groups is the Crab nebula and its pulsar. Although there seems to be more or less a general agreement on the flux of the pulsar, reports on the total Crab emission in the 1-10 MeV band differ by as much as a decade. Let me start with the pulsar.

Figure 1 shows different light curves of the Crab pulsar at low gamma-ray energies in comparison to the neighbouring X-ray and high-energy gamma-ray range. The lower curve shows excellent statistics; it was measured at X-ray energies between 100 and 400 keV by Kurfess [1] with an omnidirectional NaI-detector. The pulsar events were identified by their periodicity. From these observations gamma-ray flux values up to 3 MeV could be derived. A nearly identical experiment was flown at about the same time by Orwig et al. [2]; their light curve in the energy interval 250 keV - 2.3 MeV is also shown.

At MeV energies there are two observations of the pulsar, both obtained with a Compton telescope; one is the telescope of the University of California at Riverside [3], which measures between 3 and 25 MeV and the other one is the telescope of the Max-Planck-Institut in Garching, which measures between 1 and 20 MeV [4].

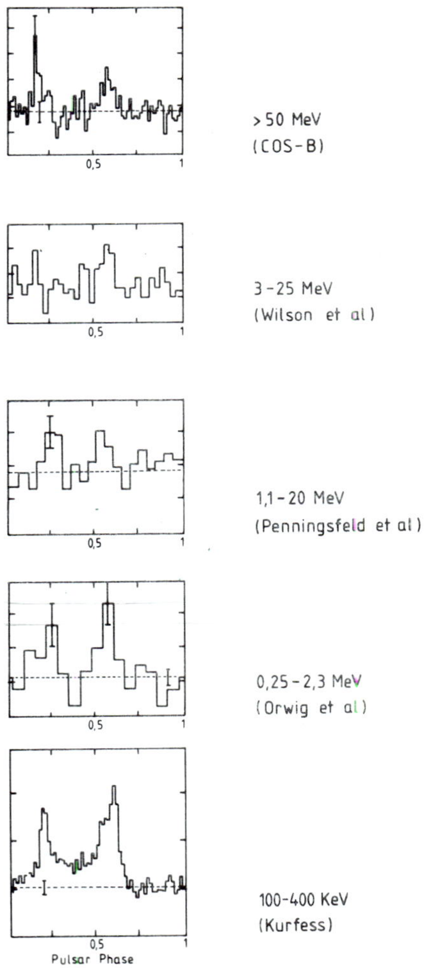


Fig. 1 Light curves of the Crab pulsar between 100 keV and 2 GeV.

The COS-B light curve at high gamma-ray energies above 50 MeV [5] is shown for comparison.

The pulse structure, consisting of a main pulse and an interpulse with a separation of about 13 ms, seems to be the same at all energies. Of course, the statistical significance is much worse at low gamma-ray energies than in the other ranges. But this fact only once again illustrates the difficulties of observations at MeV energies.

It is worth noting that the flux between both pulses is consistent with zero at all energies except between 100-400 keV, where a very clear excess flux above zero was observed.

In Figure 2 the energy spectrum of the Crab pulsar is shown. Data from various experimental groups are compared ranging from X-ray energies at 40 keV to gamma-ray energies at 2 GeV. All data are more or less consistent with a single power law over the whole range, which is proportional to $\sim E^{-2.2}$.

Let me now come to the total Crab emission including that of the nebula. Whereas a pulsar can be identified by its periodicity, it is much more difficult to detect a continuous source. In order to resolve a continuous source against the background radiation it is necessary to have directionally sensitive detectors. In most cases directionality in low-energy gamma-ray astronomy was obtained by shielding or occultation. In this case "source-on" - "source-off" observations at different times are necessary to be able to subtract the "background only" counts from a "source region" counting rate. This procedure is not without problems, because the background must be the same in both cases, in order to obtain a correct result. These difficulties may be the reason for the large discrepancies in the total gamma-ray fluxes that have been reported from the Crab between 1 and 10 MeV.

The difficulty does not exist in case of the Compton telescope. The Compton telescope - when used with all its capabilities - has imaging properties in the sense that within a large field of view much smaller regions of the sky can be resolved. Different regions of the sky are therefore observed at the same time - as in the case of spark-chamber experiments in high-energy gamma-ray astronomy.

Since we have now - for the first time - made use of the full imaging properties of the Compton telescope when analysing the Crab region, I would like to discuss this in more detail here.

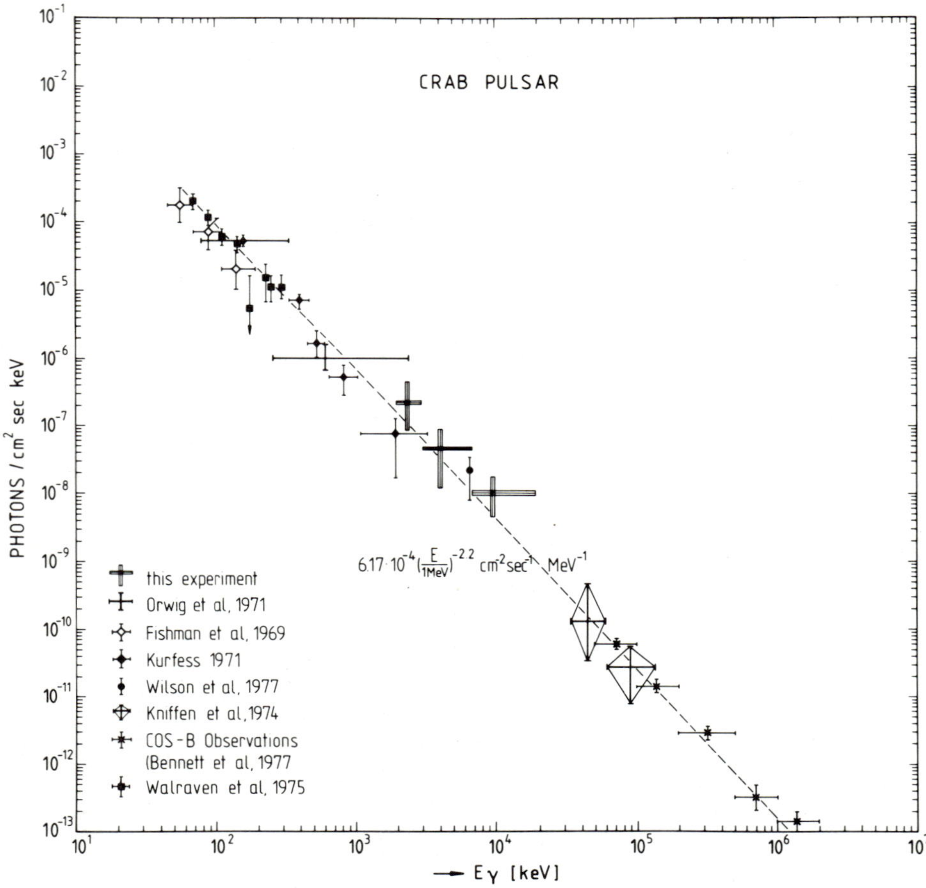


Fig. 2 Photon number spectrum of the Crab pulsar (adapted from ref. 4).

The principle of measurement of the Compton telescope is illustrated in Figure 3. It consists of two large-area scintillation detectors, an upper one of liquid scintillator and a lower one of NaI. Each detector is subdivided into optically separated cells. By measuring the energy losses $E_e'$ and $E_e''$ in both detectors and by measuring the locations in both interactions, the direction of the scattered gamma ray and the scattering angle $\varphi$ at the position of the first interaction can be determined. If only events within a certain range of scattering angles between $\varphi$ and $\varphi + \Delta\varphi$ are accepted from a celestial point source, the directions of the scattered gamma rays form an image circle of width $\Delta\varphi$ in the plane of the lower detector around the direction of the source. The source can therefore be identified by detecting its image circles. For this purpose all events are grouped into different subclasses of scattering angles $\varphi_i$. For each class the directions of the scattered gamma rays are projected onto the celestial sphere. If then a cross correlaction is performed between the event rates in different sky bins and the response of the telescope, the correlation function, which we call C(r), will show a maximum at the position of the source direction $r_{source}$. For each celestial bin within the field of view the correlation function C(r) has to be calculated. If there is no

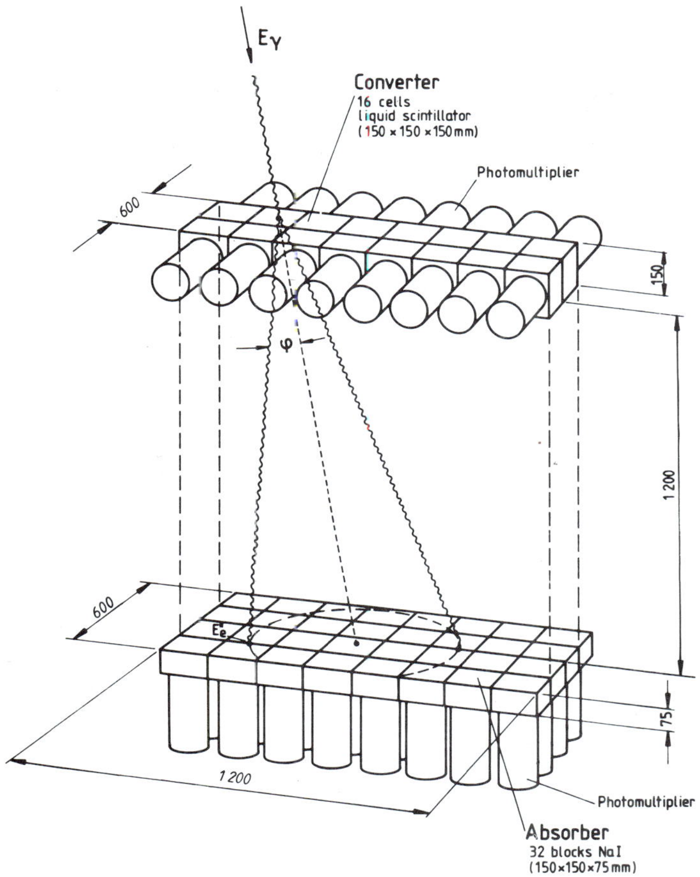


Fig. 3 Schematic drawing of the MPI Compton telescope.

source within the field, the function C(r) will fluctuate around zero with the standard deviation σ. The function C(r)/σ therefore describes the significance of the correlation.

The result of such correlation calculations is shown in Figure 4 for a field of view of the sky containing the Crab. On the abscissa the right ascension is plotted and on the ordinate the declination. The Crab position is at $\alpha = 82.5^{\circ}$ and $\delta = 21.7^{\circ}$. If special event selection criteria are applied to the data of the Compton telescope, the field of view is typically 40° FWHM. During the balloon flight the Crab was within the field of view for about 3 hours. The contours in the diagram are lines of constant C(r)/σ values, that means lines of the same significance. The position of the Crab stands out very clearly in this diagram. The position of the highest significance is at $\alpha = 83.5^{\circ}$ and $\delta = 24.5^{\circ}$, about 3° away from the true Crab position. The significance falls to half its maximum value after 6° - 8°, which is consistent with the detector resolution for a point source [6]. The excess counting rate at the position of the Crab was converted into a gamma-ray flux in order to obtain an energy spectrum of the total Crab emission. The result is shown in Figure 5 together with a compilation of spectra from other groups between 10 keV and 2 GeV. All data except those between 1 and 10 MeV can be fitted

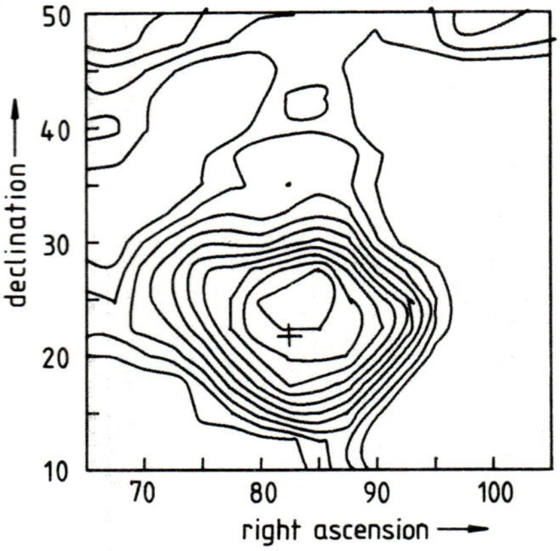


Fig. 4 Results of cross correlations for a field of view of the MPI Compton telescope containing the Crab. The energy interval of accepted events is 3 - 7 MeV. The lines mark those regions of the sky where the significance of the correlation function has constant values. The position of the Crab is marked by + (from ref. 4).

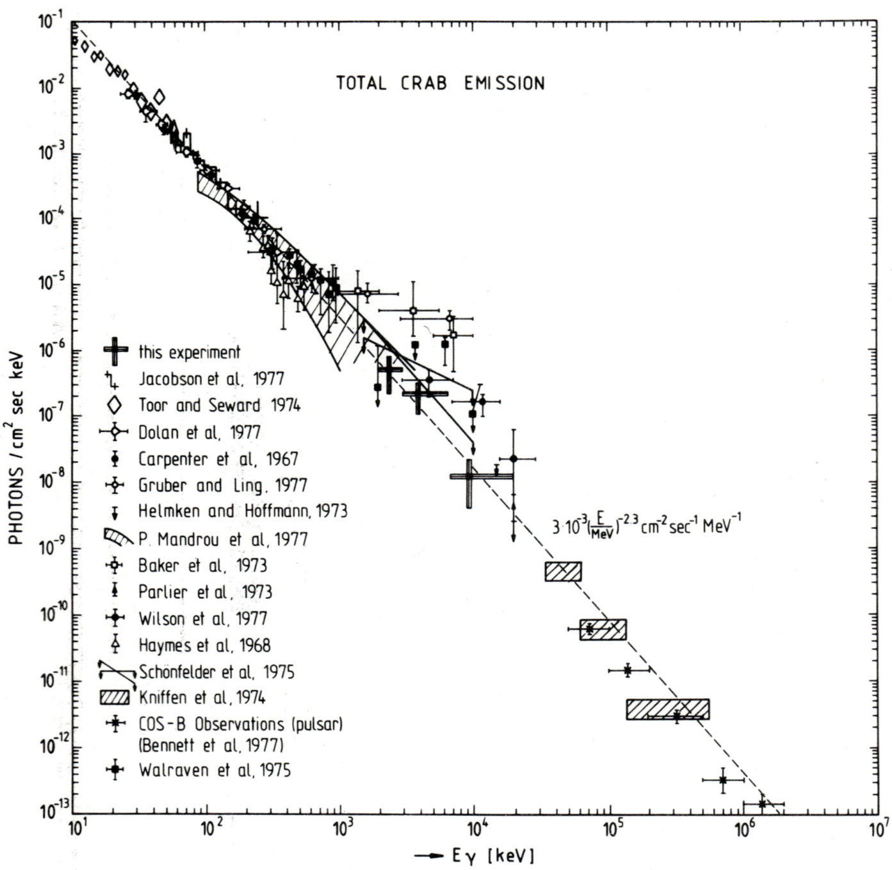


Fig. 5 Photon number spectrum of the total Crab emission (adapted from ref. 4).

by a power law $\sim E^{-2.3}$. This figure clearly illustrates the discrepancies between different measurements between 1 and 10 MeV, as mentioned above. Our new data points - the hatched crosses - are consistent with the single power law fit $\sim E^{-2.3}$.

Apart from the Crab there are only few galactic sources which have been investigated at low gamma-ray energies. Rather high fluxes have been reported for Cyg X-2 [7] and Cyg X-1 [8]. But none of these observations could be confirmed unambiguously by subsequent observations. The spectrum of Cyg X-1 is shown in Figure 6. The high flux values between 1 and 10 MeV measured by Baker et al. [8] could not be confirmed by Mandrou et al. [9] nor by Schönfelder and Lichti [10]. The same applies for the high flux values reported from early balloon experiments around 50 MeV; they also could not be confirmed by SAS-II and COS-B.

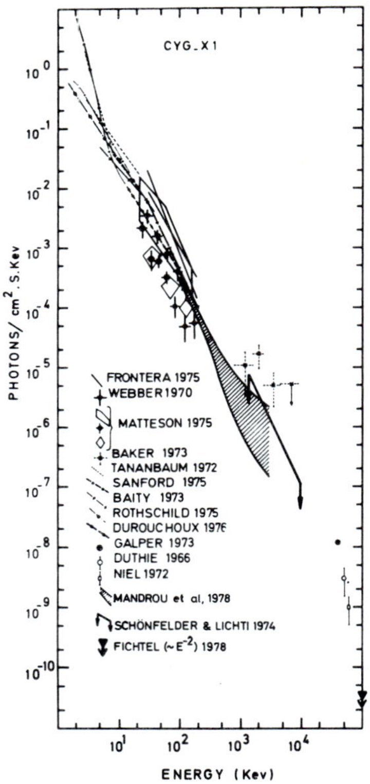


Fig. 6 Photon number spectrum of Cyg X-1 (adapted from ref. 9). The upper limit at 100 MeV was determined by SAS-II [11].

It would be of special interest to find out whether the brightest gamma-ray sources at high energies around 100 MeV, namely the Vela pulsar and the Geminga source (CG195+4) can be seen at lower energies. So far, no observations of the Vela pulsar have been made below 10 MeV. The gamma-ray flux determined during a balloon observation by Albats et al. [12] between 10 and 30 MeV is consistent with an extrapolation of the COS-B spectrum. However, the pulsar light curve only shows one peak in contrast to the double-peak structure found by SAS-II and COS-B.

An attempt by the Rice group [13] to detect the Geminga source at low gamma-ray energies was not successful. The upper limits are shown in Figure 7 and are compared with the spectrum

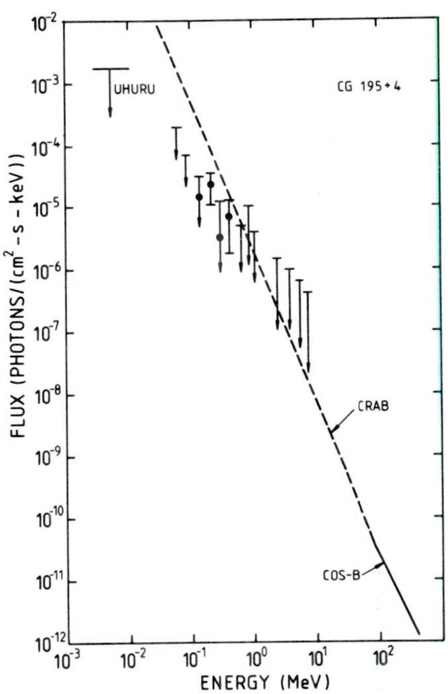


Fig. 7 Upper limits to the energy spectrum of the Geminga source.

between 50 and 500 MeV determined by COS-B.

## EXTRAGALACTIC GAMMA-RAY SOURCES

So far, gamma-ray emission has been reported from three different kinds of external galaxies.

1. Continuum gamma-ray emission between 50 and 500 MeV from the quasar 3C273 was observed by the COS-B experiment [14]. The energy spectrum is shown in Figure 8 together with X-ray measurements of Ariel V [15] and HEAO-A4 [16] between 1 and 100 keV. So far, no measurements exist between these energy ranges. The gamma-ray luminosity between 50 and 500 MeV is $3 \times 10^{46}$ erg $s^{-1}$.

2. Gamma-ray lines of low significance at 1.6 MeV and 4.5 MeV have been reported from the radio galaxy Cen A by the Rice group [17]. However, no significant continuum flux could be determined above 1 MeV. At high gamma-ray energies around 100 MeV the source could also not be detected.

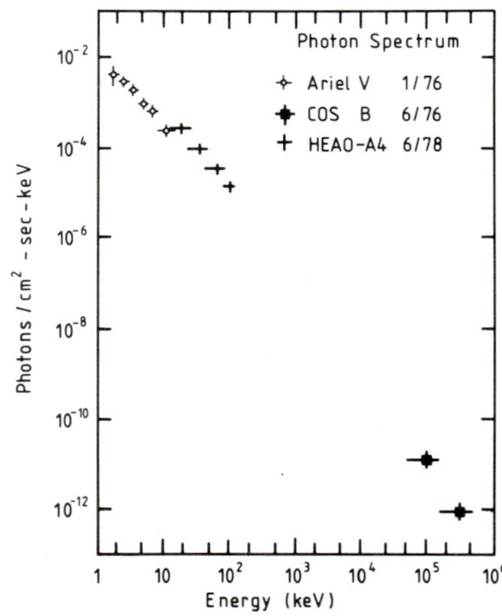


Fig. 8 Photon number spectrum of 3C273 between 50 and 500 MeV [14] and between 1 and 100 keV [15], [16].

3. Continuum gamma-ray emission from the Seyfert galaxy NGC 4151 has been reported at low gamma-ray energies from balloon observations of the Milano/Southampton group [18] and from our measurements with the Compton telescope [19]. The energy spectrum of the Seyfert galaxy is shown in Figure 9 together with data at X-ray energies.

The flux values derived from the measurement with the Compton telescope have to be considered as preliminary. An enhanced gamma-ray counting rate was observed during the transit of NGC 4151 through the field of view of the telescope which is $40^{o}$ wide (FWHM). Two basic assumptions were made in order to obtain the flux values shown in Figure 9: first, it was assumed that the observed excess counting rate is due to NGC 4151; second, it was assumed that no further source within the broad field of view of $40^{o}$ contributes to the counting rate. In the light of these assumptions - strictly speaking - the derived spectral points have to be considered as upper limits only. The tentative identification of the source with NGC 4151 is mainly based on the shape of the derived gamma-ray spectrum, which is consistent with an extrapolation from X-ray energies up to 3 MeV. When the correlation method - as described for the Crab - will be applied to this region of the sky, it will be possible to locate the source more accurately (see note added after the Symposium).

At the gamma-ray symposium at Erice in May 1979 the final results from the balloon observation of the Milano/Southampton collaboration (MISO) were presented. In this experiment an excess counting rate was observed in a $3^{o} \times 20^{o}$ field of the sky, which contains NGC 4151. The final energy spectrum derived from their data is included in Figure 9. It should be noted that the integrated significance of all their data points above 2 MeV is only about 2 $\sigma$.

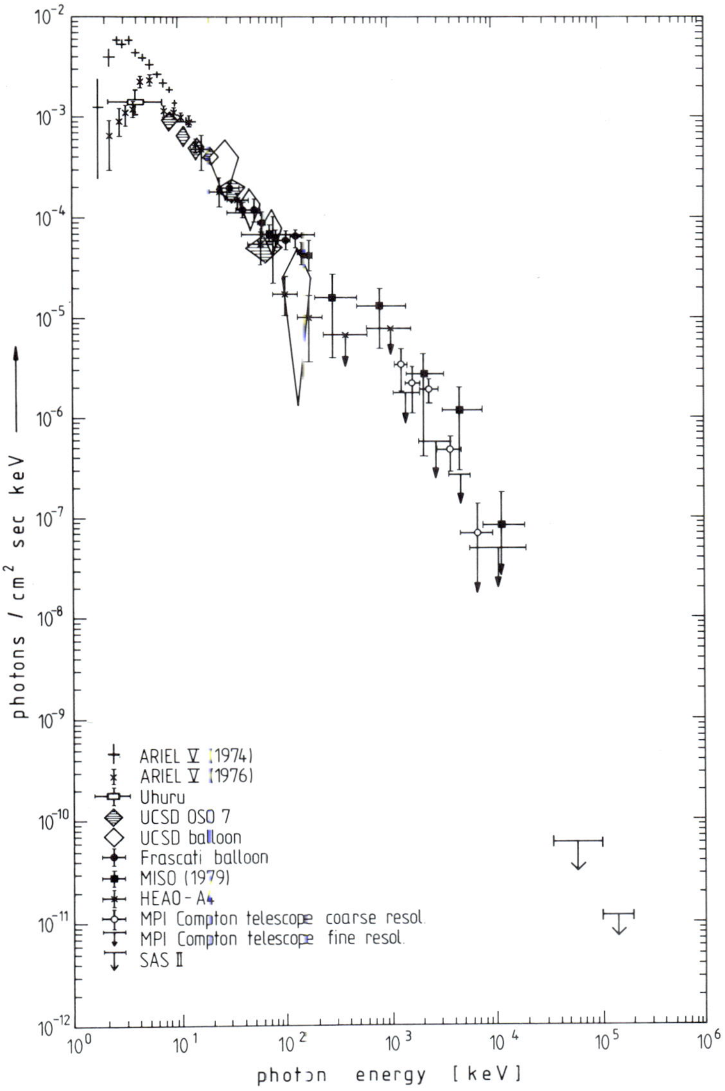


Fig. 9 Photon number spectrum of NGC 4151 (X-ray data adapted from ref. 20). The gamma-ray observations are described in the text. See also note added after the Symposium.

At the same meeting preliminary results from HEAO-A4 observations were presented by Matteson. Their spectral points up to 230 keV and two upper limits up to 1.7 MeV are also included in Figure 9.

The recent upper limits of SAS-II between 35 and 200 MeV [11] indicate that the spectrum must steepen considerably somewhere at low gamma-ray energies.

If the observation of NGC 4151 at low gamma-ray energies is real, its gamma-ray luminosity is extremely high, namely $10^{45}$ erg/s below 3 MeV compared to

$7 \times 10^{42}$ erg/s at X-ray energies between 2 and 10 keV. At least at X-ray energies this powerful energy source seems to be concentrated in a very small volume. An increase in the 2 - 10 keV flux by a factor of 1.7 over a time scale of ≈ 3 days was observed in 1976 by Elvis [21]. An increase in the 2 - 60 keV band by a factor of 2 within 1.5 days was recently observed by OSO-8 [22], and from UHURU observations [23] flares of only typically 1000 s duration in the 2 - 10 keV band were reported very recently. Furthermore, during a balloon flight of the gamma-ray detector of the Rice group [24] no source at all could be detected at the position of NGC 4151. The 2 $\sigma$ upper limits between 100 and 200 keV are about a factor of 2 lower than the preliminary HEAO-A4 points shown in Figure 9. From all these variations an upper limit on the dimension of the X-ray source of $2 \times 10^{13}$ to $2 \times 10^{15}$ cm can be derived.

One would now be interested to know if further Seyfert galaxies can be expected to be detected as gamma-ray emitters in the near future.

If all X-ray Seyferts with X-ray luminosities equal to or higher than NGC 4151 would have the same gamma-ray luminosity as NGC 4151, then only 2 of the 16 X-ray Seyferts listed by Elvis et al. [25] would at least have 0.3 times the flux of NGC 4151 and would therefore be just detectable with present instruments.

If there were proportionality between X- and gamma-ray luminosity, 6 of the 16 objects would be at least 0.3 times as strong as NGC 4151. However, proportionality between X- and gamma-ray luminosity cannot exist for all X-ray Seyferts because the resulting integrated flux would be higher than the observed diffuse cosmic flux [26].

## FUTURE ASPECTS

Let me finally try to illustrate how I see the future development in the field of low-energy gamma-ray sources. For this purpose a summary of the intensities of a variety of sources over a wide range of energies is given in Figure 10. The abscissa ranges from X-ray energies at 1 keV to gamma-ray energies at 10 GeV. On the ordinate the photon number flux $cm^{-2}s^{-1}MeV^{-1}$ is multiplied by $E^2$. In this case all sources which have a power-law photon-number spectrum proportional to $\sim E^{-2}$ will appear as a horizontal line. On the left side the flux levels of various X-ray sources are plotted from the strongest source Sco X-1 to the UHURU detection limit. On the right side around 100 MeV the gamma ray fluxes of the COS-B sources are marked. Some of the controversial high fluxes at low gamma-ray energies are also indicated. The spectra of the pulsed and total emission from the Crab are shown over the whole energy range.

What is remarkable in this diagram is that there are many X-ray sources which have no counterparts at high gamma-ray energies, even though this might be expected from a simple extrapolation of the X-ray spectra. Especially several extragalactic sources like Cen A, NGC 4151 and 3C273 have very flat X-ray spectra; however, they have only very low intensities at high gamma-ray energies. It therefore seems possible that extragalactic gamma-ray emitters are more likely to be found at low than at high gamma-ray energies.

The sensitivity levels of the next generation of low-energy gamma-ray experiments, namely the actively shielded phoswich detector of NRL [27] and the Compton telescope COMPTEL [28] on board of the American Gamma Ray Observatory GRO which is to be launched in 1985 are also indicated in Figure 10. It is clear from this figure that after GRO the low and medium energy gamma-ray range will certainly no longer be an unexplored field of astronomy as it is now. Then a break-through in this field should be achieved which may be comparable to the one made by UHURU at X-ray energies.

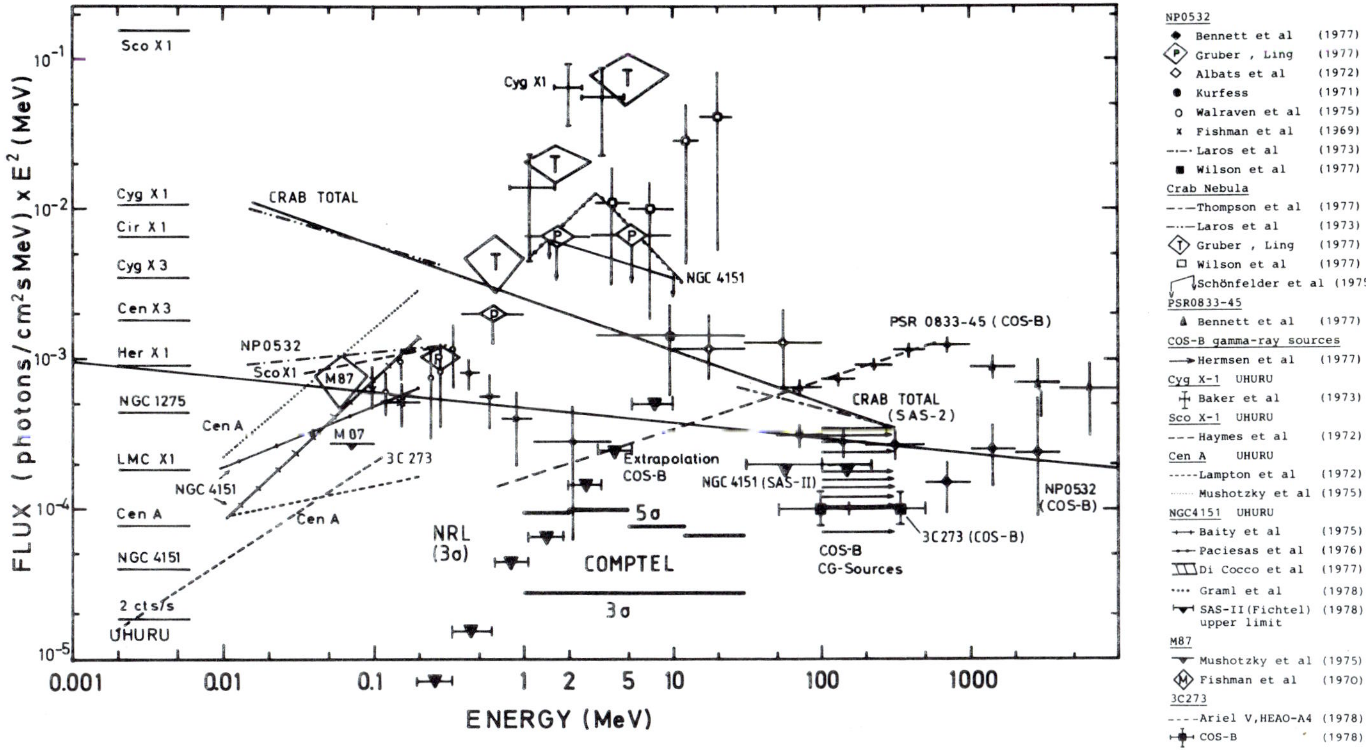


Fig. 10 Compilation of the intensities of a variety of X- and gamma-ray sources (adapted from ref. 28). On the ordinate the number of photons $cm^{-2}s^{-1}MeV^{-1}$ is multiplied with $E^2$. The sensitivity of the two low and medium-energy gamma-ray experiments on board of GRO are indicated.

Note added after the Symposium:

Immediately after the COSPAR-Symposium a correlation calculation was performed with the data of the Compton telescope for the region of the sky around NGC 4151. No enhanced gamma-ray emission was found within a circle of $6^o$ radius around NGC 4151. The 2 $\sigma$ upper limits derived from the analysis in this "fine resolution mode" are also included in Figure 9.

References

1. J.D. Kurfess, Astrophys. J. 168, L39 (1971).
2. L.E. Orwig, E.L. Chupp and D.J. Forrest, Nat. Phys. Sci. 231, 171 (1971).
3. R.B. Wilson, B. Dayton, S.H. Moon, J.M. Ryan, A.D. Zych and R.S. White, Proc. 15th Internat. Cosmic Ray Conf., Plovdiv, 1, 24 (1977).
4. F.P. Penningsfeld, U. Graser and V. Schönfelder, to be publ. in Proc. 16th Internat. Cosmic Ray Conf., Kyoto (1979).
5. K. Bennett, G.F. Bignami, G. Boella, R. Buccheri, W. Hermsen, G. Kanbach, G.G. Lichti, J.L. Masnou, H.A. Mayer-Hasselwander, J.A. Paul, L. Scarsi, B.N. Swanenburg, B.G. Taylor and R.D. Wills, Astron. Astrophys. 61, 279 (1977).
6. F. Graml, F.-P. Penningsfeld and V. Schönfelder, Proc. 12th ESLAB Symposium, ESA SP-124, 301 (1977).
7. A. Dean, G. Gerardi, C. De Martinis, G.F. Monastero, A. Russo and L. Scarsi, Astron. Astrophys. 28, 131 (1973).
8. R.E. Baker, R.R. Lovett, K.J. Orford and D. Ramsden, Nature 245, 18 (1973).
9. P. Mandrou, M. Niel, G. Vedrenne and A. Dupont, Astrophys. J. 212, 704 (1977).
10. V. Schönfelder and G. Lichti, Astrophys. J. 192, L1 (1974).
11. C.E. Fichtel, NASA Technical Memorandum 79640 (1978).
12. P. Albats, G.M. Frye, G.B. Thomson, V.D. Hopper, D.B. Mace, J.A. Thomas and J.A. Staib, Nature 251, 400 (1974).
13. R.C. Haymes, C.A. Meegan and G.J. Fishman, preprint, submitted to Astron. Astrophys. (1978).
14. B.N. Swanenburg, K. Bennett, G.F. Bignami, P. Caraveo, W. Hermsen, G. Kanbach, J.L. Masnou, H.A. Mayer-Hasselwander, J.A. Paul, B. Sacco, L. Scarsi and R.D. Wills, Nature 275, 298 (1978).
15. P. Sanford, Annals N.Y. Acad. Sci. 302, 386 (1977).
16. F.A. Primini, B.A. Cooke, C.A. Dobson. S.K. Howe, A. Scheepmaker, W.A. Wheaton, W.H.G. Lewin, W.A. Baity, D.E. Gruber, J.L. Matteson and L.E. Peterson, preprint, submitted to Nature (1979).
17. R.C. Hall, C.A. Meegan, G.D. Walraven, F.T. Djuth, D.H. Shelton and R.C. Haymes, Proc. 14th Internat. Cosmic Ray Conf., München, 1, 84 (1975).
18. G. Di Cocco, G. Boella, F. Perotti, R. Stiglitz, G. Villa, R.E. Baker, R.C. Butler, A.J. Dean, S.J. Martin and D. Ramsden, Nature 270, 319 (1977).
19. F. Graml, F.-P. Penningsfeld and V. Schönfelder, Proc. Esrange Symposium, Ajaccio, ESA SP-135, 287 (1978).
20. G. Auriemma, L. Angeloni, B.M. Belli, A. Bernardi, D. Cardini, E. Costa, A. Emanuele, F. Giovanelli and P. Ubertini, Astrophys. J. 221, L7 (1978).
21. M. Elvis, Mon.Not.Roy.Astron.Soc. 177, 7 (1976).
22. R.F. Mushotzky, S.S. Holt and P.J. Serlemitsos, NASA Technical Memorandum 79551 (1978).
23. H. Tananbaum, G. Peters, W. Forman, R. Giacconi and C. Jones, Astrophys. J. 223, 74 (1978).
24. C.A. Meegan and R.C. Haymes, preprint (1979).
25. M. Elvis, T. Maccacaro, A.S. Wilson, M.J. Ward, M.V. Penston, R.A.E. Fosbury and G.C. Perola, Mon.Not.Roy.Astron.Soc. 183, 129 (1978).

26. V. Schönfelder, Nature 274, 344 (1978).
27. J.D. Kurfess, H. Friedman, M. Ulmer, W.N. Johnson and G.H. Share, GRO-Proposal (1978).
28. V. Schönfelder, B.N. Swanenburg, J.A. Lockwood, B.G. Taylor, G. Kanbach, G. Metzner, J.A.M. Bleeker, A.J.M. Deerenberg, W. Hermsen, W.R. Webber, K. Bennett and R.D. Wills, GRO-Proposal (1978).

# THE DISCRETE SOURCES OF HIGH ENERGY GAMMA-RAY RADIATION

R. Buccheri

*MPI für extraterrestrische Physik, 8046 Garching, Federal Republic of Germany*
*(On leave from LFCTR, 90123 Palermo, Italy)*

ABSTRACT

The observational material concerning gamma-ray sources at energies above ~ 50 MeV is reviewed and then discussed in order to gain insight about their physical nature. Emphasis is given to pulsars for which experimental and theoretical arguments prove that gamma-ray emission plays a major role in their evolution. The contribution of pulsars to the observed galactic gamma-ray emission is finally estimated.

## INTRODUCTION

The problem of the existence of discrete sources of high energy gamma rays has been long debated during the early seventies when several experimenters, using spark chambers aboard at stratospheric balloons, claimed the detection of localized emission from the galactic centre [1] and [2], from the Cygnus-Cassiopea region [3], from Sagittarius [4] and from the anticentre region [5].

In fact the sensitivity of the experiments cited was low due to several reasons including the low angular resolution of the spark chambers, the short integration times and the large amount of atmospheric gamma-ray background. The statistical significance of the effects found was then poor and their reality was considered dubious.

The results of OSO-III [6] demonstrated the existence of an anisotropic flux of cosmic gamma rays exhibiting a concentration of intensity around the galactic equator with a broad maximum towards the galactic centre but did not show evidence for discrete sources. As a consequence of this and in agreement with the current theoretical ideas concerning cosmic gamma rays [7] and [8] the observed celestial distribution was explained in terms of three components, i.e. i) a general galactic component produced by $\pi^o$-process, ii) an isotropic and presumably extragalactic component and iii) a galactic-centre component in the equatorial region between $l = -30^o$ and $l = +30^o$ for which the possible contribution from discrete sources was not excluded.

The first firm evidence of the presence in the sky of gamma-ray discrete sources came from SAS-II which was able to discriminate three of them. The first one, in the galactic anticentre region, identified with the Crab pulsar PSR 0531+21 after the time analysis of the detected photons, definitely confirmed the previous discovery of gamma-ray pulsed emission from that pulsar made by Parlier et al. [9]. The second one, at $l \sim 263^{\circ}$ and $b \sim -2^{\circ}$, was identified with the Vela pulsar PSR 0833-45 while the third one at $l \sim 195^{\circ}$ and $b \sim 5^{\circ}$ is unidentified up to now([10] and references therein). Besides these sources SAS-II claimed the discovery of two more gamma-ray pulsars (PSR 1747-46 and PSR 1818-04) on the basis of pure time analysis [11] and [12].

The detection of these few sources did not change much about the interpretation of the origin of the measured gamma radiation: presenting a review of the observational results in gamma-ray astronomy Fichtel [13] pointed out that "on the basis of its intensity, energy spectrum and angular distribution and on the lack of quantitatively successful alternatives, the galactic emission appears to be best explained as resulting primarily from cosmic ray interactions with the interstellar matter, principally atomic and molecular hydrogen". Conversely more caution was used when interpreting the radiation from the regions away from the galactic plane as truly diffuse as opposed to the sum of many point sources.

A considerable advance in the comprehension of this ambiguity is presently coming from COS-B, an experiment which, due mainly to its unexpectedly long life, has collected up to now more than 66000 gamma-ray photons (to be compared with the 8000 of SAS-II). In a recent paper The Caravane Collaboration [14] presented a list of 24 discrete sources up to now detected by COS-B, to which I will refer throughout this paper. The problem of the nature of the gamma-ray sources and that of their contribution to the measured gamma radiation received a great impulse and several papers contributing to this argument have recently appeared in the literature (see for example [15], [16], [17], [18], [19], [20], [21], [22] and [23]).

At the same time a long list of papers proposing identifications with objects known at longer wavelengths has revealed the difficulty of meaningful associations due mainly to the large error boxes of the gamma-ray sources (see for example [24], [25], [26], [27], [28], [29], [30], [31] and [32]).

## THE LIST OF THE SOURCES

The up-to-date list of gamma-ray sources includes 25 objects. Two of them, referred to as CG 185-5 and CG 263-2 after the publication of the COS-B catalogue [33], merit a special attention for their characteristics are known in fine details. Their intensity is so high that detection was possible using independently pure spatial or pure temporal information and then no doubt exists about their identification with the two youngest known radio pulsars PSR 0531+21 and PSR 0833-45. Table 1 shows their main physical parameters as measured by COS-B [34] together with information concerning the associated radio pulsars. It is interesting to stress that for both objects the emission appears to be quite entirely pulsed with a similar temporal structure of two narrow phase peaks separated by $\sim 0.4$ of their period. The main difference between the two sources

TABLE 1 Physical Parameters of CG 185-5 and CG 263-2

| | CG 185-5 | CG 263-2 | |
|---|---|---|---|
| Pulsed flux ($ph/cm^2 s$) | $(5.9 \pm 0.8) \times 10^{-6}$ | $(15.6 \pm 0.8) \times 10^{-6}$ | 1/ |
| Unpulsed flux ( " ) | $< 1.7 \times 10^{-6}$ | $< 1.6 \times 10^{-6}$ | 1/2/ |
| Pulsed fraction | > 0.78 | > 0.91 | 1/2/ |
| Pulsar distance (pc) | 2000 | 500 | |
| Luminosity (ergs/s) | $1.9 \times 10^{35}$ | $4 \times 10^{34}$ | 1/3/ |

1/ Valid for E > 50 MeV

2/ 2σ upper limit

3/ A beaming factor of 0.184 is included (see [36]).

concerns their spectral behaviour: for CG 185-5 the differential spectrum is well fitted with a power law extending down to soft X-rays while a harder shape is observed in the case of CG 263-2 with the low energy side of the spectrum deviating from a power law and going below the visibility limit in the X-ray range. Differences in the spectral behaviour of the two phase peaks have also been observed in the case of CG 263-2 [35].

Sources Detected using Spatial Information

Out of the remaining 23 sources, 20 have been found by using only spatial information. The method of search used by the Caravane Collaboration consisted in the cross-correlation of the observed gamma-ray intensity along the galactic equator and away from it with the spatial profile of a point source as folded by the experiment response. Due to the limited angular resolution of the instrument, discrete sources with angular extent of the order of a few degrees are expected to appear like point sources. For what concerns the significance assigned to the correlation coefficient, it was normalized with different criteria when analysing different regions of the sky: due in fact to the anisotropy of the radiation, a given value of the correlation coefficient which is quite significant in the regions away from the galactic plane might be not significant at all in the galactic centre region where the intensity is much higher [33].

In this way the Caravane Collaboration was able to isolate, besides CG 185-5 and CG 263-2, 20 statistically significant enhancements of the gamma-ray intensity which have been ascribed to sources. For 13 of them a spectral classification was possible by comparing the detected counts in two energy ranges. The other 7 sources are located in more complex regions and a similar classification was less obvious. Table 2 shows the list of these 20 objects together with their position in galactic coordinates, intensity (relative to the intensity of CG 185-5) when published and spectral information where available.

TABLE 2 Sources Detected Using Spatial Information

| $\ell$ | b | Relative Intensity | Spectrum | Name |
|---|---|---|---|---|
| 65.5 | 0.0 | $0.30^{+0.10}_{-0.05}$ | hard | CG 64+0 |
| 75.0 | 0.0 | $0.45^{+0.15}_{-0.10}$ | | CG 75+0 |
| 78.5 | 1.5 | $1.00^{+0.30}_{-0.20}$ | | CG 78+1 |
| 95.5 | 4.5 | | hard | |
| 106.0 | 1.5 | | | |
| 121.0 | 3.5 | $0.20^{+0.10}_{-0.05}$ | hard | CG 121+3 |
| 135.5 | 1.5 | $0.30^{+0.15}_{-0.10}$ | medium | CG 135+1 |
| 176.0 | -7.0 | $0.50^{+0.10}_{-0.10}$ | soft | CG 176-7 |
| 189.0 | 1.0 | $0.40^{+0.10}_{-0.10}$ | | CG 189+1 |
| 195.5 | 4.5 | $0.90^{+0.20}_{-0.20}$ | hard | CG 195+4 |
| 219.0 | -0.5 | | medium | |
| 270.0 | -1.0 | | soft | |
| 284.0 | -1.0 | | hard | |
| 288.5 | -0.5 | | hard | |
| 291.0 | 65.0 | $0.25^{+0.08}_{-0.08}$ | medium | CG 291+65 |
| 295.5 | 0.5 | | medium | |
| 312.0 | -1.5 | $0.45^{+0.15}_{-0.15}$ | | CG 312-1 |
| 327.5 | -0.5 | $0.35^{+0.15}_{-0.15}$ | | CG 327-0 |
| 333.5 | 0.0 | $0.70^{+0.20}_{-0.20}$ | | CG 333+0 |
| 353.0 | 16.0 | | hard | |

Several identifications have been attempted with different classes of astronomical objects. Among them the identification of CG 291+65 with quasar 3C273 seems to be the most reliable because of the relative quietness of that region of the sky, while the association of CG 353+16 with the ρ-oph dark cloud must wait for verification as to the mechanism invoked [37] to supply the lack of gamma rays deriving from the standard mechanism of Black and Fazio [38].

For what concerns the other attempted identifications it is difficult at the moment to give positive answers; the error boxes associated with the positions of the gamma-ray sources are of the order of one square degree and so include many possible objects. More reliable associations must be tried by looking for spectral and/or temporal peculiarities consistent with the observed features [39]. Moreover when trying a correlation with a specific class of galactic objects a datum which has to be taken heavily into account is the thickness of the latitude distribution, equal to $\langle |b| \rangle \sim 2^{\circ}$ in the case of the gamma-ray sources.

## Sources Detected Using Temporal Information

The relative importance of the various mechanisms producing gamma rays and the context in which they act is presently a matter of discussion. Certainly much work has been done in the last five or six years in order to establish reasonable values for the amount and the distribution of cosmic rays needed to explain the observed gamma-ray flux in terms of interaction with the interstellar matter. On the other side the theoretical possibility existed that gamma-ray emission could be the result of the acceleration of high energy particles in the atmospheres of compact objects like galactic nuclei, neutron stars and black holes. With this in mind a search for time variability, typical of compact objects, was undertaken by SAS-II and COS-B. It was reasonably thought that the good time resolution of the experiments could provide a powerful tool for detecting short time variability getting rid, to a certain extent, of the low angular resolution. A systematic search for gamma-ray emission from radio pulsars was a natural policy if one considers the large number of radio pulsars detected (now amounting to more than 300) and for which the data can be made available by radio laboratories.

Table 3 lists the gamma-ray pulsars detected up to now together with their main features: to these PSR 0531+21 and PSR 0833-45 must be added as discussed above.

Indeed indications for periodicity have appeared so far in the literature concerning PSR 1818-04 [12], PSR 1742-30 [40], CG 195+5 [10] and [41] and Cyg X-3 [42], but since we are waiting for a definite confirmation they have not been taken into account. For the three objects listed in Table 3 the detection was possible on the basis of their temporal behaviour (a pulsation with the period of the corresponding radio pulsar) and a spatial check, made a posteriori, revealed the compatibility with the presence of a point source. It must be noted however that even if the confidence on their reality is sufficiently high (at least as high as the confidence on the reality of many of the sources detected via spatial information) variations in their characteristics could in principle be expected as a result of the methods used for their detection. In the case of the Crab and Vela pulsars the law giving the variation of the pulsar

TABLE 3 Gamma-Ray Pulsars

| | PSR 0740-28 | PSR 1747-46 | PSR 1822-09 |
|---|---|---|---|
| longitude | 243°.8 | 345°.0 | 21°.5 |
| latitude | -2°.4 | -10°.2 | 1°.4 |
| distance (pc) | 2600 | 740 | 650 |
| period (s) | 0.167 | 0.742 | 0.768 |
| luminosity (ergs/s) | $2.9\times10^{34}$ | $6.2\times10^{33}$ | $9\times10^{33}$ 1/ |
| energy range (MeV) | 150-500 | > 35 | 50-150 |

1/ Same beaming factor as for Table 1 included

period with time during the whole gamma-ray observation was provided by radio laboratories (J. Rankin, Arecibo and R. Manchester, CSIRO) so that their identification cannot be discussed. In the cases of Table 3 the law P = P(t) was only available at an epoch not coinciding with the gamma-ray observation and an extrapolation was necessary which could have given rise to differences between the radio and the gamma-ray pulsar period due to unpredictable events like glitches or to errors in the pulsar position. The result can be a small change in the light curve of the gamma-ray pulsar (possible case of PSR 1747-46 for which the extrapolated radio error was used) or a discrepancy in the periods of the two energy ranges (possible case of PSR 0740-28 and PSR 1822-09 for which a procedure of scanning of the period was used).

## THE NATURE OF GAMMA-RAY SOURCES AND THE PULSAR CONTRIBUTION TO THE OBSERVED GALACTIC GAMMA-RAY EMISSION

As to the nature of the gamma-ray sources the classes of astronomical objects able to emit gamma-rays with the observed characteristics are not very many. Possible interstellar clouds bombarded by cosmic rays should be significantly more massive than expected and should show radio emission. Alternatively one could think of possible mechanisms enhancing the cosmic ray intensity [37] and [43]. Among compact objects, accreting neutron stars do not show experimental evidence of gamma-ray emission apart from the claimed 4.8 h pulsation of Cyg X-3 [42] not confirmed by COS-B [44]. The mechanism of production of gamma rays from accreting neutron stars [18] must then wait for verification.

A mechanism has been proposed for the production of gamma rays by quasars and radio galaxies [21] which could explain the emission from 3C273. However being applicable to extragalactic sources this model cannot fit the narrow latitude distribution of the gamma-ray sources.

For the time being, waiting for the reality of the new emission mechanisms envisaged, only pulsars seem, as a class of galactic objects, to have the characteristics to explain well at a good extent the observed features of the gamma-ray sources. As a matter of fact the only firmly identified sources are pulsars and the circumstance that they increase with age their efficiency in converting rotational energy into gamma rays increases their importance as gamma-ray emitters. Moreover, the common temporal structure of the COS-B gamma-ray pulsars suggests a privileged role of the gamma-ray emission in the evolution of the pulsars. It is then natural to try a correlation between gamma-ray sources and pulsars. More than that we can also think in terms of unresolved gamma-ray pulsars as contributing to the galactic gamma-ray background.

By comparing the observed source intensities with those expected from pulsars it is possible to obtain the following condition on the gamma-ray luminosity of the pulsar in order to be associated with a gamma-ray source:

$$\frac{\Phi_\gamma}{\Phi_{Cr}} = f = \frac{L_\gamma}{L_{Cr}} \frac{d_{Cr}^2}{d^2} \qquad (1)$$

where the $\Phi$'s are observed intensities and d distances and where the assumption of equal spectra is made. An estimate of the distance d can be obtained on the basis of the thickness $\langle|b|\rangle$ of the latitude distribution of the gamma-ray sources. If we assume as reasonable a scale height of 150 pc, the value $\langle|b|\rangle = 2^\circ$ corresponds to an average distance d = 3 Kpc. Assuming $0.2 \leq f \leq 1$, as derived from Table 2, and substituting the value of 2 Kpc for $d_{Cr}$ the condition (1) becomes:

$$2.25 \leq \frac{L_\gamma}{L_{Cr}} \leq 0.45 \qquad (2)$$

This condition is represented in Figure 1 which shows $\frac{L_\gamma}{L_{Cr}}$ against the pulsar age as derived from a fitting of the available experimental data. From this figure we see that young pulsars (younger than 4000 years) are able to satisfy the relation (2). This limiting age for pulsars to appear as gamma-ray sources is obviously increased for distances smaller than 3 Kpc (evident case of PSR 0833-45) and/or for spectra softer than the PSR 0531+21's spectrum (evident case of PSR 1822-09).

Indeed no such radio pulsars have been found inside the error boxes of the COS-B sources but there are at least two possible reasons for this lack. One of them reflects the selection effects of the radio surveys for which young pulsars (periods less than 0.3 s) are much more difficult to detect and one is statistical in nature (pulsars younger than 4000 years are a small fraction of the total). A third reason could come from theoretical arguments: If, in fact, the geometry of emission in gamma rays is similar to a fan beam [45] we can expect more gamma-ray pulsars than radio pulsars to be visible from the earth.

As to the latitude distribution of young pulsars the experimental radio data are not presently so numerous as to allow a correct evaluation. It is anyway not unreasonable to think that young pulsars have a narrower distribution than old ones when the evidence of proper motion is taken into account. A better comprehension of this point is expected to come from future more intensive searches of young pulsars.

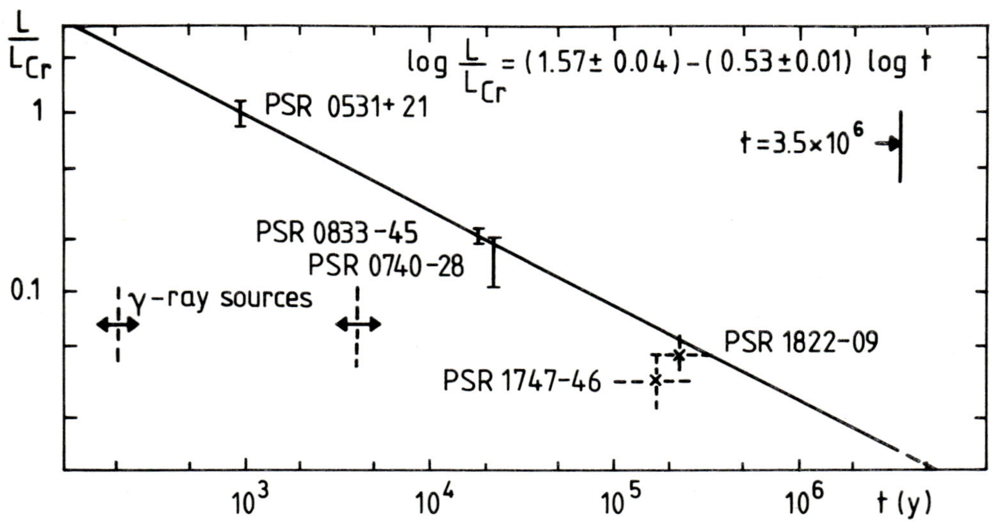


Fig.1 $L_\gamma/L_{Cr}$ versus age for gamma-ray pulsars. The ages of PSR 0531+21, PSR 0833-45 and PSR 0740-28 are assumed to equal the ages of the corresponding supernova remnants (see [46] and [47]). For the other two pulsars no association with SNR is available: they have been included in the figure for comparison by using the "characteristic age" defined by $\tau = 0.5\ P/\dot{P}$.

In Figure 1 is shown a power law relation fitting the experimental data which can be used to evaluate the contribution of resolved and unresolved gamma-ray pulsars to the galactic gamma-ray emission. If we assume that pulsars are uniformly distributed with respect to their age the contribution, till a certain age t, can be computed from the integral

$$L_P = \frac{1}{T}\int_0^t L_\gamma(t)dt = \frac{L_{Cr}}{T} \times 79 \times t^{0.47} \qquad (3)$$

where $\frac{1}{T}$ is the pulsar's birth rate. If we substitute for T the value of T = 27 years as quoted by Hills [48] we obtain, at t = 3.5 $10^6$ yr, a contribution $L_P$ = 6.6 $10^{38}$ i.e. the total gamma-ray emission of the galaxy (as quoted by Strong and Worrall [49] and corrected for the contribution from energies below 100 MeV). This result cannot be considered satisfactory since we know that, for example, $\pi^o$ processes have a significant role in the production of gamma rays: We have to conclude that the power law fitting the data does not extend till $t = 3.5\ 10^6$ yr. On the other hand it is easy to convince ourselves that, if we believe in the experimental data, the fitting is surely valid till $t = 10^6$ yr which results in $L_p = 3.7\ 10^{38}$ i.e. more than 50 % of the total luminosity.

Further measurements of old gamma-ray pulsars are then essential to establish when the pulsars switch off in gamma rays and then if their importance as gamma-ray emitters is even higher than quoted above.

I wish to express my gratitude to many colleagues of the Caravane Collaboration for interesting discussions about the arguments of the paper.

## References


1. G.M. Frye, Jr., P.A. Albats, A.D. Zych, J.A. Staib, V.D. Hopper, W.R. Rawlinson, and J.A. Thomas, Nature 231, 372 (1971).
2. G.H. Dahlbacka, P.S. Freier, C.J. Waddington, Astrophys. J. 180, 371 (1973).
3. R. Browning, D. Ramsden, and P.J. Wright, Nature Phys. Sci. 235, 128 (1972).
4. G.M. Frye, J. Staib, A. Zych, V. Hopper, W. Rawlinson, and J. Thomas, Nature 223, 1320 (1969).
5. J.P. Leray, J. Vasseur, J. Paul, B. Parlier, M. Forichon, B. Agrinier, G. Boella, L. Maraschi, A. Treves, R. Buccheri, A.Cuccia, L. Scarsi, Astron. Astrophys. 216, 443 (1972).
6. W.L. Kraushaar, G.W. Clark, G.P. Garmire, R. Borken, P. Higbie, V. Leong, and T. Thorsos, Astrophys. J. 177, 341 (1972).
7. S. Hayakawa, Progress on Theor. Phys. 8, 517 (1952).
8. P. Morrison, Nuovo Cimento 7, 858 (1958).
9. B. Parlier, B. Agrinier, M. Forichon, J.P. Leray, G. Boella, L. Maraschi, R. Buccheri, N.R. Robba, and L. Scarsi, Nature Phys. Sci. 242, 117 (1973).
10. R.C. Hartman, C.E. Fichtel, D.A. Kniffen, R.C. Lamb, D.J. Thompson, G.F. Bignami, H. Ögelman, M. Özel, T. Tümer, GSFC Symp. NASA CP-002, 15 (1976).
11. D.J. Thompson, C.E. Fichtel, D.A. Kniffen, R.C. Lamb, H.B. Ögelman, Astrophys. Lett. 17, 173 (1976).
12. H. Ögelman, C.E. Fichtel, D.A. Kniffen, D.J. Thompson, Astrophys. J. 209, 584 (1976).
13. C.E. Fichtel, Space Science Review 20, 191 (1977).
14. H.A. Mayer-Hasselwander, K. Bennett, G.F. Bignami, R. Buccheri, N. D'Amico, W. Hermsen, G. Kanbach, F. Lebrun, G.G. Lichti, J.L. Masnou, J.A. Paul, K. Pinkau, L. Scarsi, B.N. Swanenburg, R.D. Wills, Proc. 9th Texas Symp., Munich 1978, in press.
15. J.J. Quenby, Nature 269, 466 (1977).
16. L. Maraschi and A. Treves, Astron. Astrophys. 61, L11 (1977).
17. R. Buccheri, N. D'Amico, E. Massaro, L. Scarsi, Proc. Int. School of Gen. Rel. Effects, MPI-PAE/Astro 138, 376 (1977).

18. G.F. Bignami, L. Maraschi, A. Treves, Astron. Astrophys. 55, 155 (1977).

19. G.F. Bignami, P. Caraveo, and L. Maraschi, Astron. Astrophys. 67, 149 (1978).

20. R.J. Protheroe, A.W. Strong, A.W. Wolfendale, and P. Kiraly, Nature 277, 543 (1979).

21. B.N. Swanenburg, Astron. Astrophys. 70, L71 (1978).

22. B.N. Swanenburg, Proc. Int. School of Gen. Rel. Effects, MPI-PAE/Astro 138, 361 (1977).

23. N. Panagia and G. Zamorani, Astron. Astrophys., in the press.

24. K.M.V. Apparao, H.V. Bradt, R.G. Dower, R.E. Doxsey, J.G. Jernigan, F. Li, Nature 271, 225 (1978).

25. M.J. Coe, J.J. Quenby, A.R. Engel, Nature 274, 344 (1978).

26. P.C. Gregory and A.R. Taylor, Nature 272, 706 (1978).

27. P.F. Julien and H.F. Helmken, Nature 272, 699 (1978).

28. R.C. Lamb, Nature 272, 429 (1978).

29. L. Maraschi, T. Markert, K.M.V. Apparao, H. Bradt, H. Helmken, W. Wheaton, W.A. Baity, and L.E. Peterson, Nature 272, 679 (1978).

30. G.H. Share, K. Wood, R.L. Kinzer, S. Shulman, J. Samimi, W.N. Johnson, D. Yentis, J. Meekins, W. Evans, D. McNutt, E. Byram, T. Chubb, and H. Friedman, XXI COSPAR Meeting, Innsbruck 1978.

31. S. Van den Bergh, Astron. J. 84, 71 (1979).

32. R.C. Lamb and D.M. Worrall, Astrophys. J. Lett., preprint.

33. W. Hermsen, K. Bennett, G.F. Bignami, G. Boella, R. Buccheri, J.C. Higdon, G. Kanbach, G.G. Lichti, J.L. Masnou, H.A. Mayer-Hasselwander, J.A. Paul, L. Scarsi, B.N. Swanenburg, B.G. Taylor, and R.D. Wills, Nature 269, 494 (1977).

34. K. Bennett, G.F. Bignami, G. Boella, R. Buccheri, W. Hermsen, G. Kanbach, G.G. Lichti, J.L. Masnou, H.A. Mayer-Hasselwander, J.A. Paul, L. Scarsi, B.N. Swanenburg, B.G. Taylor, and R.D. Wills, Astron. Astrophys. 61, 279 (1977).

35. G. Kanbach, K. Bennett, G.F. Bignami, R. Buccheri, P. Caraveo, N. D'Amico, W. Hermsen, G.G. Lichti, J.L. Masnou, H.A. Mayer-Hasselwander, J. Paul, B. Sacco, B.N. Swanenburg, and R.D. Wills, Astron. Astrophys., preprint.

36. R. Buccheri, K. Bennett, G.F. Bignami, N. D'Amico, W. Hermsen, D.J.H. Huizing, G. Kanbach, G.G. Lichti, J.L. Masnou, H.A. Mayer-Hasselwander, J.A. Paul, B. Sacco, B.N. Swanenburg, and R.D. Wills, Astron. Astrophys., in the press.

37. G.F. Bignami and G.E. Morfill, Astron. Astrophys., preprint.

38. H.J. Black and G.G. Fazio, Astrophys. J. 195, L23 (1973).

39. E. Massaro and L. Scarsi, Nature 274, 347 (1968).

40. G. Kanbach, K. Bennett, G.F. Bignami, G. Boella, M. Bonnardeau, R. Buccheri, N. D'Amico, W. Hermsen, J.C. Higdon, G.G. Lichti, J.L. Masnou, H.A. Mayer-Hasselwander, J.A. Paul, L. Scarsi, B.N. Swanenburg, B.G. Taylor, and R.D. Wills, Proc. 12th ESLAB Symp., ESA SP-124, 21 (1977).

41. J.L. Masnou, K. Bennett, G.F. Bignami, R. Buccheri, P. Caraveo, N. D'Amico, W. Hermsen, G. Kanbach, G.G. Lichti, H.A. Mayer-Hasselwander, J.A. Paul, B.N. Swanenburg, and R.D. Wills, Proc. 12th ESLAB Symp. ESA SP-124, 33 (1977).

42. R.C. Lamb, C.E. Fichtel, R.C. Hartman, D.A. Kniffen, and D.J. Thompson, Astrophys. J. Lett. 212, L63 (1977).

43. M.A. Forman, G.F. Bignami, G.E. Morfill, and H.J. Völk, this Symp.

44. K. Bennett, G. Bignami, W. Hermsen, H.A. Mayer-Hasselwander, J.A. Paul, and L. Scarsi, Astron. Astrophys. 59, 273 (1977).

45. E. Massaro, M. Salvati, and R. Buccheri, Mon.Not.Roy.Astron.Soc., in the press.

46. E.B. Jenkins, J. Silk, and G. Wallerstein, Astrophys. J. Suppl. Series 32, 681 (1976).

47. D. Morris, D.A. Graham, J.H. Seiradakis, W. Sieber, P. Thomasson, and B.B. Jons, Astron. Astrophys. 73, 46 (1979).

48. J.G. Hills, Astrophys. J. 221, 973 (1978).

49. A.W. Strong and D.W. Worrall, J. Phys. A. 9, 823 (1976).

# DISCRETE GAMMA-RAY SOURCES: THEORY

M. Salvati

*Laboratorio di Astrofisica Spaziale, Frascati, Italy*

ABSTRACT

Having specified the notion of a "discrete gamma-ray source", we recall that the available sample of such sources is largely dominated by a galactic population, with only rare, less firmly established extragalactic emitters.

Among the galactic objects, pulsars emerge conspicuously: not only are radio pulsars found to emit gamma rays, but - more important - formerly unidentified gamma-ray sources are beginning to be associated with radio pulsars. We present a brief account of the various emission models and pattern geometries: the so-called curvature-radiation mechanism and the wide-beam option appear to be favored.

We also elaborate upon the possibility that pulsars (or a subclass thereof) are a sizeable fraction of the galactic discrete sources, and find no major inconsistency.

In the same spirit alternative models are examined: these include accreting black holes - a scheme which could conceivably be scaled up to account for gamma rays from active galactic nuclei - and dense clouds irradiated by cosmic-ray protons.

As for extragalactic sources, we only outline the essentials of the recently proposed self-Compton models, and mention the geometrical constraints associated with the photon-photon opacity.

INTRODUCTION

In the following presentation, we will restrict the notion of "discrete gamma-ray sources" to objects like the ones detected by COS-B. The instrumentation and its performance have been described elsewhere [1], and only a few basic numbers need be repeated

energy range ∿ 30 MeV - 3 GeV
angular resolution ∿ 3° - 6°
time resolution ∿ 1 ms
minimum detectable flux ∿ $1.5 \times 10^{-6}$ phot (> 50 MeV) $cm^{-2}$ $s^{-1}$

The minimum flux corresponds to $6.5 \times 10^{35}$ erg $s^{-1}$ at a typical galactic distance of 3 kpc, and to $7.2 \times 10^{44}$ erg $s^{-1}$ at a typical extragalactic distance of 100 Mpc; here the average photon energy equals 250 MeV, the same as for the Crab pulsar [2].

The restriction to the COS-B discrete sources is not a very heavy one. The gamma bursts are a different kind of phenomenon, and to them a different session of the Symposium will be devoted. The MeV sources known to us at the time of writing include only the Crab and NGC 4151: but the former is a conspicuous member of the COS-B list, and the latter falls into the category of Active Galactic Nuclei, represented in this paper by 3C273. Also the three major SAS-II sources have been confirmed.

Prior to new findings presented during this Symposium, the largest published list was the one given by Mayer-Hasselwander et al. [3]: it contains 24 objects, including PSR 0531, 0833, 0740, 1822. One must then add further detections, which have been presented in formal meetings [4] , but have not been published yet. The total number is in excess of 30. Only very few objects lie at substantial galactic latitudes and one, in particular, has been associated with 3C273. The large majority of sources, instead, cluster around b = 0°, and offer clear evidence of galactic membership.

On the other hand, the proximity to the galactic plane, when associated with the poor angular resolution, makes it impossible to establish reliable identifications by means of the usual positional criterion. Various attempts have been made to correlate the gamma-ray sources with SNR's [5] or H II regions [6], but the abundance of candidate counterparts along the plane strongly diminishes the significance of the results [7].

The only clearcut criterion remains the time structure of the detected signal. More precisely, the typical time interval between source photons amounts to several minutes, and the statistics would be too low for a conventional Fourier analysis; so one is limited to checking the signal for the period of a known radio pulsar lying within the acceptance cone [8]. As we said, there are four established identifications in the COS-B list, while the detection of PSR 1747 by SAS II is still undisputed, pending the final analysis of the new data. One source out of seven being associated with a previously known radio pulsar is already a remarkable piece of information; its value is better assessed if the following considerations are taken into account.

Several gamma-ray pulsars have been detected as a result of an ad-hoc search, where the radio period was used as a "filter" in order to extract from the background particularly weak signals. Nevertheless, the pulsars in Crab and Vela would have been detected as non-periodic sources by any "blind" survey of the gamma-ray sky. The case of PSR 0611 is a very instructive one: the source to which it might be associated appeared under the name of CG 189+1 in the first COS-B catalogue [9], and only at a later time was it realized that the pulsar could account for the observed flux if it lay at the distance of SNR IC443, rather than the DM distance [10]; the records are currently being searched for periodicity.

In summary, pulsars are not a marginal component of the gamma-ray- source population; in a later part of our paper we will try to estimate how much room they leave for other kinds of source, but one would like first to collect some ideas on how and why a pulsar emits the observed gamma flux.

For this purpose, some additional information is available:

a) Not only very young and rapid objects (like PSR 0531), but also oldish ones having periods of several hundreds of milliseconds (like PSR 1822) are able to produce gamma rays: one has to account for "special" as well as for "typical" pulsars in any comprehensive theory.

b) The gamma-ray luminosity is a surprisingly slow function of the period; the canonical way of expressing this involves the efficiency of energy conversion from star rotation into gamma rays on the one side, and, on the other, the age $P/2\dot{P}$: the former increases with the latter, and takes on values close to 1 for ages larger than about $10^5$ yr [11]. In fact, under the assumption of maximum efficiency a priority list of candidate gamma-ray pulsars has been compiled [10], and all the entries for which the analysis has been completed have indeed been detected.

c) The absence of failures in the search referred to above also implies that the gamma-ray beam sweeps a solid angle not smaller than, and inclusive of, the radio beam.

d) The available light curves are very similar to each other [11], showing always two main peaks whose phase separation has values around 0.4.

When these points are combined with the current ideas on the magnetosphere of neutron stars, the problem of modeling the gamma-ray emission becomes a well-posed one, and some satisfactory progress can be made.

## PULSARS

According to a widely accepted notion, pulsars are rotating magnetized neutron stars, whose rotational energy is extracted by large-scale electromagnetic torques, and is then available for any exotic process either close to the star, or very far away. The neutron star magnetosphere is characterized by a strongly anisotropic resistivity tensor (large across $\underline{B}$, small along it), and by large potential differences originating from electrical induction. The tendency is toward a Goldreich-Julian type of structure [12], where charged particles taken from the star surface are so distributed as to make $\underline{E} \cdot \underline{B} \approx 0$.

In order to maintain such a distribution, the charges must corotate rigidly with the star and glide along $\underline{B}$ at an arbitrary speed; no gliding is necessary on the so-called closed lines, along which the corotation velocity never exceeds c; but the open lines must carry a stream of particles from the star to infinity, or vice versa, depending on the sense of the winding up.

The available power is just the electromotive force between the magnetic axis and the last open line, multiplied by the current flowing around the circuit

$$\begin{aligned} \mathrm{EMF} &\approx 7 \times 10^{12}\ B_{12}\ P_1^{-2}\ \text{volt} \\ I &\approx 3 \times 10^{11}\ \dot{P}_{-15}\ P_1^{-1}\ B_{12}^{-1}\ \text{A} \quad . \end{aligned} \tag{1}$$

$B_{12}$ is the surface magnetic field in units of $10^{12}$ G, $P_1$ the period in s, $\dot{P}_{-15}$ the period derivative in $10^{-15}$ s $s^{-1}$. As long as carriers of both signs are available without hinderance, the investigators agree that the energy flux coincides with the Poynting-vector flux: the carriers themselves are energetically irrelevant until they get a long way outside the critical surface: there they supposedly start crossing the lines of force, and extract energy from the e.m. field.

The scenario was appealing so long as no pulsar was known with an output comparable to the rotational energy loss: the theoreticians were so to speak scared by the enormous electric fields available for acceleration, and put their efforts in reducing both the peak particle energy and the pulsed luminosity. Now, instead, we must account for the high efficiency discussed under Ib), that is, we must leave unscreened a substantial fraction of the potential (1) within the light cylinder: across the potential drop the energy will be transfered to the particles, at variance with the Goldreich-Julian magnetosphere, and then to pulsed radiation.

The neutron star magnetosphere will no longer be force-free if at least one of the carrier species is not allowed to reach its asymptotic distribution, or if it cannot do so in a short enough relaxation time [13]. Three basic possibilities have been envisaged:

a) The positive ions are bound so strongly to the neutron star crust, that the induced electric field is not sufficient to extract them in the required amount [14]. Above (half of) the polar cap a potential gap is developed, which grows until any stray particle can initiate an electro-optical cascade: the emission and absorption mechanisms adequate to the given environment are respectively curvature radiation and magnetically induced pair production. Then the positron component of the cascade can replace the ions and close the electric circuit. The expected $\Delta V$ is

$$\Delta V \approx 6 \times 10^8\ B_{12}^{-1/7}\ P_1^{-1/7}\ \rho_1^{4/7} \tag{2}$$

Under the additional hypothesis that $\rho_1$, the line's curvature radius in cm, is that of a dipole field, with no higher-order contamination, Salvati and Massaro [15] have solved the transfer problem for the gamma rays: they found that the magnetic pair production is important, but certainly does not suppress the whole flux. Furthermore, acceptable spectral shapes and pulse widths are easily obtained.

The problems with this approach are of a different nature:

aa) It is not clear whether the ion work function is really so large; temperature effects, especially important in young objects, and the possible presence of elements lighter than iron could drastically affect the picture (see for instance Flowers et al. [16] ).

ab) If the model is invoked for a high-efficiency radiator, then it is not obvious how the star disposes of the angular momentum corresponding to the rotational energy loss [17] . This criticism applies to any scheme where a large fraction of the total power output is radiated well inside the light cylinder; a possible answer could be that a comparably large power fraction is lost well outside the light cylinder, accounting for all the angular momentum which has been left behind.

ac) Assuming a very tangled magnetic configuration, i.e. putting $\rho_1$ equal to the stellar radius $\approx 10^6$ in Eq. (2), results in a typical $\Delta V$ of order $10^{12}$ volt; the current is determined by the spin-down rate and the overall EMF, Eq. (1), so the power available at the polar gap turns out to be insufficient with respect to the observed gamma-ray luminosities. The difficulty is somewhat eased if the field is taken to be purely dipolar down to the stellar surface - $\rho_1$ larger - but the situation remains unsatisfactory. This criticism, too, has a wider scope than the particular model which is being discussed: any attempt to establish a large potential gap very close to the neutron star would fail at the same $\Delta V$ , insufficient for our purposes, since the sparking threshold is relatively low in the large B field expected there [13] .

ad) With an emission region located just above the polar caps, the most natural radiation pattern should consist of two, roughly opposite, pencil beams. But the data collected at point Id) above would require that all pulsars have the radio and the gamma-ray beams in their equatorial planes; not even one oblique rotator, just a totality of orthogonal rotators. More over, the 0.4 phase interval between gamma-ray pulses would remain unexplained.

A variant which avoids the most controversial point aa) involves space charge effects: in the original Goldreich-Julian computations the particles' inertia is neglected, but a certain potential drop is necessary in order to accelerate them to the required velocity. The acceleration region develops just above the polar caps, and for Fe II ions its strength is estimated to be [18]

$$\Delta V \approx 9 \times 10^{11} B_{12}^{1/2} P_1^{-1} \text{ volt} \tag{3}$$

Of course, all the comments other than aa) remain valid, in particular the difficulties associated with the conversion efficiency and the radiation pattern apply to this variant as well. Furthemore, if for a given parameter set the space charge $\Delta V$ exceeds that of Eq. (2), the system will find it more convenient to break the vacuum rather than to accelerate the positive carriers, and the two models will merge.

In summary, this scheme has several virtues and might perhaps account for some particular portions of the gamma-ray light curves. The bulk of it, however, must be looked for somewhere else.

b) According to the models of the second class, a large $\Delta V$ can be produced along the open field lines which curve towards the spin axis [19] . With the crucial assumption that no trapped particles exist on those lines, and the moving ones all have the same sign and the same speed $\approx$ c, the proof is immediate. Because of charge conservation, $n \propto B$; but the Goldreich-Julian density is proportional to $\underline{\Omega} \cdot \underline{B}$ and decreases at a lower rate in case of favorable curvature. The deficiency of charges at large distances results in a potential drop which can be comparable to the total available EMF. At variance with case c) below, $\Delta V$ has such a sign as to pull in the direction of the outflow, and the electric circuit can keep going under any circumstances. No significant acceleration occurs if the considered lines are the site of "premature" electro-optical cascades, since then the plasma is not charge-separated, and the force-free density is easily maintained. On the other hand, the model embodies interesting geometrical properties, apparently adequate to match the observational requirements Id) [20] .

c) Similar considerations lead to a picture proposed by Holloway [21] and Cheng et al. [22] ; the following discussion applies to some extent also to b), and compensates for the lack of details there.

In a Goldreich-Julian magnetosphere the sign of the charge density along a given line changes together with the direction cosine relative to the spin axis; so a stream of a given species may happen to flow across a distribution of the opposite species, rigidly bound to the line and participating in its rotation. If the trapped particles are lost - for instance, because of centrifugal forces - they cannot be supplied anew by the star, since the relevant line carries particles with the wrong sign.

Several scenarios have been proposed at this point. The $\Delta V$ grows until the total EMF is balanced, the current flow is stopped, and the pulsar braking is limited to the magnetic dipole radiation; no gamma rays should then be emitted. Or else the $\Delta V$ becomes large enough to suck the appropriate kind of charges from beyond the light cylinder, and the circuit is maintained by a single-component plasma.

If the pulsar is very fast, a discharge of the type discussed in a) will occur at a $\Delta V <$ EMF. Of course, the dependence on the pulsar parameters will be different from Eq. (2); in particular, the typical $\Delta V$ will be much larger, since in a weaker B field a harder photon is needed in order to exceed the pair-production threshold

$$\Delta V \simeq 2 \times 10^{16} \; P_1^{4/3} \; B_{12}^{-1/3} \; \lambda \text{ volt}$$

$$\lambda = \frac{e\Delta V}{mc^2 \gamma} \simeq 2 \times 10^{-6} \; \gamma^{-1} (\Delta V/\text{volt})$$

The parameter $\lambda$ is obviously $\geq 1$, and becomes >> 1 if the radiation reaction is important.

A further possibility might arise from the presence of keV radiation in sufficient amounts [13] . The target quanta would have a frequency higher than the local Larmor frequency, and the pair-production threshold would be reached at a value of $\Delta V$ lower than that given by Eq. (4)

$$\Delta V \simeq 10^{13} \; P_1^{-1/3} \; \nu_{x5}^{-1/3} \; \lambda \text{ volt} \tag{5}$$

The target frequency $\nu_{x5}$ is normalized to 5 keV. A possible drawback is the implied X-ray luminosity, which may not be easy to account for

$$L_x \simeq 10^{32} \; P_1^{2/3} \; \nu_{x5}^{1/3} \; \lambda^{-1} \text{ erg s}^{-1} \tag{6}$$

At any rate, if in a way or another the system manages to keep the circuit going, the objections aa) and ab) are bypassed. As for the radiation efficiency, this is certainly increased, since the same current $\dot{\varepsilon}_{rot}$/EMF of Eq. (1) now experiences wider potential jumps than in case a).

Also the geometrical problems outlined in ad) are offered a solution. The 0.4 phase lag is reproduced as the average peak separation, if random inclinations are assumed for the magnetic axis and the line of sight, and if the gamma-radiation pattern coincides with the magnetic equatorial plane [23 ] . Such a wide pattern is plausible only when the active region is close to the light cylinder. Furthermore, no single-peaked light curves are expected, and the solid angle filled up by the gamma rays is found to be $\pi^2$ sr, in accordance with the "visibility" of such objects stressed at paragraph Ic).

If we look at the preceding discussion with an optimistic attitude, we might perhaps take it as evidence that a pulsar can develop accelerating regions with the right properties in the right places. It still remains to convert the particle energy efficiently into gamma rays, and the relevant radiation processes are briefly reviewed in the following.

Since Sturrock [24] , the authors agree that curvature emission must be the primary process, because the large magnetic field suppresses any significant transverse component of the particle motion. The simple-minded approach involves ivalent magnetic field

$$.7 \times 10^{3} \ \gamma \ \rho_1^{-1} \qquad G \qquad (7)$$

he large-angle synchrotron formulae; we give typical
wer and radiation length in a parametrized form adequate to
e the curvature radius is of the order of the light cylinder radius

$$\begin{aligned} \nu &\approx 2\text{x}10^{-21} \ \gamma^{3} \ P_1^{-1} && \text{MeV} \\ \dot{\varepsilon} &\approx 2\text{x}10^{-28} \ \gamma^{4} \ P_1^{-2} && \text{erg s}^{-1} \\ \ell &\approx 10^{32} \ \gamma^{-3} \ P_1^{2} && \text{cm} \end{aligned} \qquad (8)$$

A high efficiency is achieved when the radiation length becomes shorter than a representative size of the active region, which in the general case is a fraction $(\Delta V/\text{EMF})^{1/2}$ of the critical distance $cP/2\pi$ [14] . This defines a minimum Lorentz factor $\gamma_{min}$, and a minimum potential drop $\Delta V_{min}$ sufficient to impart $\gamma_{min}$

$$\Delta V_{min} \approx 1.4\text{x}10^{13} \ B_{12}^{1/7} \ \text{volt} \qquad (9)$$

For a potential difference in excess of the above one, the particles reach an equilibrium between gains and losses, so that the low-energy relationship $mc^2\gamma = e\,\Delta V$ is changed into

$$\gamma \approx 6 \times 10^{5} \ B_{12}^{1/8} \left(\frac{\Delta V}{\text{volt}}\right)^{1/8}$$

$$\lambda = \left(\frac{\Delta V}{\Delta V_{min}}\right)^{7/8} \qquad (10)$$

Eqs. (9) and (10) are unavoidable for any efficient mechanism working close to the light cylinder, except for minor modifications when the acceleration region reduces to the thin outer gap of Cheng et al. [22] . So, on the one hand we require that the expected potential drop - either EMF, or Eq. (4), or Eq. (5) - be larger than Eq. (9); on the other, however, we should use Eq. (10) in a self-consistent derivation of the $\Delta V$ which will finally be established. For instance, even a $\Delta V \approx 10^{15}$ volt in a pulsar with $P_1 = 0.1$ would be insufficient for initiating vacuum breakdown, because of the lower $\gamma$'s allowed by the radiative losses.

A final note of caution refers to the possible occurrence of significant inverse

Compton losses against mm photons (if the targets had higher frequencies, the Klein-Nishina cutoff would be operative). The mm luminosity needed for an appreciable effect may be estimated as $(B_{eq}^2/8\pi)\, c 4\pi R^2 \gtrsim 10^{31}$ erg s$^{-1}$, a value so high that the envisaged possibility becomes very remote.

If we were interested in case a), we should go on discussing the behaviour and the contribution of the secondary particles [15], characterized by a much lower energy and a much larger pitch angle than the primary ones. However, in case c) it is not obvious how such a secondary population could arise: there, pair production by primary photons occurred mostly above the (very thin) accelerating layer, and the pair members could not but live with the birth energy. Here, instead, the site for pair production is the accelerating region itself, and the newly born particles experience the same $\Delta V$ as their progenitors. This point is especially relevant to the connection between the gamma-ray and the radio luminosities.

At any rate, the above question - and all the problems which remain open - do not imply that pulsars are inexplicable objects: our knowledge, however gross, is certainly a working one, and in reaching it the contribution of the gamma-ray observations has been of paramount importance.

## OTHER SOURCES?

Besides pulsars, a certain number of different objects have been considered for gamma-ray emission, although such a line of work suffers from the lack of unquestionable observational support.

Accreting black holes of several solar masses could convert perhaps one tenth of the inflow energy into medium-to-soft gamma rays [25]. In a turbulent environment the relative motion of magnetized cells gives rise to induced electric fields, and the acceleration of runaway particles becomes plausible. The gamma rays are produced with different efficiencies and different spectra because of synchrotron and inverse-Compton emission from the electrons, and pion emission from the protons.

If a typical COS-B source radiates about $10^{35}$ erg s$^{-1}$, and the efficiency of the process is not fictitiously inflated, then one needs more than $10^{36}$ erg s$^{-1}$ in the accretion flow. This appears feasible only in a close binary system, perhaps not an extreme case like the canonical X-ray binaries, but certainly something qualitatively similar. In fact, the same scenario has been invoked in connection with a supposed gamma-ray emission from Cyg X-3 [26].

The number of such sources in the Galaxy, instead of being the black-hole production rate multiplied by the Galaxy lifetime, is only the production rate of black holes in close binary systems multiplied by the lifetime of the accretion stage. A gamma-ray black hole being a young object implies that the production rate of interest is limited by the rate of SN explosions at the present time; more precisely, it cannot be a major fraction thereof, since we know that the birthrate of radio pulsars is already difficult to account for.

Alternative models resort to the interaction of cosmic rays with interstellar matter. This mechanism of producing gamma rays has been pointed out a long time ago [27,28] , and has been investigated mostly in connection with the diffuse galactic background [29,30] . It goes together with a number of intricate questions, as the distribution of the interstellar matter, the extrapolation of the local cosmic flux, the correlation or anticorrelation of the two ingredients [31] . Such topics will be covered by successive authors, in charge of the specific subject.

Here we just recall that gamma-ray telescopes have a comparatively poor angular resolution, and that a clump of interstellar matter emitting because of the above process might be interpreted as a point source. This idea lies behind the various attempts which have been made to correlate gamma-ray sources with SNR's or H II regions [5,6] . More recently, Montmerle and Cesarsky [32] have pointed out that a good correlation exists between gamma-ray sources and enhancements in the gamma-ray background, on the one side, and supernova remnants interacting with the dense environment of OB associations (SNOB's) on the other. Preliminary estimates of the bremsstrahlung flux from the observed radioelectrons are already encouraging; still better results are obtained if a sensible proton-to-electron ratio is assumed.

A more specific model is suggested by Abdulwahab et al. [33] , who identify a cluster of CG sources in the anticenter region with density enhancements irradiated by a proton burst from SN 1054. Several authors, on the basis of the positional coincidence with a COS-B source, have investigated the gamma-ray emissivity of the $\rho$ Oph cloud complex, and have invoked a local production of high-energy particles by a variety of acceleration mechanisms [34,35] .

It will be noted that in all the available examples an enhanced interstellar density is supplemented with a nearby source of cosmic rays. In order to produce $10^{35}$ erg $s^{-1}$ by means of the local cosmic-ray flux, one would need $10^{7}$ solar masses of target nuclei, that is, a conspicuous object, at the upper end of the molecular cloud sequence. Moreover, such a cloud would extend over several tens of parsecs, and, at a distance of 1 kpc, it would perhaps not give a point image.

Independently of any theoretical consideration on the emission mechanisms, the role of non-pulsar gamma-ray sources can be assessed by subtracting the known contributions to the total galactic luminosity. The latter quantity is estimated [36] to be 6-7 x $10^{38}$ erg $s^{-1}$; to the original 5 x $10^{38}$ erg $s^{-1}$ the power radiated between 50 and 100 MeV has been added, under the assumption that the spectrum has a number index equal to 2, in order to achieve consistency with the COS-B bandwidth.

The contribution of the diffuse component is taken to be of the order of 2 x $10^{38}$ erg $s^{-1}$; in estimating this number, we tried to be as simple-minded as possible, ignoring all the complications alluded to in the preceding. We took the yield per target atom [37] , assumed that the local cosmic-ray flux is representative of the whole Galaxy, and multiplied by a gas mass of 5 x $10^{9}$ $M_{\odot}$, as given by Mezger [38] .

As for the discrete sources, let us recall that in a disk population with a negligible thickness and a distribution independent of the galactocentric distance a simple relation holds between the observable number of sources $N_{obs}$ and their luminosity integrated over the whole disk $L_{gal}$. Call n(L) the luminosity function, $\phi$ the minimum detectable flux, S the area of the Galaxy; then

$$n(L) = 0 \quad \text{if} \quad L < L_{min}, \quad L > L_{max}$$

$$N_{obs} = \int_0^\infty 2\pi r \, dr \int_{4\pi r^2 \phi}^\infty n(L) \, dL = \int_0^\infty n(L) \, dL \, .$$

$$\int_0^{(L/4\pi\phi)^{1/2}} 2\pi r \, dr = \int_0^\infty n(L) \, dL \, \frac{L}{4\phi} \qquad (11)$$

$$L_{gal} = \int_0^\infty n(L) \, dL \, L \, S = N_{obs} \, 4 \, \phi \, S$$

For a fixed $N_{obs}$, the total population luminosity increases when the density increases toward the galactic center, or when the thickness has a non-negligible value h; the former statement is obvious, the latter follows from a selection effect against low-luminosity objects: when the visibility distance becomes smaller than h, the sampled volume is a sphere rather than the (larger) circumscribed cylinder.

In our case, $\phi$ may be set at 0.3 times the Crab level, since very few sources are catalogued with a lower flux even if the quoted detection threshold is somewhat better. One gets $L_{gal}$ 1.7 x $10^{37}$ $N_{obs}$ erg $s^{-1}$; so the pulsar contribution to the galactic luminosity is at least 0.8 x $10^{38}$ erg $s^{-1}$; all other classes of source could yield a few times as much, certainly less than ten times in any case, in agreement with there being six non-pulsar-identified sources out of seven in the detected sample.

Our final point will be a daring suggestion: in the discussion about the pulsar magnetosphere and the gamma-radiation pattern, we offered evidence for a beaming factor of 80%; from the radio data, i.e. pulse widths, frequency of double-peaked light curves, and so on, a beaming factor of 20% is derived [39]. If indeed out of four gamma-ray pulsars only one is a radio pulsar visible from Earth, then the estimates of Buccheri et al. [10] should be multiplied by 4: one would expect 40 detectable sources and a total source luminosity not less than 7 x $10^{38}$ erg $s^{-1}$. These numbers could appear excessively large, but one should recall the serious uncertainties affecting all the results.

A further discrepancy is brought about by the latitude distribution. Assuming that pulsars and gamma-ray sources are one and the same thing, and recalling that pulsars extend to several 100 pc above the galactic plane, one finds an average $|b|$ of 10° - 11°, while the known sources show a much flatter distribution, with a $<|b|>$ around 2° [4].

It must be noted, however, that also the detected gamma-ray pulsars have a $<|b|>$ much smaller than expected; this could be evidence that similar selection effects against high-latitude objects are at work both for pulsars and general sources; or else, the gamma-ray pulsars are a subset characterized by a lower typical age and a stronger concentration toward the galactic plane.

A viewpoint of this type, intermediate between the all-pulsar and the mysterious-object options, has been adopted by Panagia and Zamorani [40] ; the unidentified COS-B sources should be special pulsars, with extreme gamma-ray luminosities and stellar progenitors in a definite mass range; the picture seems to meet several positional and stellar-evolutive constraints, and - as we said - also the $<|b|>$ problem would be solved.

## ACTIVE GALACTIC NUCLEI

The source at 291+65 has been tentatively associated with the quasar 3C273; if the identification is correct, the observed flux corresponds to a gamma-ray luminosity of $2 \times 10^{46}$ erg $s^{-1}$, which is to be compared with somewhat smaller $L_x$ and $L_{opt-IR}$ [41] . The simple fact that gamma rays are observed, plus the assumption that the source volume is the same at all wavelengths, imposes an interesting constraint on the size [42] : the optical depth to photon-photon pair producing collisions between gamma and X-ray quanta is equal to 1 when

$$R \gtrsim L_{x46} \, \nu_{x5}^{-1} \text{ pc} \qquad (12)$$

The dimensions implied by Eq. (12) offer a strong argument against purely thermal, accretion models [43] : it is already difficult to get to 100-MeV energies, but at any rate one would expect such energies at the inner edge of an accretion disk, a few Schwarzschild radii away from the central object: this should be at least three orders of magnitude less than the sizes we are talking about.

One is then compelled toward non-thermal models for the gamma-ray emission, which - we would like to stress - is not a minor fraction of the total output. Once non-thermal mechanisms must be invoked, theoretical economy criteria require that the whole spectrum be explained in such a framework. A clue is provided by the near-equality of the luminosities in the various frequency bands, which is immediately suggestive of a cascade of inverse Compton scatterings with an amplification factor close to 1.

Th... ...n is met when the energy density in the large scale electroma-
...of the order of the energy density contributed by the synchro-
...d hence of the overall radiation energy density. The source
... a converter of Poynting-vector flux into observable radiation,
...e which has long been debated in connection with pulsars and

Spinars, indeed, are presently enjoying a new period of fame, since they can embody at the same time the electrodynamics of pulsars, and the conversion efficiency of black holes. A basic difference with respect to the pulsar case is the role of inverse Compton losses: in a typical pulsar, the gamma-ray luminosity is vastly in excess of any lower-frequency output, and can reach the total Poynting-vector flux; this immediately rules out inverse Compton as a significant loss mechanism if the same particles have to account for the low- and the high-frequency emission: so large an amplification factor would entail a target-photon power already larger than the maximum available one.

In the case of 3C273, however, the nice picture has a problem: requiring a Poynting-vector flux of several $10^{46}$ erg $s^{-1}$, a synchrotron frequency in the near IR, and the first and second Compton orders in the X-ray and gamma-ray ranges, respectively, one finds $R \approx 10^{-4}$ pc. Such a small size is ruled out by the same argument used against thermal disks. A possible remedy is to make only one Compton jump from optical-IR to gamma rays, so that $R \approx 2$ pc. The X-rays, then, must be accounted for by a different mechanism, involving perhaps a different population of relativistic particles, or even an accretion flow with canonical properties.

Further information can be gained from the overall shape of the spectrum. Upper limits to the gamma-ray flux from NGC 4151 have been obtained by SAS-II and COS-B, and approximately fall upon the extrapolation of the $E^{-3}$ law found by Schonfelder above the MeV region [45] . If the break there is interpreted as the onset of the photon-photon pair production opacity, and if indeed the $E^{-3}$ spectrum extends over a considerable range, then a homogeneous spherical model is ruled out. The target photons fill up the space outside the source as well as inside, and the geometrical dilution factor comes into play only at about one source radius above the surface. So, even gamma rays emitted exactly on the surface must deal with an optical depth comparable to those produced at the center. The expected cutoff is exponentially steep, unless gamma rays of different frequencies are emitted at different distances from the center.

As a final point we will remark that the spectrum of 3C273 appears to be rather flat at hard X-rays [46 ] , and has a slope -2.7 in the gamma-ray range. Observational results around 1 MeV would be most helpful, in view of the considerations developed for NGC 4151.

It is a pleasure to acknowledge discussions with Drs. Panagia and Massaro; Drs. Massaro, Buccheri, Scarsi and Scharlemann critically read parts of the manuscript.

## REFERENCES

1. L. Scarsi et al.: Proc. 12th ESLAB Symp., ESA SP-124, 3 (1977).
2. G. Kanbach et al.: Proc. 12th ESLAB Symp. ESA SP-124, 21 (1977).
3. H. A. Mayer-Hasselwander et al.: Proc. IX Texas Symp. on Relativistic Astrophysics, Munich, in the press (1978).

4. B. Swanenburg: EPS Study Conference "Gamma-ray Astronomy after COS-B", Erice (1979).
5. R. C. Lamb: Nature 272, 429 (1978).
6. A. W. Strong: Nature 269, 394 (1977).
7. E. Massaro, L. Scarsi: Nature 274, 346 (1978).
8. R. Buccheri et al.: Proc. 12th ESLAB Symp., ESA SP-124, 309 (1977).
9. W. Hermsen et al.: Nature 269, 494 (1977).
10. R. Buccheri et al.: Nature 274, 572 (1978).
11. R. Buccheri et al.: Astr. Ap., in the press (1979).
12. P. Goldreich, W. Julian: Ap. J. 157, 869 (1969).
13. M. Ruderman: Proc. IX Texas Symp. on Relativistic Astrophysics, Munich, in the press (1978).
14. M. Ruderman, P. Sutherland: Ap. J. 196, 51 (1975).
15. M. Salvati, E. Massaro: Astr. Ap. 67, 55 (1978).
16. E. Flowers et al.: Ap. J. 215, 291 (1977).
17. R. H. Cohen, A. Treves: Astr. Ap. 20, 305 (1972).
18. W. Fawley et al.: Ap. J. 217, 227 (1977).
19. E. T. Scharlemann et al.: Ap. J. 222, 297 (1978).
20. W. Fawley: preprint (1978).
21. N. Holloway: Nature Phys. Sci. 246, 6 (1973).
22. A. Cheng et al.: Ap. J. 203, 209 (1976).
23. E. Massaro et al.: M.N.R.A.S., in the press (1979).
24. P. A. Sturrock: Ap. J. 164, 529 (1971).
25. L. Maraschi, A. Treves: Ap. J. Lett. 218, L 113 (1977).
26. P. Mészàros et al.: Nature 268, 420 (1977).
27. S. Hayakawa: Progr. Theoret. Phys. 8, 571 (1952).
28. P. Morrison: Nuovo Cimento 7, 858 (1958).
29. C. E. Fichtel et al.: Ap. J. 208, 211 (1976).
30. F. W. Stecker: The Structure and Content of the Galaxy and Galactic Gamma Rays, NASA GSFC CP-002, 315 (1977).
31. C. J. Cesarsky et al.: Proc. 12th ESLAB Symp., ESA SP-124, 153 (1977).
32. T. Montmerle, C. J. Cesarsky: this volume.
33. M. Abdulwahab, P. Morrison: Ap. J. Lett. 221, L 33 (1978).
34. M. Cassé, J. A. Paul: this volume.
35. G. F. Bignami, G. Morfill: submitted to Astr. Ap. (1979).
36. A. W. Strong, D. M. Worral: J. Phys. A 9, 823 (1976).
37. F. W. Stecker: Ap. J. 185, 499 (1973).
38. P. G. Mezger: Astr. Ap. 70, 565 (1978).
39. J. H. Taylor, R. N. Manchester: Ap. J. 215, 885 (1977).
40. N. Panagia, G. Zamorani: Astr. Ap. 75, 303 (1979).
41. B. N. Swanenburg et al.: Nature 275, 298 (1978).
42. G. Cavallo, M. J. Rees: M.N.R.A.S. 183, 359 (1978).
43. A. Cavaliere, P. Morrison: Ap. J., in the press (1979).
44. F. Pacini, M. Salvati: Ap. J. 225, L 99 (1978).
45. V. Schonfelder: Nature 274, 344 (1978).
46. D. M. Worral et al.: Ap. J., in the press (1979).

# HIGH-ENERGY GAMMA-RAY SOURCES OBSERVED BY COS-B

R. D. Wills[+++], K. Bennett[+++], G. F. Bignami[**], R. Buccheri[+], P. Caraveo[**], N. D'Amico[***], W. Hermsen[*], G. Kanbach[+], G. G. Lichti[+++], J. L. Masnou[++], H. A. Mayer-Hasselwander[+], J. A. Paul[++], B. Sacco[***] and B. N. Swanenburg[*]
(the Caravane collaboration)

[*] *Cosmic-Ray Working Group, Huygens Laboratorium, Leiden, The Netherlands*
[**] *Laboratorio di Fisica Cosmica e Tecnologie Relative del CNR, Milan, Italy*
[***] *Istituto di Fisica, Università di Palermo, Palermo, Italy*
[+] *Max-Planck-Institut für extraterrestrische Physik, Garching bei München, Federal Republic of Germany*
[++] *Service d'Electronique Physique, Centre d'Etudes Nucléaires de Saclay, France*
[+++] *Space Science Department of ESA, ESTEC, Noordwijk, The Netherlands*

## ABSTRACT

COS-B has now detected 29 localised sources of high-energy (> 100 MeV) gamma-rays. An attempt has been made to eliminate selection effects from the study of their distribution in the sky, by excluding those whose intensities are below the threshold for detection in the brightest part of the Galaxy. The 18 sources left show, in addition to their concentration near the Galactic equator, a marked concentration in the first and fourth quadrants of longitude. The energy spectra of four of the sources, in the range 100 MeV to 3 GeV, have been derived and are shown to be markedly dissimilar.

## INTRODUCTION

The ESA satellite COS-B was launched into an eccentric earth orbit (100 000 km apogee) on 9 August 1975. It carries a single spark-chamber experiment for gamma-ray astronomy, the construction, operation and performance of which have been described elsewhere |1,2|. Figure 1 shows the region of the sky studied so far in the search for discrete sources of high-energy gamma radiation. It represents most of the data acquired in the first three years of operation. Study of these data is continuing together with the analysis of later observations. The satellite and experiment are still functioning well and it is foreseen to continue operations at least until the end of 1979.

## DETECTION OF SOURCES BY SPATIAL ANALYSIS

Previous reports |3,4,5,6,7| of detection of sources by COS-B have been based on results derived by differing analysis techniques. The present work is confined to results obtained by the systematic analysis of the spatial distribution of measured gamma-ray directions. The data used were obtained in about 30 separate observations, in each of which the pointing direction of the instrument was maintained constant for about one month. The fields of view covered by these observations generally overlap. In order to improve the statistical accuracy of the measurements data from such overlapping observations have been merged, without regard to the epoch of recording. It is intended that the investigation of long-term variability of gamma-ray sources will form the subject of a later, more detailed, analysis.

The merged data have been analysed using a cross-correlation method: the frequency distribution of arrival directions of gamma rays was correlated with the distribution expected from a point source (point-spread function) as determined by calibration and confirmed by flight data on the strong source PSR0833-45 |2|. A gamma-ray source is defined as a significant excess of counts compatible with the instrument point-spread function. A truly point-like object and a feature extending up to about 1 degree are not distinguished. The combined characteristics of the method, the instrument and the gamma-ray sky imply that the significance of a source will depend on its intensity and energy spectrum and on the intensity and energy spectrum of the background against which it is seen.

Figure 1 shows the positions of 29 sources which have been detected above 100 MeV by this method. Preliminary positions, error circles and fluxes are given in Table 1. Two excesses reported earlier as sources at (270.0, - 1.0) and (176.0, - 1.0) are not included because their energy spectra are so soft that the enhancements they show are not significant above 100 MeV. The source CG 189+1 is omitted because its enhancement is not compatible with a single point-spread function as used in the present (improved) analysis technique.

Because of the variation over the sky of the gamma-ray background (especially that from the disc of the Galaxy) the sample is subject to strong selection effects. In Figure 1 is indicated which of the sources exceed a minimum flux ($\sim 1.3 \times 10^{6}$ photon $cm^{-2}$ $s^{-1}$ above 100 MeV) which generally permits detection throughout the Galactic plane. This subset is believed to be complete in the region $90^{\circ} < l < 300^{\circ}$, $-15^{\circ} < b < +15^{\circ}$. It is unlikely to be complete in the region $300^{\circ} < l < 90^{\circ}$, especially at low latitudes ($|b| \lesssim 2.5^{\circ}$) where the Galactic emission is intense and highly structured |7|.

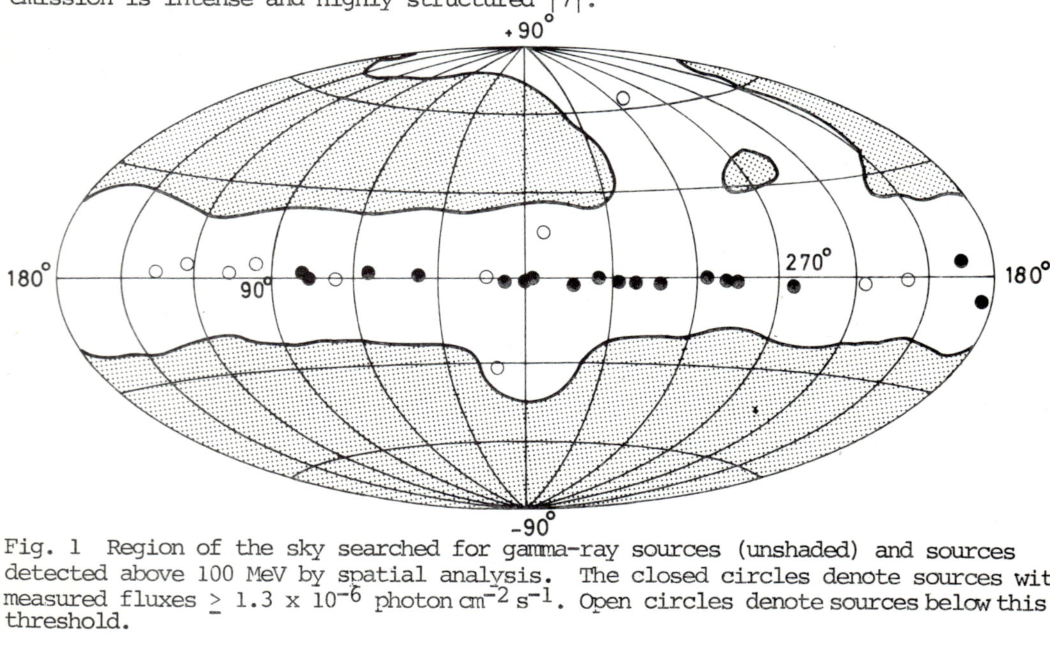


Fig. 1 Region of the sky searched for gamma-ray sources (unshaded) and sources detected above 100 MeV by spatial analysis. The closed circles denote sources with measured fluxes $\geq 1.3 \times 10^{-6}$ photon $cm^{-2}$ $s^{-1}$. Open circles denote sources below this threshold.

TABLE 1 Gamma-Ray Source Parameters

| $l^{II}$ (deg) | $b^{II}$ (deg) | error radius (deg) | flux > 100 MeV (photon $cm^{-2}$ $s^{-1}$) | $l^{II}$ (deg) | $b^{II}$ (deg) | error radius (deg) | flux > 100 MeV (photon $cm^{-2}$ $s^{-1}$) |
|---|---|---|---|---|---|---|---|
| 6.7 | - 0.5 | 1.0 | $2.4 \times 10^{-6}$ | 235.5 | - 1.0 | 1.5 | $1.0 \times 10^{-6}$ |
| 10.5 | -31.5 | 1.5 | 1.2 | 263.6 | - 2.5 | 0.3 | 13.2 |
| 13.5 | 0.5 | 1.0 | 1.0 | 284.0 | - 1.0 | 1.0 | 2.7 |
| 36.5 | 1.5 | 1.0 | 1.9 | 288.5 | - 0.5 | 1.3 | 1.6 |
| 54.2 | 1.7 | 1.0 | 1.3 | 289.3 | 64.6 | 0.8 | 0.6 |
| 66.0 | 0.0 | 0.8 | 1.2 | 295.5 | 0.5 | 1.0 | 1.3 |
| 75.0 | - 0.5 | 1.0 | 1.3 | 312.0 | - 1.3 | 1.0 | 2.1 |
| 77.8 | 1.5 | 1.0 | 2.5 | 321.0 | - 1.2 | 1.0 | 1.3 |
| 95.5 | 4.0 | 1.5 | 1.1 | 327.5 | - 0.5 | 1.0 | 2.2 |
| 106.0 | 1.5 | 1.5 | 1.0 | 333.5 | 0.0 | 1.0 | 3.8 |
| 121.0 | 4.0 | 1.0 | 1.0 | 342.5 | - 2.5 | 1.0 | 2.0 |
| 135.0 | 1.5 | 1.0 | 1.0 | 353.0 | 16.0 | 1.5 | 1.1 |
| 184.5 | - 5.8 | 0.4 | 3.7 | 356.5 | 0.3 | 1.0 | 2.6 |
| 195.1 | 4.5 | 0.4 | 4.8 | 359.5 | - 0.5 | 1.0 | 1.8 |
| 218.5 | - 0.5 | 1.3 | 1.0 | | | | |

Notes:
- error radius defines approximately the 90% confidence error circle
- fluxes are approximate, since an $E^{-2}$ spectrum is assumed for simplicity.

## ENERGY SPECTRA OF SOURCES

In a few cases, where a source is strong relative to the ambient background and where data have been acquired during multiple observations, it has been possible to estimate the intensity of the source in separate limited ranges of energy. The spectra of four sources, corrected for the effects of the instrumental energy resolution, are compared in Figure 2. In order to emphasise the differences the differential photon flux N(E) ($cm^{-2}$ $s^{-1}$ $GeV^{-1}$) has been multiplied by $E^2$ so that an $E^{-2}$ spectrum would appear parallel to the abcissa. None of these spectra look alike, though at least two of them (Crab, Vela) belong to similar objects. Differences between the spectra of the total radiation from the Crab and Vela sources and those of the pulsars PSR0531+21 and PSR0833-45 |8| are discussed elsewhere |9|.

## DISCUSSION

Only two of the gamma-ray sources presented here can be identified with complete certainty. These are the Crab and Vela pulsars whose timing signature is unequivocal |8|. A third source has been identified with high confidence with the quasar 3C273 |5|. This identification is strengthened by the second observation by COS-B in June/July 1978, the combination of the two data sets yielding the significantly reduced error box given in Table 1. The source at (353.0, + 16.0) has been associated with the ρ Oph cloud complex |7|. Since the measured intensity is an order of magnitude greater than that predicted by Black and Fazio |10| for production by cosmic-ray interactions with the gas of the cloud, improved models have been proposed to support the identification |11,12|.

In all other cases, including the strong source CG 195+4 discovered by the SAS-2 experiment |13|, the question remains open. The problems of identifying gamma-ray sources, either individually or as a class (or number of classes) have been discussed by Swanenburg |6|, Massaro and Scarsi |14| and Buccheri |15|, who cite numerous examples of proposed identifications. One clear common property of the

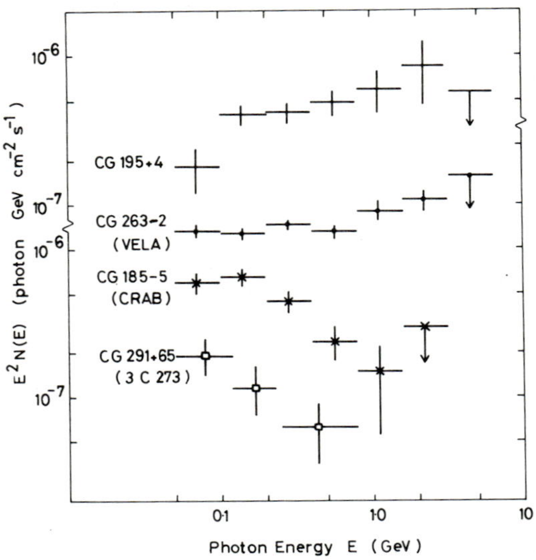


Fig. 2 Spectra of four gamma-ray sources (differential photon number spectra multiplied by the square of the photon energy).

majority of the sources is their Galactic nature, which is evidenced by the distribution of their Galactic latitudes. The mean absolute value of the latitudes of the 26 sources in Table 1 within 15° of the equator is $<|b|> = 1.5°$, suggesting typical distances in excess of 2 kpc, assuming a scale height of about 70 pc. Other features of their distribution are a warp, corresponding to the 'hat-brim' distortion of the Galactic plane, and some concentration within the solar circle, indicated by the longitude distribution of the 18 stronger sources.

Taking a value of $2 \times 10^{-6}$ photon (> 100 MeV) $cm^{-2}$ $s^{-1}$ for the typical flux of a source one can conclude that their gamma-ray luminosities are in the range $10^{35}$ to $10^{36}$ erg $s^{-1}$. The absence in the Uhuru catalogue |16| of obvious counterparts for the majority of sources indicates that their X-ray luminosities cannot exceed one tenth of this value. This places severe constraints on models of the sources since they are required to emit the major part of their energy as high-energy gamma-rays.

Acknowledgement

GGL acknowledges the receipt of an ESA fellowship.

References

1. G.F. Bignami, G. Boella, J.J. Burger, P. Kierle, H.A. Mayer-Hasselwander, J.A. Paul, E. Pfeffermann, L. Scarsi, B.N. Swanenburg, B.G. Taylor, W. Voges and R.D. Wills, Space Sci. Instrum. 1, 245 (1975).
2. L. Scarsi, K. Bennett, G.F. Bignami, G. Boella, R. Buccheri, W. Hermsen, L. Koch, H.A. Mayer-Hasselwander, J.A. Paul, E. Pfeffermann, R. Stiglitz, B.N. Swanenburg, B.G. Taylor and R.D. Wills, Proc. of the 12th ESLAB Symposium, ESA SP-124, 3 (1977).
3. K. Bennett, G.F. Bignami, M. Bonnardeau, R. Buccheri, W. Hermsen, G. Kanbach, G.G. Lichti, H.A. Mayer-Hasselwander, J.A. Paul, L. Scarsi, R. Stiglitz, B.N. Swanenburg and R.D. Wills, Astron. Astrophys. 56, 469 (1977).
4. W. Hermsen, B.N. Swanenburg, G.F. Bignami, G. Boella, R. Buccheri, L. Scarsi, G. Kanbach, H.A. Mayer-Hasselwander, J.L. Masnou, J.A. Paul, K. Bennett, J.C. Higdon, G.G. Lichti, B.G. Taylor and R.D. Wills, Nature 269, 494 (1977).

5. B.N. Swanenburg, K. Bennett, G.F. Bignami, P. Caraveo, W. Hermsen, G. Kanbach, J.L. Masnou, H.A. Mayer-Haswelwander, J.A. Paul, B. Sacco, L. Scarsi and R.D. Wills, Nature 275, 298 (1978).
6. B.N. Swanenburg, in: Stars and Star Systems, Reidel, Dordrecht, 1979, p. 1.
7. H.A. Mayer-Hasselwander, K. Bennett, G.F.Bignami, R. Buccheri, N. D'Amico, W. Hermsen, G. Kanbach, F. Lebrun, G.G. Lichti, J.L. Masnou, J.A. Paul, K. Pinkau, L. Scarsi, B.N. Swanenburg and R.D. Wills, Proc. of the 9th Texas Symposium in Annals of the New York Academy of Sciences, in press (1979).
8. K. Bennett, G.F. Bignami, G. Boella, R. Buccheri, W. Hermsen, G. Kanbach, G.G. Lichti, J.L. Masnou, H.A. Mayer-Hasselwander, J.A. Paul, L. Scarsi, B.N. Swanenburg, B.G. Taylor and R.D. Wills, Astron. Astrophys. 61, 279 (1977).
9. G.G. Lichti et al., this volume.
10. J.H. Black and G.G. Fazio, Astrophys. J. 185, L7 (1973).
11. G.F. Bignami and G. Morfill, Astron. Astrophys., in press (1979).
12. M. Cassé and J.A. Paul, this volume.
13. D.J. Thompson, C.E. Fichtel, R.C. Hartman, D.A. Kniffen and R.C. Lamb, Astrophys. J. 213, 252 (1977).
14. E. Massaro and L. Scarsi, Nature 274, 346 (1978).
15. R. Buccheri, this volume.
16. W. Forman, C. Jones, L. Cominsky, P. Julian, S. Murray, G. Peters, H. Tananbaum and R. Giacconi, Astrophys. J. Suppl., 38, 357 (1978).

# COS-B MEASUREMENTS OF THE SPECTRA OF THE CRAB AND VELA GAMMA-RAY SOURCES

G. G. Lichti+++, R. Buccheri***, P. Caraveo**, G. Gerardi***, W. Hermsen*, G. Kanbach+, J. L. Masnou++, H. A. Mayer-Hasselwander+, J. A. Paul++, B. N. Swanenburg* and R. D. Wills+++(the Caravane collaboration)

*Cosmic Ray Working Group, Huygens Laboratory, Leiden, The Netherlands
**Laboratorio di Fisica Cosmica e Tecnologie Relative del CNR, Milan, Italy
***Laboratorio di Fisica Cosmica e Tecnologie Relative del CNR, Università di Palermo, Italy
+Max-Planck-Institut für extraterrestrische Physik, Garching bei München, Federal Republic of Germany
++Service d'Electronique Physique, Centre d'Etudes Nucléaires de Saclay, France
+++Space Science Department of the European Space Agency, ESTEC, Noordwijk, The Netherlands

ABSTRACT

COS-B measurements above 50 MeV of the pulsed and total energy spectra of the Crab and Vela gamma-ray sources are reported here. It is found that the pulsed fraction of the Crab is consistent with 100 % above 400 MeV while below this energy the pulsed fraction lies between 45 % and 65 %. The emission from Vela is pulsed at least to 90 % above 50 MeV.

INTRODUCTION

Since the discovery [1, 2, 3] and the more detailed measurements [4, 5] of high-energy gamma-ray emission from pulsars many attempts have been made to understand how these gamma rays are created [see e.g. 6, 7, 8] . It is now widely accepted that gamma rays are produced in general by the interaction of highly energetic particles with the strong magnetic fields of the pulsar magnetospere either via curvature radiation or via synchrotron radiation. Because of the high energy content of the emitted photons the gamma-ray emission plays a fundamental role in pulsar physics and in particular its spectrum may yield unspoiled information about the primary particles and the magnetic field structure close to the neutron star [9].

Detailed energy spectra in the gamma-ray energy range have been published so far only for the Crab and Vela pulsars [10]. At that time each source was looked at by COS-B only once and the data of these two observations were used for that analysis. Since then COS-B observed both pulsars again and the analysis is more advanced, especially in the suppression of instrumental background events. New results on the pulsed and total energy spectra of their emission above 50 MeV are presented.

## THE ENERGY SPECTRUM OF THE CRAB GAMMA-RAY SOURCE

The Crab was observed by COS-B from 1975 August 18 till September 16, and again from 1976 September 29 till November 1. The data of both observations were used for this analysis.

The statistics allowed us to construct light curves for several different energy ranges from which the emission spectrum of the pulsed component could be derived, using the phase intervals defined in reference [10]. The spectrum was deconvolved using the known energy dispersion of the experiment [11]. The resulting pulsed spectrum is shown in figure 1 for four energy ranges. Above 800 MeV only 2$\sigma$ upper limits could be given. The spectrum can be very well represented by a single power law of the form

$$F(E) = (2.0 \pm 0.2) \cdot 10^{-7} \cdot E^{-(2.2 \pm 0.2)} \text{ photon cm}^{-2} \text{ s}^{-1} \text{ GeV}^{-1}$$

where the errors (1$\sigma$), quoted throughout the paper, are those due to statistics only. This is in good agreement with the measurement of SAS-2 [12].

Performing a spatial analysis using a cross correlation method the total emission spectrum from the Crab could be derived for five energy intervals. Between 1.6 GeV and 3.2 GeV only a 2$\sigma$ upper limit can be given. The spectrum is shown in figure 2. It is compared with the pulsed spectrum given by the dashed line. It is seen that the spectra are in good agreement above 400 MeV. This result is consistent with the emission above this energy being totally pulsed. In the energy range between 50 and 400 MeV the total spectrum lies significantly above the pulsed spectrum.

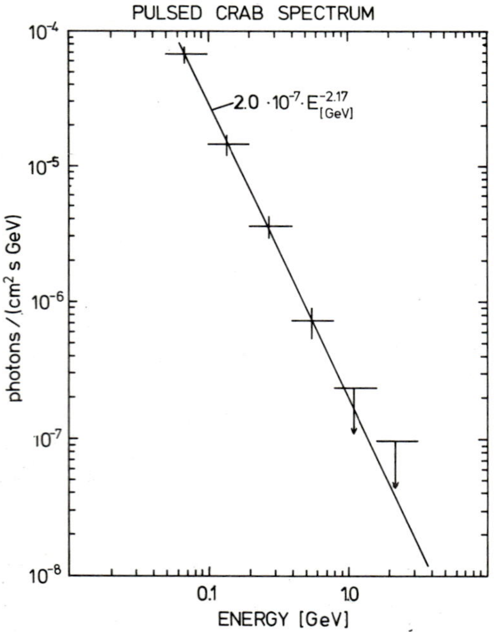


Fig. 1: Energy spectrum of the pulsed emission from the Crab pulsar

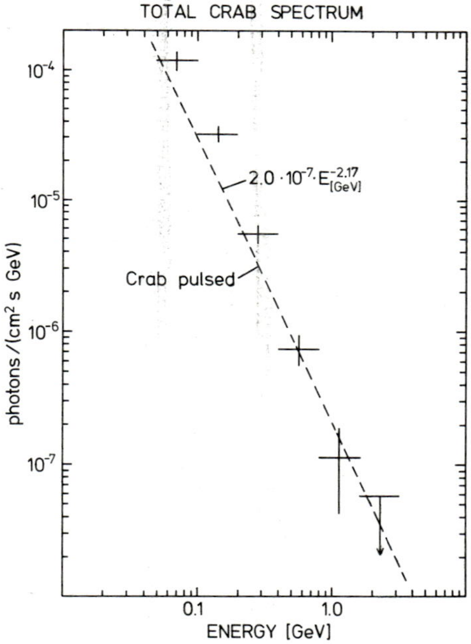


Fig. 2: Energy spectrum of the total emission from the Crab gamma-ray source

The pulsed emission contributes only between 45 % and 65 % to the total emission and in average the pulsed fraction in this energy range is 55 ± 7 %.

## THE ENERGY SPECTRUM OF THE VELA GAMMA-RAY SOURCE

Vela was observed by COS-B from 1975 October 20 to November 27. For this observation the deconvoluted energy spectrum of the pulsed emission has been constructed from the pulsed counts derived from light curves obtained for eight different energy intervals. The resulting energy spectrum in the range from 50 MeV to 3 GeV is shown in figure 3. Between 3 and 6 GeV only a 2σ upper limit was obtained. The spectrum can be fitted reasonably well with a single power law of the form

$$F(E) = (1.46 \pm 0.07) \cdot 10^{-6} \cdot E^{-(1.89 \pm 0.06)} \quad \text{photon cm}^{-2}\ \text{s}^{-1}\ \text{GeV}^{-1}$$

The differences between this spectrum and that published by Bennett et al. [10] are mainly due to a different definition of the pulsed phase intervals in the light curve (details will be published elsewhere [13]). The integrated pulsed flux above 100 MeV is $\sim 1.2 \cdot 10^{-5}$ cm$^{-2}$ s$^{-1}$ and is in agreement with the updated SAS-2 result [14].

From the spatial analysis of the arrival directions of the measured gamma rays using the same cross correlation method as above the energy spectrum of the total emission from the Vela gamma-ray source was determined. The result is shown in figure 4. For six energy ranges flux values could be derived, while in the interval from 3.2 to 6.4 GeV only a 2σ upper limit can be given. In the range from 50 MeV to 3.2 GeV the spectrum can be fitted again with a single power law of the form

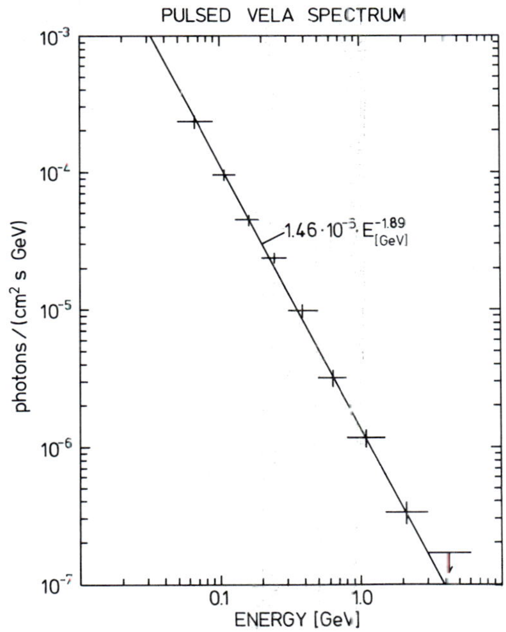


Fig. 3: Energy spectrum of the pulsed emission from the Vela pulsar

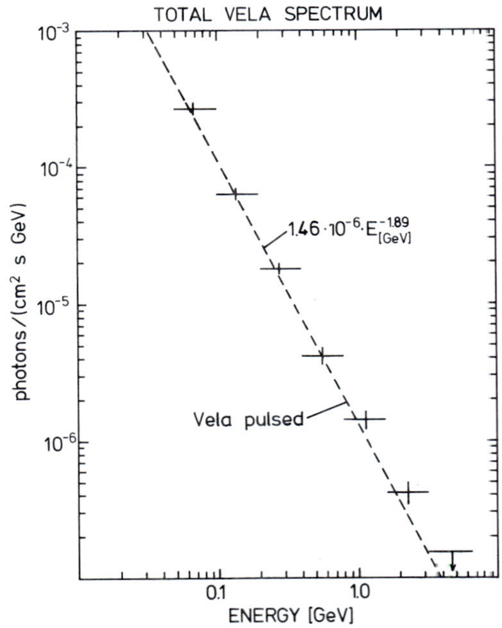


Fig. 4: Energy spectrum of the total emission from the Vela gamma-ray source

$$F(E) = (1.60 \pm 0.15) \cdot 10^{-6} \cdot E^{-(1.90 \pm 0.08)} \text{ photon cm}^{-2} \text{ s}^{-1} \text{ GeV}^{-1}$$

To derive this fit only the indicated statistical errors were used. Possible systematic uncertainties are somewhat larger for the total than for the pulsed spectrum.

The total spectrum is compared with the fit to the pulsed spectrum indicated by the dashed line (figure 4). Within the statistical uncertainties the two spectra are identical. A more precise evaluation based on a $\chi^2$ test shows that Vela is pulsed at least to 90 % above 50 MeV.

## DISCUSSION AND CONCLUSIONS

Detailed energy spectra of the pulsed and total emission from the Crab and Vela gamma-ray sources have been measured by COS-B. An unpulsed emission from the Crab was detected for the first time up to an energy of about 400 MeV. Such an emission was already observed up to 100 MeV by SAS-2 [12]. A comparison of the total emission with measurements in hard X-rays [15] shows good agreement within the statistical uncertainties. The energy spectrum of the pulsed emission from the Crab pulsar can be fitted with a single power law which extrapolates well to measurements in the X-ray energy region [15]. This suggests that a common physical process may be responsible for the production of X and gamma rays.

The pulsed and total energy spectra of the Vela gamma-ray source can both be fitted with single power laws. The comparison of the two spectra shows that they are identical within the statistical uncertainties. Since no steady emission was measured from this source the great majority of the observed gamma rays must originate near the pulsar itself. The extrapolation of the pulsed gamma-ray spectrum towards X-ray energies is at least a factor 4 above the upper limits of the pulsed flux around 30 keV [16]. Thus a break in the energy spectrum of the emission from the Vela pulsar is needed and therefore different production mechanisms might be responsible for the production of X and gamma rays.

## REFERENCES

1. J. Vasseur, J. Paul, B. Parlier, J.P. Leray, M. Forichon, B. Agrinier, G. Boella, L. Maraschi, A. Treves, R. Buccheri and L. Scarsi, Nature 226, 534 (1970).
2. R. Browning, D. Ramsden and P.J. Wright, Nature Phys. Sci. 232, 99 (1971).
3. P. Albats, G.M. Frye jun., A.D. Zych, O.B. Mace, V.D. Hopper and J.A. Thomas, Nature 240, 221 (1972).
4. D.A. Kniffen, R.C. Hartman, D.J. Thompson, G.F. Bignami, C.E. Fichtel, T. Tümer and H. Ögelman, Nature 251, 397 (1974).
5. D.J. Thompson, C.E. Fichtel, D.A. Kniffen and H.B. Ögelman, Astrophys. J. 200, L 79 (1975).
6. D.J. Thompson, Astrophys. J. 201, L 117 (1975).
7. H. Ögelman, S. Ayasli and A. Hacinliyan, Symp. on "The Structure and Content of the Galaxy and Galactic Gamma Rays", Goddard Space Flight Center, NASA CP-002, 109 (1976).
8. P.E. Hardee, Astrophys. J. 216, 873 (1977).
9. M. Salvati and E. Massaro, Astron. & Astrophys. 67, 55 (1978).
10. K. Bennett, G.F. Bignami, G. Boella, R. Buccheri, W. Hermsen, G. Kanbach, G.G. Lichti, J.L. Masnou, H.A. Mayer-Hasselwander, J.A. Paul, L. Scarsi, B.N. Swanenburg, B.G. Taylor and R.D. Wills, Astron. & Astrophys. 61, 279 (1977).
11. L. Scarsi, K. Bennett, G.F. Bignami, G. Boella, R. Buccheri, W. Hermsen, L. Koch, H.A. Mayer-Hasselwander, J.A. Paul, E. Pfeffermann, R. Stiglitz,

B.N. Swanenburg, B.G. Taylor and R.D. Wills, Proc. of 12th ESLAB Symposium, Frascati, ESA SP-124, 3 (1977).
12. D.J. Thompson, C.E. Fichtel, R.C. Hartman, D.A. Kniffen and R.C. Lamb, Astrophys. J. 213, 252 (1977).
13. G. Kanbach, K. Bennett, G.F. Bignami, R. Buccheri, P. Caraveo, N. D'Amico, W. Hermsen, G.G. Lichti, J.L. Masnou, H.A. Mayer-Hasselwander, J.A. Paul, B. Sacco, B.N. Swanenburg and R.D. Wills, in preparation.
14. D.A. Kniffen, C.E. Fichtel, R.C. Hartman, R.C. Lamb and D.J. Thompson, Proc. of 12th ESLAB Symposium, Frascati, ESA SP-124, 45 (1977).
15. M.S. Strickman, W.N. Johnson and J.D. Kurfess, Astrophys. J. 230, L 15 (1979).
16. G.R. Ricker, M. Gerassimenko, J.E. McClintock, S.G. Ryckman and W.H.G. Lewin, Astrophys. J. 186, L 111 (1973).

SESSION 2

# Discrete Gamma-ray Sources—II

*Chairman:* **B. V. SREEKANTAN**

# GAMMA RAY EMISSION FROM SUPERNOVA REMNANTS

R. Cowsik and S. Sarkar

*Tata Institute of Fundamental Research, Homi Bhabha Road, Bombay 400 005, India*

## ABSTRACT

Recent radio, optical and X-ray observations of galactic supernova remnants (SNR) have considerably increased our knowledge of their physical parameters and have encouraged considerable theoretical speculation regarding the dynamics of their evolution. In this work we point out that gamma-ray astronomy can be an additional powerful tool in clarifying some important features of SNR. To illustrate this claim, we obtain a lower limit to the magnetic field in the radio-emitting shell of the young remnant Cassiopeia-A, by considering the generation of bremsstrahlung gamma rays by the relativistic electrons that emit the observed synchrotron radiation. We also study the process of gamma-ray generation through inverse-Compton scattering of Lyman-alpha photons from H II regions and OB associations, that are generically associated with supernova remnants.

## INTRODUCTION

Supernovae are definitely an important factor in determining the physical conditions in the interstellar medium and the abundances of the heavy elements in the galaxy as well as probably being involved in the generation of galactic cosmic rays [1]. The remnants of supernovae are identified as extended sources of non-thermal radiation at radio frequencies and they usually exhibit shell-like structures that are expanding outwards. The radio emission is recognized to be due to the synchrotron process and early attempts [2, 3] at explaining the strength and temporal evolution of the radio luminosity were mainly concerned with estimating the relativistic electron density and the magnetic field in the remnant in idealized approximations. More recently, the evolution of SNR has been discussed in terms of a more elaborate fluid dynamical model [4], in which the physical conditions at the boundary region between the ejected matter and the surrounding interstellar medium are explored in more detail. In this picture, it is suggested that the interstellar magnetic field is significantly amplified (to equipartion values) in the region of radio emission, by turbulence generated through magnetohydrodynamic instabilities. The suggestion that such regions of amplified field would accelerate particles to high energies by the Fermi mechanism [5] has been investigated to yield an analytic solution for the evolution of the synchrotron luminosity that satisfactorily explains all the radio data on the young remnant Cassiopeia A, as well as the general evolutionary characteristics of galactic SNR [6]. The strength of the magnetic field is obviously an important parameter in all these theories.

Current ideas regarding stellar nucleosynthesis and the evolution of supernova progenitors [7] lead us to expect that the ejected material in SNR would comprise several solar masses of matter that is considerably enriched in heavy elements and that relict H II regions would be associated with the remnants. Optical and X-ray observations of the remnant Cassiopeia-A [8, 9] provide definite evidence for these expectations. It is then interesting to consider the generation of gamma rays by the processes of bremsstrahlung in the ambient matter and inverse-compton scattering of the Lyman-alpha photons from the associated H II region, by the radio emitting electrons. The strength of the gamma-ray emission would be proportional to the number density of the relativistic electrons (knowing the matter and photon densities) whereas the strength of radio emission would depend on both the relativistic electron density and a positive power of the local magnetic field. The observed gamma ray flux would then yield an upper limit to the relativistic electron density, which when coupled with the radio data would yield a lower limit to the magnetic field in the remnant.

## A LOWER LIMIT TO THE MAGNETIC FIELD IN CASSIOPEIA-A

To illustrate the technique, we choose the SNR Cassiopeia-A on which extensive radio [10, 11], optical [8, 12] and X-ray [9, 13] data is available and deduce from these data the mass and composition of the ejected matter in the radio-emitting shell. We then calculate the expected integral fluxes of gamma rays for various values of the local magnetic field, and compare these with the experimental upper limit on the gamma ray emission from this remnant, as deduced by us from the COS-B data [14]. As can be seen in Fig. 1, a minimum magnetic field value of $H = 8 \times 10^{-5}$ gauss is required to fit the COS-B upper limit on the integral flux at 100 MeV. Simple compression of the interstellar field at the remnant boundary [3] would yield a value for the field of only $1.2 \times 10^{-5}$ gauss and such a small field strength is clearly ruled out by the observations. Further, there is no evidence for a pulsar in this remnant, to which the derived high magnetic field can be attributed [7]. This then represents definite evidence for turbulent enhancement of the interstellar magnetic field as in the model of Gull [5]. Consideration of the gamma ray generation by inverse-compton scattering of Lyman-alpha photons from the associated H II region further improves the field limits as shown in Fig. 2. The density of Lyman-alpha photons is, however, a rather uncertain parameter and while we cannot obtain explicit values of the gamma-ray flux, the indications are that this process is as important as the process of electron bremsstrahlung. Further details of the calculations and results obtained are given in ref. [15].

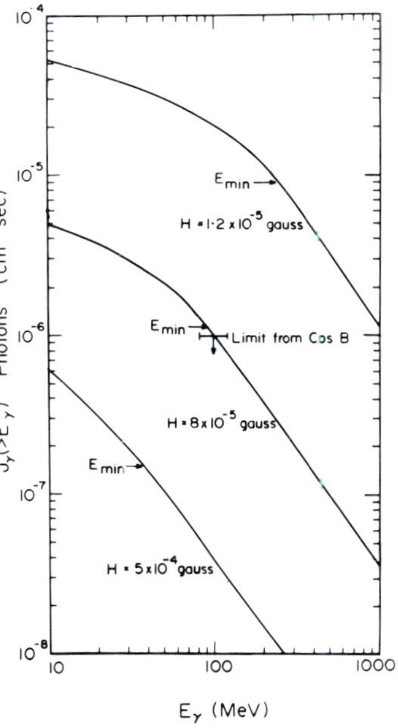


Fig. 1. Integral flux of gamma rays from Cassiopeia-A due to bremsstrahlung in the ejected matter, for various values of the magnetic field, H, in the radio-emitting shell.

(i) H = 1.2 x $10^{-5}$ G; corresponding to compression of the interstellar field at the remnant boundary.

(ii) H = 8 x $10^{-5}$ G; the minimum field that satisfies the COS-B upper limit at 100 MeV.

(iii) H = 5 x $10^{-4}$ G; corresponding to equi-partition between the particle and magnetic field energy densities.

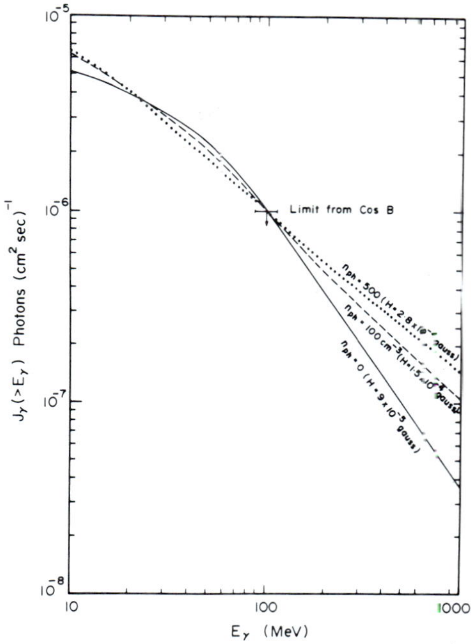


Fig. 2. The expected shape of the integral gamma ray spectrum for various values of $n_{Ph}$, the Lyman alpha photon density. The processes considered are the inverse Compton scattering of these photons and bremsstrahlung in the ejected and swept up interstellar matter.

## REFERENCES


1. L. Woltjer, A. Rev. Astron. Astrophys., 10, 129 (1972).
2. I.S. Shklovskii, Sov. Astron. A.J., 4, 243 (1960).
3. H. Van der Laan, Mon Not. R. Astr. Soc., 124, 125 (1962).
4. S.F. Gull, Mon. Not. R. Astr. Soc., 161, 47 (1973).
5. J.S. Scott and R.A. Chevalier, Astrophys. J. (Letts.), 197, L 5 (1975).
6. R. Cowsik, Astrophys. J., 227, 856 (1979).
7. W.T. Arnett, Astrophys. J., 195, 727 (1975).
8. S. Van den Bergh & W.W. Dodd, Astrophys. J., 162, 485 (1970).
9. S.H. Pravdo, R.H. Becker, E.A. Boldt, S.S. Holt, R.E. Rothschild, P.J. Serlemitsos & J.H. Swank, Astrophys. J. (Letts.), 206, L 41 (1976).
10. J.M.M. Baars and A.P. Hartsuijker, Astr. Astrophys., 17, 172 (1972).
11. I. Rosenberg, Mon. Not. R. Astr. Soc., 151, 109 (1970).
12. R.A. Chevalier, Astrophys. J., 208, 826 (1976).
13. P. Gorenstein, F.R. Harnden & W.H. Tucker, Astrophys. J., 192, 661 (1974).
14. W. Hermsen et al., Caravanne Collaboration, Nature, 269, 494 (1977).
15. R. Cowsik & S. Sarkar, Mon. Not. R. Astr. Soc., to be published (1979).

# NEW EVIDENCE FOR LINKS BETWEEN SNOBs AND GALACTIC γ-RAY SOURCES

T. Montmerle and C. J. Cesarsky

*Section d'Astrophysique, Centre d'Etudes Nucléaires de Saclay, France*

ABSTRACT

On the basis of the most recent COS-B data, it is shown that SNOBs are likely to be a class of γ -ray sources. Indeed, 13 SNOBs out of 32 are now seen to coincide with sources, within the angular resolution of COS-B, and 9 to 11 unidentified γ-ray sources out of 26 can be associated with SNOBs. The relativistic electrons present in the SNR generate γ-rays via the inverse Compton effect on the photon field of the associated HII region, and via bremsstrahlung on the surrounding dense medium. While the contribution of the first process is shown to be generally small, bremsstrahlung can account for the observed γ-ray flux. However, in the absence of spectral information, an important $\pi^{\circ}$ decay contribution cannot be ruled out.

## I. SNOBs AND COS-B γ-RAY SOURCES

In recent work, Montmerle [1, 2] defined SNOBs as supernova remnants (SNR) physically linked with OB associations (or HII regions as their tracers), and presumably interacting with them and/or with the surrounding molecular complexes. Evidence for an actual link is established to various degrees of certainty, and this has led one to define different "coincidence classes" : (a) "Class A", directly observed evidence for coincidence, (b) "Class B", indirect evidence, based mainly on distance criteria (for details, see ref. 1). We introduce here (c) "Class X", i.e., SNOBs in which the presence of a SNR is not confirmed, but suspected on various grounds (see § 3, and Tables 3 and 4, of ref. 2).

It has been shown that links probably exist between SNOBs and galactic γ-ray emission, as seen from the COS-B data. The evidence is based on angular coincidences between SNOBs and γ-ray sources or peaks in the COS-B γ-ray map of the galactic disk, with the result that out of 17 unidentified sources, 5 to 7 could be associated with SNOBs (of class A only), and that out of 32 SNOBs (all classes), 18 to 21 were seen, either as a source or as a peak, in the γ-ray data (see ref. 2 and Fig. 1).

Now at the recent Erice workshop "Gamma-Ray Astronomy After COS-B", nine new COS-B sources close to the galactic plane were reported ( [4 ] ; see Table 1). Out of these, four are seen to have SNOB counterparts (see Table 1). Thus updated figures are as follows : out of 26 unidentified γ-ray sources close to the galactic plane, 9 to 11 may be associated with SNOBs ; conversely, 13 out of 32 SNOBs (all classes) can be now associated with sources, and a further 7 with peaks of γ-ray emission. It is remarkable to note that the ratio (SNOB candidates)/(unidentified sources) has remained approximately constant (between 1/3 and 1/2) as new COS-B sources were being discovered (see the catalogues released at Frascati, Kiel and Erice).

Table 1

New COS-B γ-ray sources$^{(*)}$

| ℓ | b | error radius | proposed SNOB identification | Coincidence class |
|---|---|---|---|---|
| 6.7 | - 0.5 | 1.0 | W28 + HII | A |
| 13.5 | + 0.5 | 1.0 | W33 | X |
| 36.5 | + 1.5 | 1.0 | - | |
| 54.2 | + 1.7 | 1.0 | - | |
| 235.5 | - 1.0 | 1.5 | - | |
| 321.0 | - 1.2 | 1.0 | RCW89 + HII | B |
| | | | (G321.9 - 0.3) + HII | B |
| 342.5 | - 2.5 | 1.0 | - | |
| 356.5 | + 0.3 | 1.0 | - | |
| 359.5 | - 0.5 | 1.0 | (G0.0 + 0.0) + HII | A |

$^{(*)}$close to the galactic plane. One further source, at ℓ = 10.5, b = - 31.5, completes the list of ten new sources reported by the Caravane collaboration at the Erice Workshop [4].

## II. INTERPRETATION

In addition to reporting new γ-ray sources at the Erice workshop, the Caravane collaboration gave fluxes above 100 MeV for sources known previously by their position only. Thus, in the SNOB context, one disposes now of 8 objects (instead of 3), the possible γ-ray flux of which is known (see Table 2). This allows more comprehensive calculations than in ref. 8, while retaining the same simplifying assumptions. (In what follows, the reader is referred to that paper for details.)

In the present paper, we fill focus on γ-rays produced by the relativistic electrons present in the SNR, either by the inverse Compton effect on the photon field of the associated HII region, or by bremsstrahlung on the ambient (dense) medium. (In the absence of precise spectral information on γ-ray sources, the third relevant process, π° decay, will not be dealt with, since it involves many unknown parameters.)

a) Inverse Compton effect. The photon density in SNOBs is not well known. Indeed, because of the presence of dust, the transfer of photons through the HII region is not fully understood. Nevertheless, it is possible to avoid a number of difficulties by estimating upper limits to the photon flux, hence to the γ-ray flux.

The basic quantity for estimating the photon field is the flux $N_c^*$ (ph. $s^{-1}$) of Lyman continuum photons from the ionizing stars and being transferred through the HII region. It can be found either from radio observations (continuum or recombination line), or from a knowledge of the characteristics of the ionizing stars. We have taken the numbers, for radio HII regions (and upper limits when they are not listed but correspond to known O and B stars) from ref. 7.

Then, assuming that the photon field completely fills the synchrotron-emitting volume, it can be shown, from ref. 8, that the γ-ray flux due to the inverse Compton effect is :

$$\Phi_c(>E_\gamma) < \frac{\delta}{|2\alpha|} S_\nu B_5^{1-\alpha} \frac{\nu^{-\alpha}}{\zeta(\alpha)} N_{49}^* \theta^{-2} d^{-2} [<\epsilon>/E_\gamma]^{-\alpha} K_d 10^{-7.23+0.48\alpha} \quad (1)$$

Table 2

Gamma-ray emission from SNOBs

| | SNOB | W28 | IC443 | Carina | RCW89 | G321.9-0.3 | Kes27 | RCW103 | G335.2 + 0.1 |
|---|---|---|---|---|---|---|---|---|---|
| (1) | $\ell$, b ($\gamma$-ray source) | 6.7 - 0.5 | 189.0+1.0 | 288.5-0.5 | 321.0-1.2 | | 327.5-0.5 | 333.5 + 0.0 | |
| | $\Phi_{\gamma,\,obs}(10^{-6}cm^{-2}s^{-1})$ | 2.4 | 1.4 | 1.6 | 1.3 | | 2.2 | 3.8 | |
| (2) | $\alpha_{SNR}$ | - 0.38 | - 0.40 | -0.60~-0.40 | - 0.34 | - 0.31 | - 0.61 | - 0.55 | - 0.46 |
| | $S_{408}$(Jy) | 460 | 230 | 80(at 30 MHz) | 94 | 18.3 | 58 | 44 | 27.1 |
| | $\theta$(') | 49 | 40 | < 25 | 25.8 | 24.2 | 21 | 9.4 | 18.6 |
| (3) | d(kpc) | 1.5 | 2.8 | 2.6 | 5.1 | 4.2 | 5.5 | 3.9 | 4.6 |
| (4) | $N^*_{49}$ (ph. $s^{-1}$) | < 8 | < 8 | 21.6 | 10.3 | 13.4 | < 8 | < 8 | < 8 |
| (5) | $\frac{\Phi_c}{\Phi_{\gamma,\,obs}} \times B_5^{1-\alpha}$ < (a) | 0.075 | 0.026 | 0.010~0.023 | 0.012 | 0.011 | 0.0054 | 0.022 | 0.0018 |
| | (b) | 0.83 | 0.32 | 0.29 ~ 0.46 | 0.11 | 0.090 | 0.26 | 0.70 | 0.033 |
| (6) | $n/B_5^{1-\alpha}$ | 54 | 60 | 576~582 | 153 | 787 | 123 | 411 | 1250 |

Notes to Table 2. (1) From ref.4. (2) From ref. 5. $\alpha$ not known for Carina ; flux at 30 MHz from ref. 6. (3) From ref. 1. (4) From ref. 7. The quoted upper limits correspond to HII regions not listed, hence not classified as giant HII regions. Note that these limits are a factor 5 below the figure used in ref. 8. (5) Upper limit to the inverse Compton contribution, obtained by maximizing the photon flux from the ionizing stars (see text, § IIa). (a) Normally expected ; (b) In the extreme case in which all Lyman photons are converted into IR radiation. (6) If $\gamma$-ray emission is due solely to bremsstrahlung (see text, § IIb).

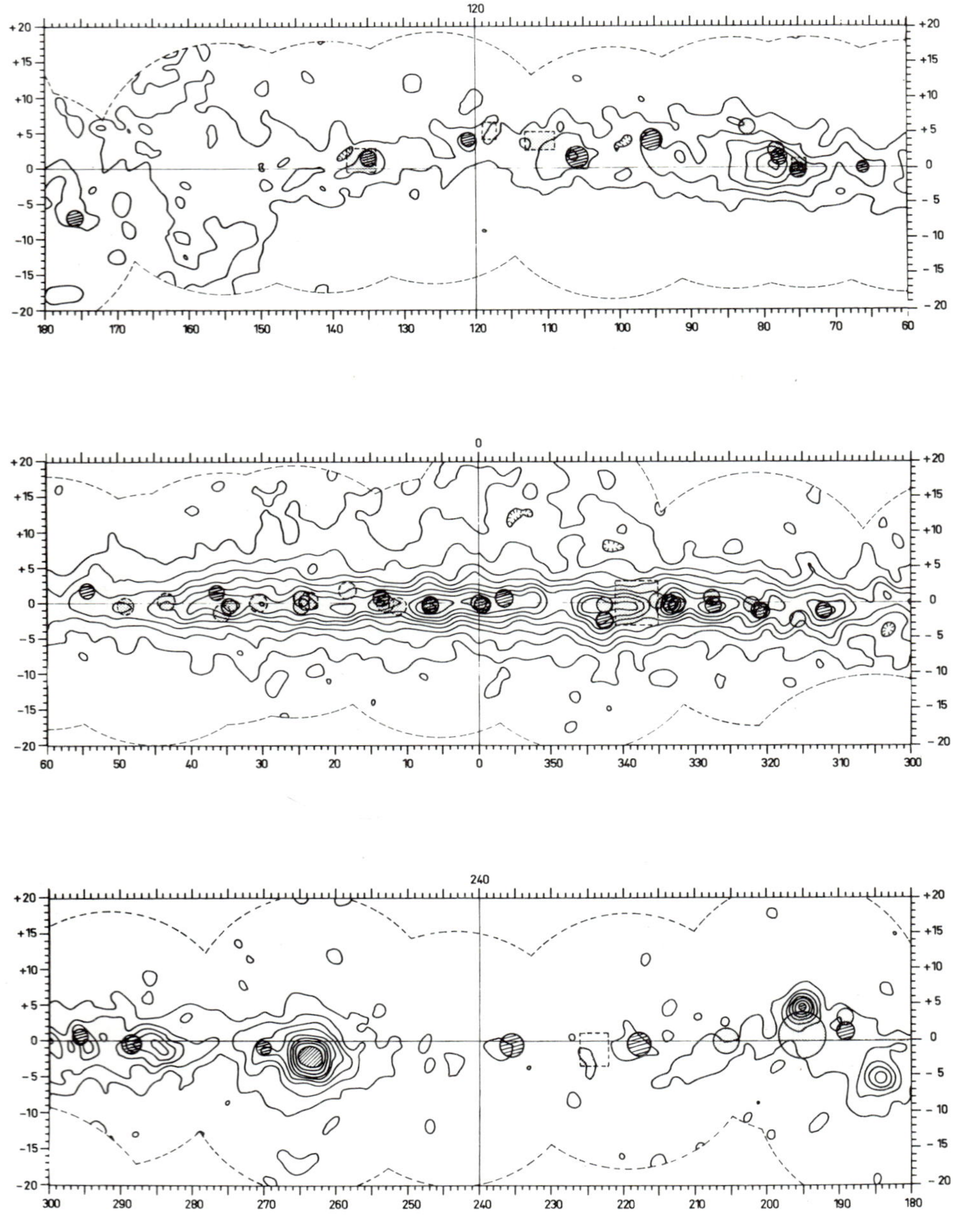


Fig. 1. The 2-dimensional map of the γ-ray sky, as derived from the COS-B data [3 ]. Hatched circles : unidentified γ-ray sources [4 ]. Circles : SNOBs of classes A and B. Dotted circles : HII regions displaying a mixture of thermal and non-thermal radio spectra. Dotted rectangles : HII regions or OB associations in which the presence of a SNR is suspected.

In equation (1) the synchrotron flux density $S_\nu$ (in Jy) is proportional to $\nu^\alpha$ ($\nu$ in GHz); the inverse Compton $\gamma$-rays above $E_\gamma$ (in units of 100 MeV) result from the interaction of relativistic electrons in a magnetic field $B_5$ (in $10^{-5}$ G) with photons of average energy $<\epsilon>$(in eV), associated with a stellar Lyman-continuum flux $N^*_{49}$ (in units of $10^{49}$ ph.s$^{-1}$) and permeating an HII region of angular diameter $\theta$ (in minutes of arc) at a distance d (kpc); the factor $\zeta(\alpha)$ comes from the theory of synchrotron radiation (see ref. 9). The factor $\delta$, on the other hand, takes into account the spectral type of the stars ($\delta = 2.7$ for O6 stars). The factor $K_d$, which is discussed below, takes into account the possibility that the photon field may be increased because of dust or by trapping. (The numerical results are given in Table 2.)

(i) If there is little dust, then a representative value of $<\epsilon>$ is $\epsilon_L = 13.6$ eV. The possibility of trapping of Ly-$\alpha$ radiation due to resonant scattering in the gas, brings a contribution $\Phi_{c,L}$ to add to eq.(1), (in which $K_d = 1$). It can be shown that the corresponding $K_d$ factor is $K_{d,L} = ag/\delta$, a being the effective photon pathlength, and g the proportion of Lyman continuum radiation absorbed by dust. Since a <10, and g $\simeq$ 0.5, one has, at most, $\Phi_{c,L} < 2\ \Phi_c$. As shown in Table 2 (case a), this is well below the observations.

(ii) If there is much dust, all Lyman-continuum photons are absorbed "on-the-spot" by the dust, hence $K_d = 1$, and re-radiated in the infrared, at an average energy $<\epsilon> = \epsilon_{IR}$ corresponding to a wavelength $\sim 50\ \mu$. This is case (b) of Table 2.

As a result, for reasonable values of the magnetic field, the upper limit to $\Phi_c$ is below observations in all cases. While more detailed calculations are still to be performed, it appears that the inverse Compton contribution to the $\gamma$-ray flux in SNOBs is generally small. The figures used here on a greater number of objects thus strengthen our previous conclusions [8].

b) Bremsstrahlung. Assuming that the $\gamma$-rays from SNOBs come entirely from the bremsstrahlung process, one can find a relation independent of the synchrotron-emitting volume V and of the distance which reads [8]:

$$n/B_5^{1-\alpha} = r\ (-2\alpha)^2\ (1/S_\nu)\ \zeta(\alpha)\ 10^{5.66+2.42\alpha}\ f_B/f_n \qquad (2)$$

In this expression, n is the matter density in cm$^{-3}$, r = $\Phi_\gamma$ (sources)/ $\Phi_\gamma$(Crab) above 100 MeV, from ref.[4]; $f_B$ and $f_n$ are the filling factors of the magnetic field and of the matter, respectively, with respect to V.

Then, for $f_B \simeq f_n$ and $B \simeq 10^{-5}$G, Table 2 shows that n must be typically in the range $\sim 10^2$ to $10^3$ cm$^{-3}$. Such values are likely to be expected in the vicinity of SNOBs; they are close to the lower limits one has from the estimates of the r.m.s. value of the electron density in some of the corresponding HII regions (W 28, see ref. 10; W 33, Carina, see ref. 11), or from the values deduced from CO or HI observations near IC 443 [12, 13, 14], for instance.

Thus, it is confirmed that bremsstrahlung appears as a prime candidate to explain the $\gamma$-ray emission from SNOBs, while one cannot preclude a contribution from $\pi^\circ$ decay.

## REFERENCES


1. T. Montmerle, Ap.J., 231, 95 (1979a).
2. T. Montmerle, in Proc. 16th Int. Cosmic Ray Conf.(Kyoto), paper OG 4-20 (1979b).
3. H. Mayer-Hasselwander et al., in Proc. 9th Texas Symposium on Relativistic Astrophysics (Munich), in press (1978).
4. G. Swanenburg et al., presented at the Workshop "Gamma-Ray Astronomy after COS-3" (Erice, Italy, May 17-23) (1979).
5. D.H. Clark and J.L. Caswell, M.N.R.A.S., 174, 267 (1976).
6. B.B. Jones, Austr. J. Phys., 26, 545 (1973).

7. L.F. Smith, P. Biermann and P.G. Mezger, Astr. Ap., 66, 65 (1978).
8. T. Montmerle and C. Cesarsky, in Proc. 16th Int. Cosmic Ray Conf. (Kyoto), paper OG 4-21 (1979).
9. L. Woltjer, Ann. Rev. Astr. Ap., 10, 129 (1972).
10. C. Goudis, Astr. Sp. Sci., 40, 91 (1976).
11. C. Goudis, Astr. Sp. Sci., 47, 109 (1977).
12. R.H. Cornett, G. Chin and G.R. Knapp, Astr. Ap., 54, 889 (1977).
13. L.K. De Noyer, M.N.R.A.S., 183, 187 (1978).
14. R. Giovanelli and M.P. Haynes, Ap.J., 230, 404 (1979).

# BALLOON-BORNE OBSERVATIONS OF NGC 4151 USING THE MISO TELESCOPE

F. Perotti*, A. Della-Ventura*, G. Sechi*, G. Villa*, G. Di Cocco**, R. E. Baker***, R. C. Butler***, A. J. Dean***, S. J. Martin***+ and D. Ramsden***

**Laboratorio di Fisica Cosmica e Tecnologie Relative del C.N.R. Via Bassini 15, Milan, Italy*
***Laboratorio T.E.S.R.E., C.N.R. Via de Castagnoli 1, Bologna, Italy*
****University of Southampton, Dept. of Physics, Southampton, England*
*+ Now at Science Research Council, Appleton Laboratory, England*

ABSTRACT

The Seyfert Galaxy NGC 4151 was observed from balloon altitudes using the MISO low energy gamma-ray telescope on 23rd May 1977. The spectrum was measured over the range 0.04 to 20 MeV and a change in the spectral index was observed at about 2MeV. The data at low energies is in good agreement with other hard X-ray data but at higher energies the data could be consistent with an exponential spectrum having an e-folding energy of 2.1 MeV.

## A BRIEF DESCRIPTION OF THE TELESCOPE

A schematic diagram of the detector which was used for this observation of NGC4151 is shown in Fig. 1. The two scintillators S1 and S2 form the central Compton-coincidence detector. The upper detector S1 is a liquid scintillator (NE 311) and provides the target material in which the incident photons are Compton scattered. Some of the scattered photons interact again in the lower detector S2 which is a large NaI(Tl) crystal. Charged particle events are vetoed by anti-coincidence detectors A1 and A3. The area of this central detector unit is 560 $cm^2$. Each of the elements is 10 cm thick and they are separated by a distance of 3 cm. This Compton-coincidence telescope is only crudely directional and it is therefore shielded from atmospheric $\gamma$-rays by a semi-active shield made by distributing 6 radiation lengths of lead between plastic scintillators. A cylindrical lead collimator inside the shield restricts the aperture to $10^o$ HWHM for $\gamma$-rays in the range 0.2 - 20 MeV. The angular resolution of the telescope is increased in one plane to $1.5^o$HWHM, by the introduction of a slatted collimator. The lower element of the detector is also surrounded by a neutron shield which is designed to reduce the flux of thermal and epithermal neutrons entering the Sodium Iodide crystal. The shield contains a mixture of boron, tungsten and lithium set in a wax moderator.

More details of this telescope system and the results of the calibration tests are reported elsewhere (1). The measured angular resolution of the telescope at 662KeV and 4.43MeV is shown in Fig 2. The telescope was pointed using an alt-azimuth mount.

## OBSERVATION OF NGC 4151

The Seyfert Galaxy NGC 4151 was one of the objects included in the observation programme followed during the first flight of the MISO telescope which was launched

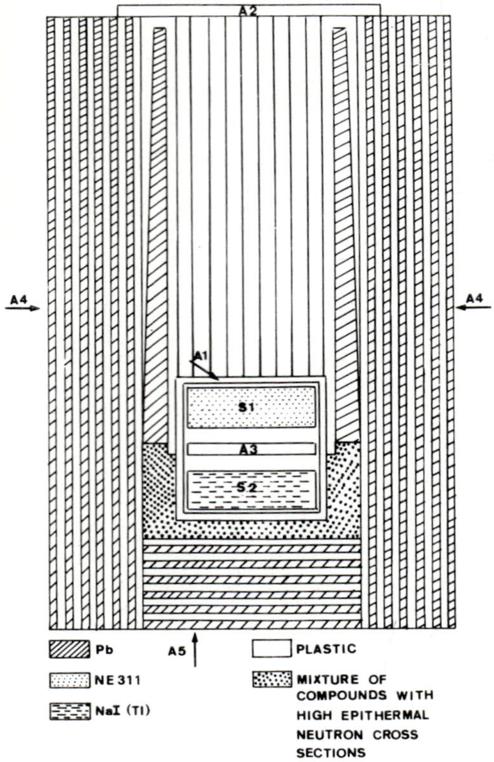


Fig 1. A Schematic View of the MISO Gamma-ray Telescope.

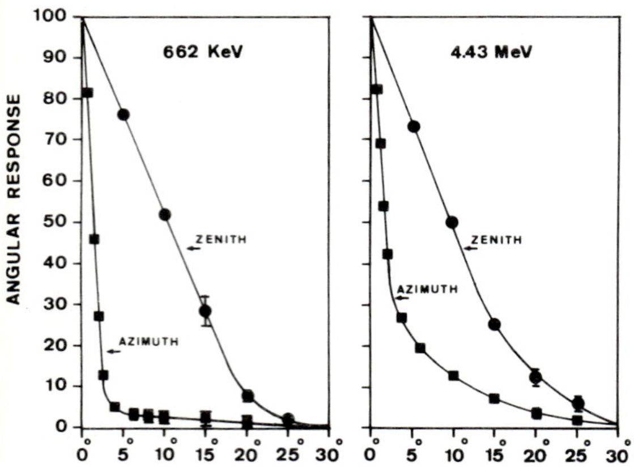


Fig 2. The Measured Angular Response of the MISO Telescope at Two Energies.

from Palestine, Texas on the 22 May 1977. The analysis of the 'quick-look' data obtained from one of the elements of the Compton-coincidence system in the form of the integral counting rate in the range 0.15MeV - 20MeV, demonstrated for the first time that there was γ-ray emission from NGC4151 (2). This integral data was, at that time, interpreted as being an extension of the X-ray spectrum which appeared to have a spectral index of $\alpha = -1$. Our observation of NGC4151 was conducted using two different values for the telescope zenith angle (18°and 7°). The mean float altitude was close to 7 mb during the observation which lasted 3 hours. The observation period was equally divided between source and background measurements. For each zenith angle setting, the observation of the source was both preceded and followed by a background measurement. During the source observation NGC4151 was allowed to drift through the field-of-view of the telescope. The background was determined by pointing towards a region of the sky clear of known X- and γ-ray sources. This direction was not close to the azimuth of NGC4151.

During the period of the source observations, the balloon altitude remained steady at a level close to 7 mb. However, when the background measurements were made, the altitude varied between 6.2 and 8 mb. The data obtained during the OFF-source periods was used to determine the dependence of the background counting rate on the residual atmospheric pressure. During these periods the zenith angle of the telescope was held at a constant value. This data was used to calculate the expected background counting rate during the ON-source period. This was subtracted from the total counting rate due to the source and the background. When the contribution from the events in each of the detectors (S1 and S2) and events involving the coincidence of both detectors (S1.S2) were taken into account, an excess γ-ray counting rate of $7.2 \pm 1.2$ counts/s was obtained. This excess was equivalent to a 6 $\sigma$ variation in the background.

From the telescope attitude sensor data it was possible to compute the angular offset between the telescope axis and the direction to NGC4151 every128 ms.

The γ-ray excess was correlated with this offset angle. The result of this correlation is shown in Fig 3. This data clearly suggests a close relationship with the angular response of the telescope ($3^\circ$ FWHM). The difference between the expected position of NGC4151 and the maximum in the observed counting rate is consistent with the absolute pointing accuracy of the telescope orientation system. The gap between $+1.2^\circ$ and $+1.8^\circ$ in Fig 3 follows from a loss of recorded data during this phase of the observation. As a consequence of the background measurement routine no data was obtained just outside the $3^\circ$ field-of-view of the MISO telescope. The counting rate profile shown in Fig 3 provides good evidence that the γ-ray excess observed during this flight came from a direction very close to that of NGC4151.

The energy-loss spectrum for the observed counting rate excess has been converted to a photon spectrum at the top of the atmosphere using a technique similar to that described by Dolan (8). This procedure takes into account the absorption of photons in the residual atmosphere above the detector, absorption in the instrument window, the re-distribution of the photon energy through Compton interaction and pair production, and the energy resolution of the detector. The systematic errors introduced in deriving a photon spectrum are estimated to be less than 10%.

The photon spectrum obtained by the MISO telescope is shown in Fig 4 along with some other relevant X-ray measurement and the γ-ray spectrum observed by Schönfelder (3). For clarity, the measurements of OSO-7 (4) and the 1976 Ariel-5 data (5) have not been shown since they are essentially identical to the more recent measurements of OSO-8 (6). Also shown are the two sigma upper limits obtained by SAS-2 at high energies (7).

In principle, a gainshift or an azimuth or latitude dependence in the background counting rate could lead to systematic errors in the determination of the intensity of the γ-ray source and it is therefore important to estimate the magnitude of these effects. The temperature of the detector and of the associated electronics was monitored throughout the flight but during the observation of NGC4151 it varied by less than $0.5^\circ C$. The outputs of both the high and low voltage power supplies were stable. The voltage varied by less than 0.1%. Moreover, in-flight calibrations made using a $^{88}Y$ source and the atmospheric annihilation line at 511 keV, set an upper limit on the gainshift of about 0.2%. The azimuth dependence of the background counting rate was found to be negligible compared with the statistical errors. This would be expected since the observations were made using small zenith angles. During the two observations of NGC4151 the geographical positions of the balloon remained practically constant ($\pm 0.2^\circ$). A consideration of these effects suggests that their combined contribution would generate an uncertainty of less than $\pm$ 1 count/sec, that is less than 15% of the measured flux.

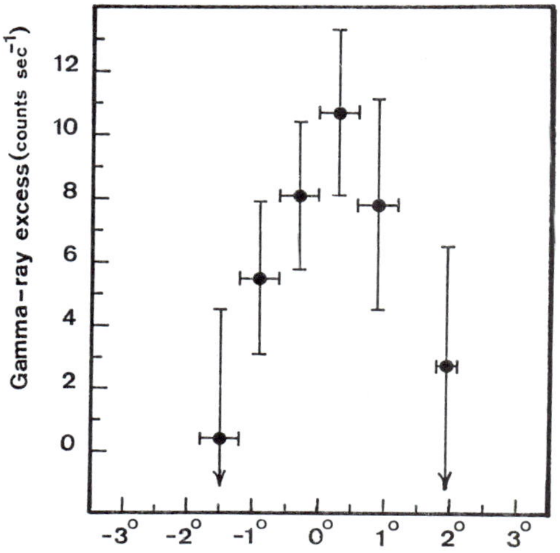

Fig 3. Correlation between γ-ray excess and angle from telescope axis to direction of NGC4151.

## DISCUSSION

The observations reported here were carried out a few days prior to the OSO-8 observation of NGC4151 which spanned the period 26 May to 5 June 1977. During that period there was no statistically significant variation in the spectral index. Also, the OSO-8 data (6) suggests that the source was in a 'high' state of activity when our measurements were made. With this in mind

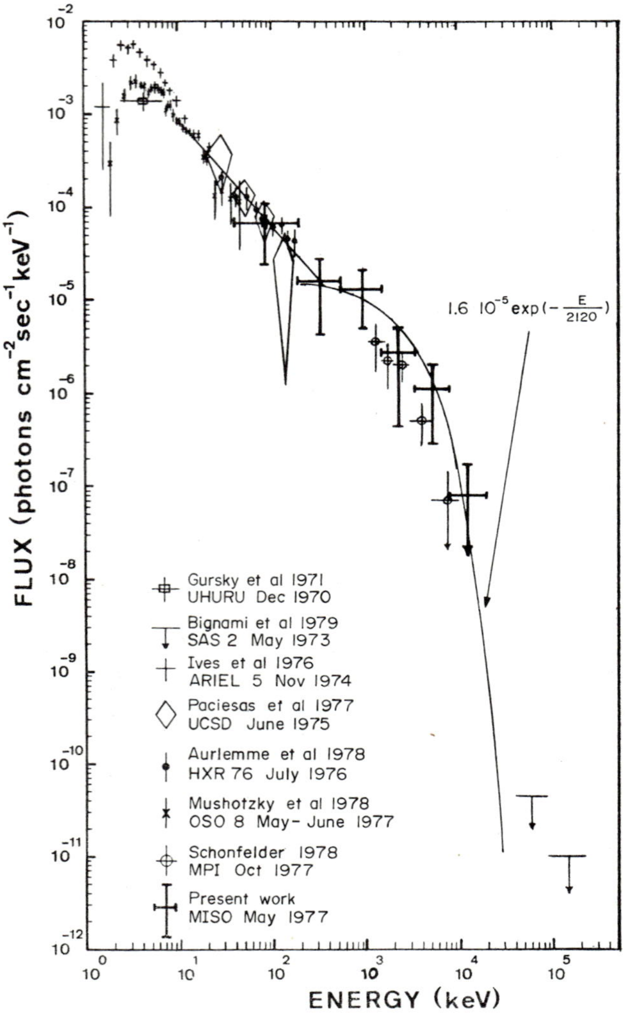


Fig 4. The Photon Spectrum of NGC 4151.

the general agreement between our data and that obtained by Schönfelder (3) may not be real and the differences may be attributable to variations in the intensity of the source. If we take into account the earlier hard X-ray data, the spectrum of NGC4151 up to about 3MeV may be described by a power law having an index of $\alpha = 0.9$. At higher energies, our data along with the upper limits set by SAS-2 (7) suggests that the spectrum steepens rapidly. Above 3MeV the data may be described by a second power law having an index of between 3.4 and 4. The reduced $\chi^2$ value for the two fits is 0.73 with 19 degrees of freedom. If one were on the other hand to fit a single power law to our data and that from the Hard X-ray observations, a reduced $\chi^2$ of 2.3 is obtained for 19 degrees of freedom. This suggests that the change in the spectral index is significant. Alternatively our measurements above 200keV may be represented by an exponential curve of the form, $dI/dE =$
$(1.6 \pm 0.7)10^{-5}\exp(-E/(2120 \pm 480)$
photons cm$^{-2}$sec$^{-1}$keV$^{-1}$. This corresponds to a reduced $\chi^2 = 0.5$ with 3 degrees of freedom. The data between 10 and 200keV would then be best fit by a power law of the form, $dI/dE =$
$(1.1 \pm 0.3)10^{-2}\ E^{-(1.1 \pm 0.1)}$
photons cm$^{-2}$sec$^{-1}$keV$^{-1}$. This corresponds to a reduced $\chi^2 = 1.1$ with 23 degrees of freedom. These two fits are shown in Fig 4.

REFERENCES


1. R.E. Baker et al. Nucl. Instr. & Meth. 158 (1979)p.595
2. G. di Cocco et al. Nature 270 (1977)p.319
3. V. Schönfelder Nature 274 (1978)p.344
4. W.A. Baity et al Ap.J.(Letters)199 (1975) L5
5. P. Barr et al Mon.Not.R.astr.Soc. 181 (1977) L115
6. R.F. Mushotzky et al. Ap.J.(Letters)225 (1977)L115
7. G.F. Bignami et al. NASA Technical Memorandum 80269
8. J.F. Dolan Astrophys.Sp.Sci.17,472

# ULTRA HIGH ENERGY GAMMA RAYS

P. V. Ramana Murthy

*Tata Institute of Fundamental Research, Homi Bhabha Road, Bombay 400 005, India*

## ABSTRACT

Though celestical sources of gamma-rays have been detected to emit gamma-rays upto energies of a few GeV in balloon and satellite borne experiments, the fluxes are so low even at these energies as to make it impossible to extend the energy spectrum to energies beyond 10 GeV using the same techniques. Ground based experiments using atmospheric Cerenkov technique make it possible to extend the gamma-ray energy spectrum to energies beyond 100 GeV. Besides the traditional celestial sources (e.g. SN remnants and pulsars) one has to also consider the possibility of Primordial Black Holes being sources of high energy gamma-rays which they generate through the Hawking process. Experiments carried out over the years at many laboratories including our own to detect high energy gamma-rays are summarised and the implications of the results discussed.

## INTRODUCTION

Gamma-ray astronomy has come into its own during the past decade, occupying a place of importance alongside the astronomies at the other wavelengths in the electromagnetic spectrum. High energy gamma-rays, being neutral, are unaffected by the interstellar and intergalactic magnetic fields and, therefore preserve the sense of direction of their travel from the source to us. This knowledge of the angular coordinates of the source coupled with evidence gathered from studies at other wavelengths of the electromagnetic spectrum helps us in most cases to identify the source. The kind of information the gamma-rays provide us on the particles, fields and processes in the distant regions in space strongly depends on the energy range of gamma-rays. The very low fluxes of gamma-rays render the experiments very difficult even at energies as low as 1 GeV. This difficulty is further enhanced as the gamma-ray energy increases; for, their flux is a decreasing function of energy. Nevertheless one pursues undauntedly the experimental studies of ultra high energy gamma-rays at energies greater than 100 GeV with the hope of understanding the topics listed above and also to gain information useful in the broader attempt of understanding the origin of cosmic rays.

The first attempt to detect celestial ultra high energy gamma-rays using atmospheric Cerenkov technique was carried out nearly 25 years ago by Jelley and Galbraith [1]. Since then there were many experiments carried out in many countries, most of them yielding only upper limits to the gamma-ray fluxes. There is strong evidence, however, that some sources e.g. the Crab pulsar, Centaurus A and Cygnus X-3 do emit high energy gamma-rays. The recent phenomenal success of the COS-B satellite in confirming and/or discovering as many as 30 discrete gamma-ray sources [2, 3] in the energy range 50 MeV to a few GeV provides an added impetus to investigate if these sources also emit gamma-rays at energies beyond 100 GeV. It is also important to undertake sky surveys for new sources of high energy gamma-rays irrespective of whether they were seen at other wavelengths.

## PRODUCTION PROCESSES

Supernova remnants, pulsars, galactic centre, galactic plane, other galaxies and primordial black holes are considered to be potential sources of high energy gamma-rays. Perhaps distinctly different mechanisms are in operation in the various sources. Theoreticians are handicapped by not knowing all the relevant parameters in predicting the high energy gamma-ray fluxes with certitude. Even the enormous and impressive theoretical activity centered on pulsars during the past decade has not led to a universally accepted theory of emission of radiation by pulsars. Many conclusions are contradictory to each other. The situation is perforce difficult often involving new concepts in the frontiers of physics. Our experience with the laboratory physics is plainly of no help in dealing with the bewildering ranges of magnetic fields and densities encountered in the pulsar setting. Rather than discuss any particular model, I will merely list below some of the physical processes important to the production of high energy gamma-rays. This subject has been discussed by several authors in the past; see, for example, Fazio [4].

Meson Production. Inelastic collisions of high energy protons with interstellar matter lead to production of neutral pions which decay into gamma-rays. Gamma-ray energy is typically an order of magnitude lower than that of the proton and the intensity proportional to the product of proton flux and the matter density.

Bremsstrahlung. High energy electrons produce high energy gamma-rays in their interaction with matter by bremsstrahlung. Gamma-ray energy is a large fraction of that of the electron. The production rate is proportional to the product of electron flux and the matter density.

Inverse Compton Process. High energy electrons interact with ambient low energy but abundant photon flux raising the photon energy to very high values by inverse Compton process. Gamma-ray energies are comparable to those of the electrons if $E \gg (mc^2)^2/\epsilon$ where $E$ and $mc^2$ are the energy and the rest mass of the electron and $\epsilon$ the energy of the ambient photon.

Synchrotron Radiation. High energy electrons emit high energy gamma-rays in their interaction with magnetic field by synchrotron process. The photon energy is usually several orders of magnitude lower than that of the electron.

Curvature Radiation. This radiation is unique to the pulsar environs. In some of the pulsar models, high energy electrons accelerated very close to the neutron star surface move out of the polar caps along the magnetic field lines. Since the electrons are forced to glide along the field lines which are curved, they suffer acceleration resulting in the emission of high energy gamma-rays. In some models the gamma-ray energies can be as high as $10^{13}$ eV.

Hawking Process. According to Page and Hawking [5], it is conceivable that hot and highly compressed regions believed to exist shortly after the 'big bang' might have collapsed into black holes. These objects are called Primordial Black Holes (PBH) as distinct from stellar black holes. The temperature of a PBH is given by $10^{26}$ $M^{-1}$ $^{o}K$ where M is the mass of the PBH in grams. The PBH in the mass range $10^{14}$ to $10^{15}$ gms would have survived until the present epoch. Their temperatures are as high as $10^{12}$ $^{o}K$ and the event horizon radii as small as one fermi or less. At such small confines quantum mechanical effects come into play and lead to particle emission by PBH through spontaneous creation of pairs of particles near the event horizon. As the particle emission progresses the mass of the PBH decreases raising its temperature which leads to even more copious production of particles. The system is thus clearly in a run-away state; for details, see Page and Hawking [5] and Carr [6]. The PBH with masses $\sim 5.10^{14}$ gm, should they exist, must be evaporating at this epoch so fast as to effectively explode. One does not yet know enough of particle physics to predict in detail the nature, energies and time-scale of particle emission. Depending on the particle physics models, it is estimated that each exploding PBH emits among other radiations either $\sim 10^{30}$ photons of energy $\sim 10^{12}$ eV over a period of 0.1 sec or $\sim 10^{37}$ photons of energy $\sim 250$ MeV over 100 nanoseconds. As noted by Page and Hawking, a definite observation of gamma-rays from a PBH would indeed be a tremendous vindication of General Relativity and Quantum Theory and would give us important information about the early Universe and strong interaction physics at high energies that probably could not be obtained in any otherway.

We close this section with a general observation. Whatever may be the ultimate model of high energy gamma-ray emission, one cannot help marvel at the apparently very high efficiencies of the particle acceleration and radiation emission processes operating in the celestial settings as compared with those achieved by humans. In Table 1 we show the production efficiencies of gamma-rays, defined as the ratio of gamma-ray luminosity of a pulsar to its mechanical energy loss rate, for 4 pulsars. Similar, though not identical, values of efficiency were published by Buccheri et al. [7]. Two assumptions are made here: (a) the ultimate energy source in a pulsar is its rotational kinetic energy and (b) the gamma-ray emission is isotropic. The efficiency, in principle, has to be always less than 1, perhaps by several orders of magnitude. It is seen that for some pulsars the principle of energy conservation seems to be violated. We must hasten to add that the effects of beaming (instead of isotropic emission) and the uncertainties in distances and moments of inertia may shift the efficiency figures shown in the last column of Table 1 lower by two orders of magnitude or up by an order of magnitude. Even then these efficiencies are

Table 1 Gamma-ray Conversion Efficiencies of Pulsars

| Pulsar | Relative* Intensity | Gamma-ray Energy Flux, KeV/cm$^2$ s $\varphi$ | Gamma-ray Luminosity $L_\gamma$ ergs/s $L_\gamma = 4\pi D^2 \varphi$ | Mechanical Energy-loss Rate, $\dot{E}$ $\dot{E} = I\omega\dot{\omega}$ ergs/s | Gamma-ray Conversion Efficiency |
|---|---|---|---|---|---|
| PSR0531+21 | 1.0 | 0.8 | $6.2.10^{35}$ | $4.7.10^{38}$ | $1.3.10^{-3}$ |
| PSR0833-45 | 3.3 | 2.6 | $1.3.10^{35}$ | $7.1.10^{36}$ | $1.8.10^{-2}$ |
| PSR0740-28 | 0.4 | 0.3 | $4.2.10^{35}$ | $1.4.10^{35}$ | 3 |
| PSR1822-09 | 0.8 | 0.6 | $5.2.10^{34}$ | $4.6.10^{33}$ | 11 |

*Values taken from Reference 3. Values of D, $\omega$ and $\dot{\omega}$ are taken from Taylor and Manchester [8].

higher 1/ than or, at the least, comparable with the terrestrial standards $\sim 10^{-3}$ at modern proton-synchorotrons achieved only by meticulous care in applying the electrical engineering technology. Perhaps a basic ingredient is missing from all the theories considered so far. In any case it is prudent for the experimental groups to be willing to be guided but not inhibited by theorists in their quest for high energy gamma-rays.

## ATTENUATION PROCESSES

Once they exit from the source region, the ultra high energy gamma-rays get attenuated very little in the interstellar or intergalactic regions. The most important process of attenuation of high energy gamma-rays is the electron pair-production as a result of their interaction with the ambient photon flux. Fig. 1, taken from Gould and Schreder [10], shows the attenuation coefficient for high energy gamma-rays as a function of energy. It is seen that the Universe is by and large transparent to high energy gamma-rays out to distances of $10^9$ ly except at energies beyond a few times $10^{14}$ eV. Even here one will be able to detect high energy gamma-rays from galactic sources out to distances $\sim 10^4$ ly.

1/ Some years ago we offered a similar argument (Chatterjee et al [9] ) to convince ourselves why one should not expect to detect ultra high energy gamma-rays from pulsars with the sensitivities then possible. Now that one has actually detected radiations from pulsars implying very high conversion efficiencies, one should perhaps be no longer inhibited by efficiency arguments in one's attempts to detect high energy gamma-rays from pulsars.

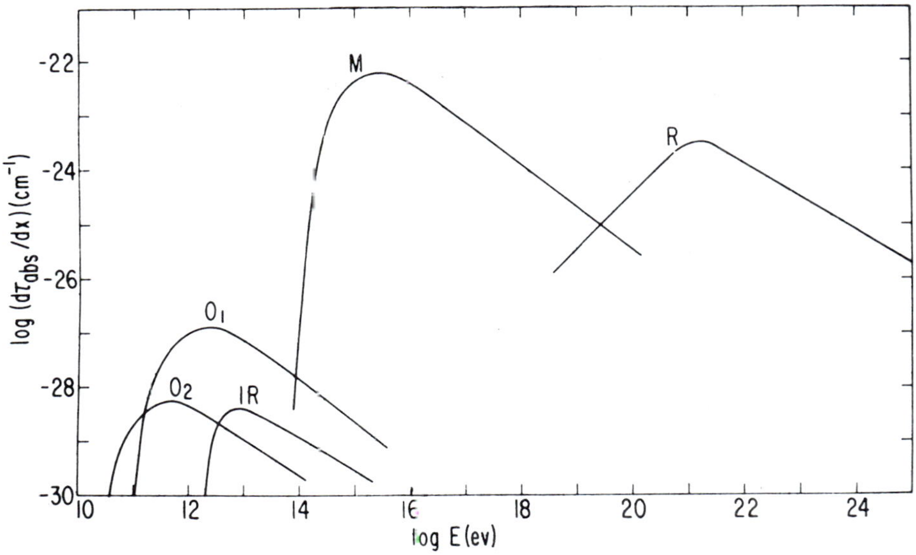


Fig. 1. Absorption coefficient vs. energy for high energy photons due to pair creation in interactions with various components of the background radiation; taken from Gould and Schreder [10]. The curve labelled M represents the thermal microwave radiation, R the non-thermal radio spectrum, IR the infrared and $0_1$ and $0_2$ the optical radiation from cool and hot stars.

An important process of attenuation of high energy gamma-rays operates in the source region very near the surface of the pulsars. Although high energy gamma-rays are produced in abundance by the curvature radiation mechanism mentioned in the previous section, they suffer enormous energy degradation in their interaction (leading to pair-production) with the intense magnetic and electric fields in the pulsar environment; for example Massaro and Salvati [11] estimate that the degradation in the high energy gamma-ray luminosity can be as much as 2 to 4 orders of magnitude at energies $\sim$ 100 GeV.

## EXPERIMENTAL TECHNIQUES

Atmospheric Cerenkov Technique. This technique is useful for gamma-ray studies in the energy range 100 to $10^5$ GeV. All the experiments on gamma-rays (in this energy range) conducted so far are ground based and that, for a good reason. We will illustrate with an example why this is so. Consider the time averaged gamma-ray integral energy spectrum from the Crab Pulsar; it is of the form $2.5 . 10^{-7} E^{-1.09}$ $cm^{-2}s^{-1}$ (with E in GeV) based on observations up to a few GeV. Even if one assumes that the spectrum does not steepen up to 100 GeV, the flux beyond 100 GeV, is $1.7 . 10^{-9} cm^{-2} s^{-1}$. Assuming a 10 hour balloon flight with a rather large area detector of 2.5 m diameter, comparable to the one used by McBreen et al [12], one receives just 3 gamma-rays during the entire flight. Though the exposure can be higher in a satellite-borne experiment, the detector will be much smaller in area, again leading to only a few gamma-rays. On the other hand, with the atmospheric Cerenkov technique, one can make use of the fact that the collection area can be several orders of magnitude higher than the physical area of the detector as will be explained below. Taking a collection area of 100 m radius and 50 hours of observation, the same flux will result in nearly $10^5$ gamma-rays from Crab pulsar. This number is large enough to be detected even in the presence of the rather large cosmic ray air shower background. The atmospheric Cerenkov technique has no competition at all for detecting celestial ultra high energy gamma-rays in the range $10^{11}$ to $10^{14}$ eV.

We will mention here only a few salient features of this technique; for more information, see a recent and detailed review article by Porter and Weekes [13]. A high energy gamma-ray on entering the earth's atmosphere initiates an electromagnetic cascade shower. The more energetic electrons in this shower which are above the Cerenkov threshold (20 to 75 MeV depending on the depth in the atmosphere) produce low intensity Cerenkov light. Though the electrons and photons in the cascade are absorbed in the atmosphere, the optical Cerenkov light penetrates the atmosphere reaching the detectors at mountain altitude or at sea-level. The Cerenkov light pool created by a 300 GeV gamma-ray, for example, may contain several million photons spread over a few nano-seconds in time and over distances of the order of several hundred metres from the axis. If the field of view is restricted to a cone of $1^o$ diameter, the photon density $\sim 7/m^2$ near the core dropping off drastically at distances beyond 100 m. A single mirror of only a few $m^2$ in area looked at by a photomultiplier set up at a suitable threshold can thus detect gamma-rays incident over $\sim 10^5$ $m^2$ in area. Detailed Monte Carlo calculations are performed by Browning and Turver [14] on the Cerenkov photon density distributions. Since the amount of Cerenkov light is very small, the experiment can be carried out only during clear moonless nights. The celestial source may be practically a point as seen from the earth; yet the Coulomb scattering of the electrons of the cascade in the atmosphere and the finite emission angle of the Cerenkov light spoil the resolution to make the source look like a diffused object of $1^o$ to $2^o$ in diameter. The weak Cerenkov light is focussed by a parabolic mirror on to a fast photo-multiplier. There is usually a mask at the photomultiplier cathode to limit the view to a cone of $1^o$ to $3^o$ in diameter.

The Night Sky Background, NSB, ($\sim 7 \times 10^7$ photons/$cm^2$ sec. ster. when the view does not include any bright star) and the atmospheric Cerenkov radiation produced by the Extensive Air Showers (EAS) initiated by charged primary cosmic rays constitute the background to the experiment. NSB is reduced first by limiting the view to a very small solid angle around the celestial object of interest and then it is almost eliminated by taking fast ($\sim$ 5 to 10 ns) coinci-dences between several photomultipliers with small integration times. NSB can also be eliminated by demanding a majority logic among the several photo-multipliers or adding the signals from the photomultipliers and then biasing off at a suitable discrimination level. With the NSB almost eliminated, the EAS is the only remaining background to the gamma-ray experiments, which is not easy to remove. In the energy range of our interest ($10^{11}$ to $10^{13}$ eV) the EAS contain too few muons and hadrons to be used as a veto signal. The lateral distribution of the Cerenkov light in the two cases, gamma-rays and EAS, do not significantly differ from each other. Grindlay [15] made use of the Cerenkov light produced by muons in the EAS (second maximum) as a veto to decrease the contribution from EAS. This technique, however, works only at energies $> 10^{12}$ eV; for, there are too few muons in the EAS to produce Cerenkov light below $10^{12}$ eV. It must be remembered that there are no standard candles (high energy gamma-ray sources) in the sky to calibrate the detector systems. Though troublesome as a background, the EAS produced Cerenkov radiation is useful in the sense that it can be used to calibrate the detector systems. There are two factors operating against the EAS background: (a) Even at energies $> 10^{12}$ eV not all the energy of the cosmic ray primaries goes into electromagnetic cascades and (b) as the energy decreases below $10^{12}$ eV, cosmic rays are more and more inefficient in producing Cerenkov light. The reasons for the latter are (i) the emission angle of pions at lower energies spreads out the Cerenkov light to larger distances lowering the photon density at a given distance and (ii) much of the energy is lost to the EM cascades by $\pi \rightarrow \mu$ decays. The progressively increasing advantage gamma-rays enjoy over cosmic rays (with decreasing energy) is clearly demonstrated in Fig. 2 taken from Weeks and Turver [16]. Since there is independent evidence that cosmic ray primaries at energies $10^{11}$ to $10^{14}$ eV are isotropic, it is hopeless to detect isotropic diffuse gamma-ray flux at these energies using this technique. One can use this technique therefore only to detect point sources by one of the two methods: (a) Drift scan, in which one compares the emission from source region with that from a neighbouring region and (b) Tracking, in which one keeps the source region all the time in the view of a detector and examines if the detected signals have any periodicity as expected in the case of pulsars.

The EAS rate is proportional to the product of the collection area, A, and to the solid angle of view, $\pi \theta^2$ ($\theta$, small) while the gamma-ray (signal) rate is proportional to A. If the experiment operates for a time T, then the standard deviation for the EAS background = $(A\,T\,\theta^2)^{\frac{1}{2}}$. The excess counts in the source direction are proportional to A T. The signal-to-noise ratio, then, is proportional to $AT/(A\,T\,\theta^2)^{\frac{1}{2}} = A^{\frac{1}{2}}\,T^{\frac{1}{2}}\,\theta^{-1}$. To sharpen the gamma-ray signal, one has to maximise the observation time T and decrease $\theta$, the opening angle of view. There is a limit, $\sim 1^o$ in diameter, below which one cannot decrease the angle; for, the fraction of Cerenkov light collected and consequently the detection efficiency falls very rapidly below this value. To maximise the collection area, A, for a given gamma-ray energy, one decreases the Cerenkov photon threshold, by using large area mirrors. This has also the more important effect of lowering the high energy gamma-ray threshold. If the source is pulsating with its period known from experiments at other frequencies beforehand, sensitivity of a given detector system is improved by a factor of 5 to 10 compared to the one for a steady source. Any search for pulsed gamma-rays can be meaningful only if one has an accurate knowledge of P and $\dot{P}$ from measurements at other frequencies at an epoch as close to the epoch of gamma-ray experiment as possible.

There have been several experiments performed in the past using this technique; see Porter and Weekes [13] for a complete survey. We will illustrate some details of this kind of experiment by citing the one at Ootacamund (Bhat et al.[17])

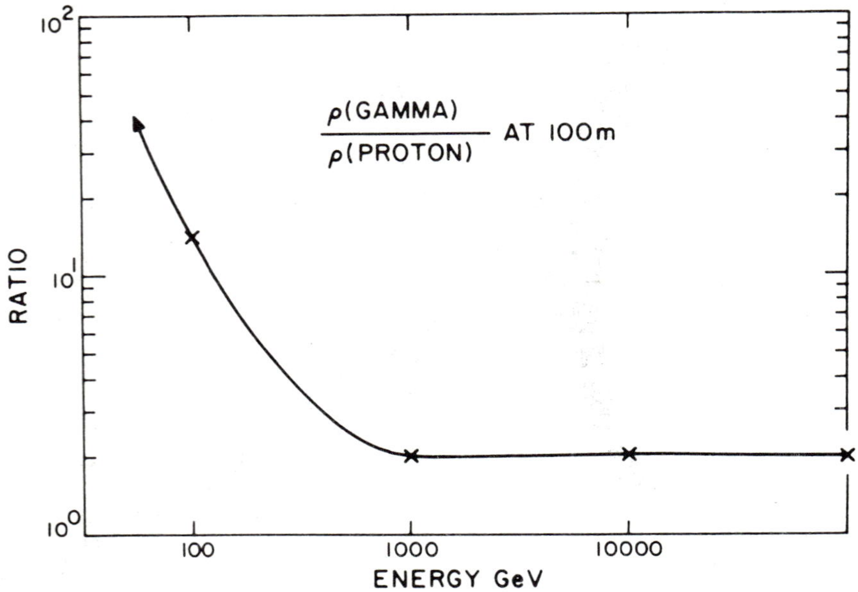


Fig. 2. The ratio of Cerenkov photon densities, ρ (GAMMA)/ ρ (PROTON), at 100 m from axis (for gamma and proton initiated cascades) is shown as a function of energy of the initiating particle; taken from Weekes and Turver [16].

to look for pulsed high energy gamma-rays from pulsars. The experiment consists of 8 large (1.5 m diam.) and 10 small (0.9 m diam) closely spaced and equatorially mounted parabolic mirrors each having a fast photomultiplier mounted at its focus; see Figures 3 and 4. Signals from 4 to 5 photomultipliers (amounting to mirror area of $\sim$ 4.8 $m^2$ for each bank) are added, discriminated and put in fast coincidence with the signals from three other similar banks.

All the mirrors are trained at and made to track the pulsar (typically for 2 to 4 hours) under investigation. The time of each event derived from an accurate clock is recorded on a magnetic tape to a precision of 0.1 ms. The event times are later analysed for the periodicity appropriate to the pulsar. The sensitivity of our experiment is at a level of $\sim 1 \times 10^{-11}$ quanta $cm^{-2}sec^{-1}$ (time averaged flux) at a gamma-ray energy threshold of 250 GeV. We will present the very preliminary results from this experiment along with those from the others in the next section.

## RESULTS

Primordial Black Holes. Results bearing on PBH from high energy gamma-ray and other experiments are summarised in Table 2. Todate there is no positive evidence that PBH do indeed exist and some are currently exploding in the manner we think they should be. One is therefore able to only set some upper limits.

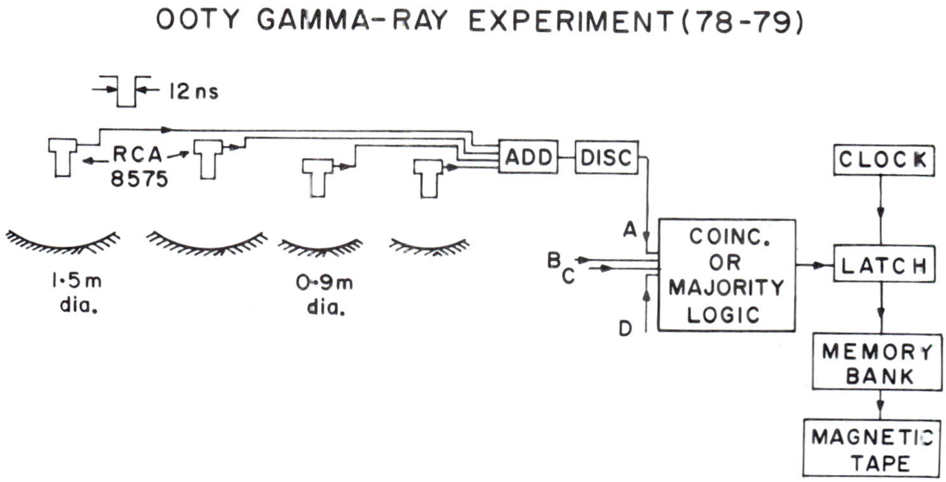


Fig. 3 Schematic of the gamma-ray experiment being carried out by Bhat et al [17] at Ootacamund, India. Total mirror area of each of the 4 banks is 4.8 $m^2$. Only bank A is shown in figure; other three banks are similar.

Fig. 4 A photograph of the gamma-ray experiment of Bhat et al [17]. (Photo by A.R. Apte)

TABLE 2 Upper Limits on PBH Explosions

| Method | Reference | Upper Limit Explosions $yr^{-1} pc^{-3}$ |
|---|---|---|
| Indirect | | |
| From diffuse gamma-ray background | Page and Hawking [5] | $0.4.10^{-6}$, if uniformly spread. 0.4, if confined to galactic halos. |
| Direct | | |
| Composite Particle Model | | |
| Gamma-ray flash $E_\gamma \sim 250$ MeV | Porter and Weekes [18] | 0.4 |
| Radio flash | Phinney and Taylor [20] | $2.10^{-9}$ |
| Elementary Particle Model | | |
| Atmosph. Cerenkov Technique ($E_\gamma \sim 1.5.10^{12}$ eV) | Porter and Weekes [21] | $3.10^4$ |
| Particle Detectors (see-level $E_\gamma \sim 5.10^{13}$ eV) | Fegan et al. [22] | $6.10^3$ |
| Particle Detectors (800 g. $cm^{-2}$ $E_\gamma \sim 5.10^{13}$ eV) | Bhat et al. [23] | $3.10^3$ |

From the observed value of diffuse gamma-ray background, Page and Hawking [5] have been able to set an upper limit of $0.4 \times 10^{-6}$ $yr^{-1}$ $pc^{-3}$ to PBH explosion rate if they are dispersed uniformly throughout the Universe. This limit could go up to 0.4 $yr^{-1}$ $pc^{-3}$, if PBH are confined to galactic halos. From the absence of coincidences between atmospheric Cerenkov light flashes (that could have been caused by a flood of 250 MeV gamma-rays on the top of the atmosphere) seen by two large area (70-80$m^2$) mirrors 400 Km apart, Porter and Weekes [18] concluded that the 99% C.L. upper limit to PBH explosions is 0.04 $yr^{-1}$ $pc^{-3}$. This observation sets a better (lower) upper limit to PBH than the diffuse gamma-ray background provided PBH are clustered in galactic halos and the composite particle model (see References 5 and 6) operates during the explosion of PBH. Rees [19] argues that collective interaction of $e^-$ and $e^+$ (produced in the PBH explosions) with an ambient interstellar magnetic field can, on some specific assumptions, generate radio bursts powerful enough to be detected from anywhere in the galaxy or even beyond. Phinney and Taylor [20] looked for such flashes and, not finding any, set an upper limit of $2.10^{-9}$ yr$^{-1}$ pc$^{-3}$. In three other experiments, evidence was sought for an unusually rapid sequence of atmospheric Cerenkov flashes (Porter and Weekes [21] ) or of small size showers (Fegan et al. [22], Bhat et al. [23] ) which could have been initiated by the high energy gamma-rays from PBH explosions in the energy range $E = 1.5.10^{12}$ eV to $10^{14}$ eV. Having seen none beyond what is expected on the basis of Poisson fluctuations in the average event rate, each of the three groups set the upper limits shown in Table 2. It is very unlikely, on the basis of Elementary Particle Model (see references 5 and 6), that continued observations on ultra high energy gamma-rays with atmospheric Cerenkov technique or with particle detectors (small size air showers) will be able to lower the limits below $10^3$ yr$^{-1}$ pc$^{-3}$.

Pulsars. Pulsars are high on the list of any high energy gamma-ray experiment for two reasons: (i) Details of models apart, pulsars are strongly believed to contain large electromagnetic fields capable of accelerating charged particles to very high energies and (ii) the distinct signature of periodicity for the source relieves the experimenter from the burden of some technical aspects - constancy in the clarity of the night sky, comparison with a neighbouring region etc. It is imperative, on the other hand, to know the latest values of pulsar parameters to fold the event times modulo the pulsar period. We recall that at least 4 pulsars emit low energy gamma-rays upto 1 GeV; these are PSR0531+21, 0833-45, 0740-28 and 1822-09 (see Reference 3). Two more are suspected low energy gamma-ray emitters: PSR1747-46 and 1818-04 (see Kniffen et al. [24]).

PSR0531+21 (Crab Pulsar). This object has been studied most extensively by the Smithsonian group at Mt. Hopkins in USA. The time-averaged flux values (in units of quanta cm$^{-2}$s$^{-1}$) observed by this and other groups are: $2.10^{-12}$ at $E_\gamma \geq 2.10^{12}$ eV by Jennings et al. [25], $4.10^{-12}$ at $E_\gamma \geq 800$ GeV by Grindlay et al [26], $4.7.10^{-13}$ at $E_\gamma \geq 2.10^4$ GeV by Erickson et al. [27] and $1.2.10^{-11}$ at $E_\gamma \geq 500$ GeV by Gupta et al. [28]. In some cases information on the phase relative to the radio peak is not available; in the cases where available, the high energy gamma-ray peak did not coincide with the radio peak. In many other experiments only upper limits could be set for the fluxes; see e.g. References 13 and 28.

Our group at the Tata Institute of Fundamental Research has been continuing the observations on the Crab pulsar each winter from 1976-77 with progressively increasing sensitivity by enlarging the mirror area. Though we saw a positive flux in 1976-77 (Gupta et al. [28] ) as quoted above, we did not see any pulsed gamma-rays from this pulsar the following two years (77-78 and 78-79) even at reduced thresholds. We set, therefore, upper limits of $4.7.10^{-12}$ at $E_\gamma \geq 375$ GeV (1977-78) and $6.10^{-12}$ at $E_\gamma \geq 250$ GeV (78-79).

PSR0833-45. Grindlay et al. [29] observed this pulsar from Australia and set an upper limit of $1.0.10^{-11}$ quanta cm$^{-2}$s$^{-1}$ at $E_\gamma \geq 300$ GeV to the flux from this object. Though our location at Ootacamund (+ 11°23') was not ideal to look at this pulsar (zenith angles $\gtrsim 56°$), we Gupta et al. [28], observed this pulsar during the winter of 76-77 with no positive flux of high energy gamma-rays. We have again observed it a few months ago (78-79, Bhat et al [17] ) at a calculated threshold of $E_\gamma$ = 500 GeV. An analysis of the event times using the pulsar parameters supplied by Dr. R.N. Manchester resulted in the phasogram shown in Figure 5. The phase separation between the main pulse (4.4 $\sigma$) and interpulse (2.5 $\sigma$) is about 0.42. The probability of observing

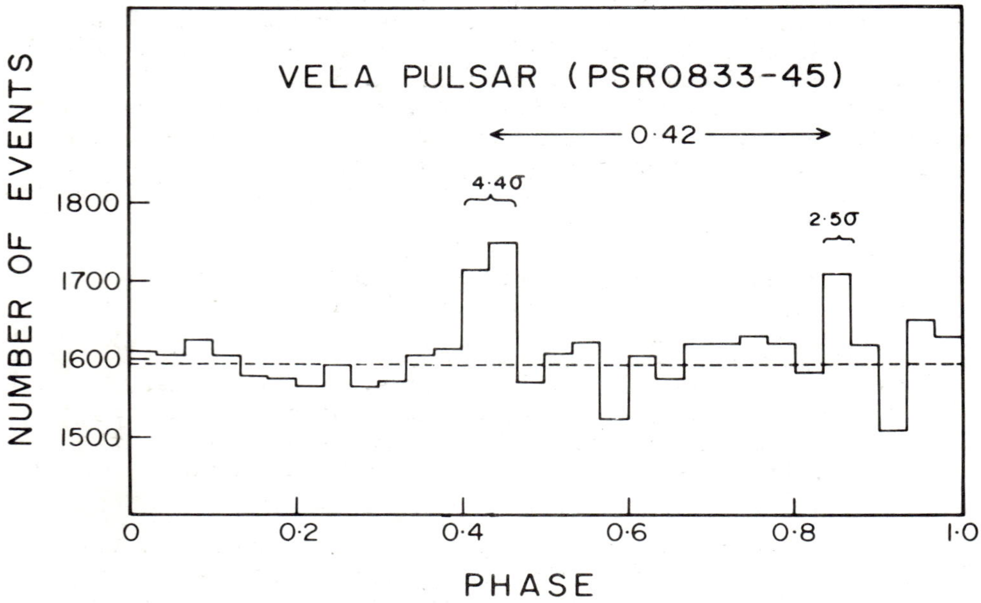


Fig. 5 Phasogram of the events from data on PSR0833-45. Phase interval between the two peaks marked M (4.4 $\sigma$) and I (2.5 $\sigma$) is 0.42.

these two peaks by pure chance is as small as $10^{-5}$. The excess counts seen in the two peaks correspond to a time-averaged flux of $(1.5 \pm 0.3).10^{-11}$ quanta $cm^{-2} s^{-1}$.

PSR0740 - 28. We, Bhat et al. [17], looked at this object a few months ago and no pulsed high energy gamma-rays were seen. A 90% C.L. upper limit to the flux of gamma-rays with $E_\gamma \geq$ 300 GeV from this object is placed at $2.7.10^{-11}$ quanta $cm^{-2}s^{-1}$.

PSR0950+ 08. Not having seen any gamma-rays from this object (nearest to earth), we place a 90% C.L. upper limit to the flux of gamma-rays with $E_\gamma \geq$ 250 GeV from this object at $8.10^{-12}$ quanta $cm^{-2}s^{-1}$.

CYG X-3. This source was seen to emit low energy gamma-rays by SAS-2 but not by COS-B. The Crimean Group in USSR found that this object emits gamma-rays at $E_\gamma \geq 2.10^{12}$ eV. Results obtained during the period 1972-75 were summarised by Stepanian et al. [30]. The object seems to be emitting high energy gamma-rays in two distinct components: (a) a periodic (P = 4.792416 hours) and constant amplitude component at a flux value of $1.3.10^{-11}$ quanta

$cm^{-2}s^{-1}$ and (b) a sporadic and highly fluctuating component at an average flux value of $1.7.10^{-11}$ quanta $cm^{-2}s^{-1}$, both at energies $E_\gamma \gtrsim 2.10^{12}$ eV. The total flux averaged over four years is $(3.0 \pm 0.8).10^{-11}$ quanta $cm^{-2}s^{-1}$. The sporadic component has a steeper spectrum than the periodic one. In August 1974, the sporadic component shot up to a flux value of $(6.2 \pm 2.1).10^{-11}$ quanta $cm^{-2}s^{-1}$ from its average value of $1.7.10^{-11}$. Weekes and Helmeken [31], however, did not find any evidence in 1976 for gamma-ray emission by this object and placed an upper limit of $6.5.10^{-11}$ quanta $cm^{-2}s^{-1}$ at $E_\gamma \geq 3.10^{11}$ eV. Stepanian and his colleagues set up a new experiment (Weekes [32]) at Tian-Shan in Siberia and again observed Cyg X-3 (1977) to emit high energy gamma-rays very clearly (5.6 sigma over the background at the expected Right Ascension) with the same period as inferred earlier. In summary, there is a clear-cut evidence for emission of high energy gamma-rays by this object albeit with some unusual fluctuations.

Centaurus A. Grindlay et al [29] observed this source from Australia and detected a flux of $(4.4 \pm 1.0).10^{-11}$ quanta $cm^{-2}s^{-1}$ at $E_\gamma \geq 3.10^{11}$ eV.

Crab Nebula. Fazio et al. [33] observed a steady flux of $(4.4 \pm 1.4).10^{-11}$ quanta $cm^{-2}s^{-1}$ at $E_\gamma \geq 2.5.10^{11}$ eV from the Crab nebula over a three-year (1969-72) period. The authors draw attention to the interesting observation that most of the effect during this observation was contributed by 3 separate periods each between 60 and 120 days after an observed spin-up of the Crab pulsar (Sept. '69, Aug. '71 and Oct. '71).

Other Sources. There are many more suspected sources - X-ray sources, pulsars, SN remnants, galaxies, clusters of galaxies, quasars etc. - which were also searched for emission of high energy gamma-rays. In almost all the cases, no high energy gamma-ray emission was seen at the sensitivity of the respective detectors; for upper limits, see Porter and Weekes [13].

## DISCUSSIONS AND CONCLUSIONS

As a result of intensive experimental efforts in the field of high energy gamma-rays, it has been possible to establish that some objects - the Crab and Vela pulsars, Crab nebula, Cygnus X-3 and Centaurus A - emit gamma-rays at energies greater than a few hundred GeV. It is also true that a large number of suspected sources do not emit high energy gamma-rays at the level of detection sensitivities. The currently reported positive fluxes of high energy gamma-rays from the Crab Nebula, if interpreted as due to inverse Compton scattering of electrons, imply that the magnetic fields must be greater than $5.10^{-4}$ gauss in the Crab Nebula. Some of the experimental results on the emission of pulsed high energy gamma-rays by pulsars (PSR0833-45, 0950+08 and even 0531+21) are tantalizing. The peaks in the phase plots are not too small to be ignored; nor are they strong enough to convince others without much effort. Continued observations with perhaps reduced thresholds and backgrounds will help establish the truth. If it is reliably established that pulsars do emit high energy gamma-rays, then the parameters behind the objections regarding the energy degradation of high energy gamma-rays in the pulsar magnetospheres have to be re-examined. Perhaps the observations will lead us to understand how far away from the neutron star surface the high energy gamma-rays originate and if there are any secondary acceleration mechanisms operating.

It is too early, with the meagre observations on hand, to proceed and build a credible model. Perhaps the present status of the field of high energy gamma-rays is comparable to that of 100 MeV gamma-rays 10 years ago. As it happened with the 100 MeV gamma-rays, the hope is that the next ten years of high energy gamma-ray studies will add significantly to our knowledge of particles, fields and processes in various celestial settings.

## ACKNOWLEDGEMENTS

It is a pleasure to thank P.N. Bhat, S.K. Gupta, B.V. Sreekantan, S.C. Tonwar and P.R. Viswanath, my colleagues on the Ooty Gamma-ray experiment, preliminary results from which are mentioned in this paper. My thanks are due to Trevor Weekes of Smithsonian Astrophysical Observatory (USA) for filling me in on the latest experimental results from the Crimean (USSR) group.

## REFERENCES

1. J.V. Jelley and W. Galbraith, J. Atmos. Terr. Phys. 6, 304 (1955).
2. W. Hermsen et al., Nature 269, 494 (1977).
3. K. Pinkau, Nature 277, 17 (1979).
4. G.G. Fazio, Ann. Rev. Astron. and Astrophys. 5, 481 (1967).
5. D.N. Page and S.W. Hawking, Astrophys. J. 206, 1 (1976).
6. B.J. Carr, Astrophys. J. 206, 8 (1976).
7. R. Buccheri et al., Nature 274, 572 (1978).
8. J.H. Taylor and R.N. Manchester, Astron. J. 80, 794 (1975).
9. B.K. Chatterjee et al., Nature Phys. Sci., 231, 126 (1971).
10. R.J. Gould and G.P. Schreder, Phys. Rev., 155, 1408 (1967).
11. E. Massaro and M. Salvati, Astron. Astrophys. 71, 51 (1979).
12. B. McBreen et al., Astrophys. J., 184, 571 (1973).
13. N.A. Porter and T.C. Weekes, S.A.O. Special Report 381 (1979).
14. R. Browning and K.E. Turver, Nuovo Cimento, 38 A, 223 (1977).
15. J.E. Grindlay, S.A.O. Special Report No. 334 (1971).
16. T.C. Weekes and K.E. Turver, Proc. of 12th ESLAB Symp., (1977).
17. P.N. Bhat et al., Private Communication (1979).
18. N.A. Porter and T.C. Weekes, M.N.R.A.S. 183, 205 (1978).
19. M.J. Rees, Nature 266, 333 (1977).
20. S. Phinney and J.H. Taylor, Nature 277, 117 (1979).
21. N.A. Porter and T.C. Weekes, Nature 277, 199 (1979).
22. D.J. Fegan et al., Nature 271, 731 (1978).
23. P.N. Bhat et al., Preprint (1979).
24. D.A. Kniffen et al., Proc. of 12th ESLAB Symp., ESA SP-124 (1977).
25. D.M. Jennings et al., Nuovo Cimento, 20B, 71 (1974).
26. J.E. Grindlay et al., Astrophys. J., 209, 592 (1976).
27. R.A. Erickson et al., Astrophys. J. 210, 539 (1976).
28. S.K. Gupta et al., Astrophys. J. 221, 268 (1978).
29. J.E. Grindlay et al., Astrophys. J. 201, 82 (1975).
30. A.A. Stepanian et al., 15th Int. Cosmic Ray Conf. Vol. 1, 135 (1977).
31. T.C. Weekes and H.F. Helmeken, Proc. of 12th ESLAB Symp., (1977).
32. T.C. Weekes, Private Communication, May 1979.
33. G.G. Fazio et al., Astrophys. J. Lett. 175, L 117 (1972).

# COSMIC $\gamma$ RAY LINE OBSERVATIONS

G. Vedrenne

*Centre d'Etude Spatiale des Rayonnements, Toulouse, France*

ABSTRACT

The present status of solar and non-solar nuclear gamma-ray line research from several 10's of keV up to several MeV is reviewed. After a brief summary of the experimental difficulties in this field, the main results are presented. Although somewhat fragmentary these results demonstrate clearly the existence of solar and cosmic nuclear lines ; future programs and especially satellite experiments such as HEAO C and GRO promise to advance our understanding of these phenomena considerably.

## INTRODUCTION

Since $\gamma$ ray lines are produced in different ways - radioactive elements which decay by gamma emission, nuclear reactions which may result from inelastic collisions between cosmic radiation and the interstellar medium, and processes of radioactive capture and annihilation - their detection is a highly efficient way of studying cosmic nuclear processes. In addition, electron transitions in very intense magnetic field regions, encountered in neutron stars, may give lines in the X-ray range. The 58 keV line, detected for the first time by Trümper et al.[1,2] in the source Her X-1, may be associated with this process and is therefore of very great interest to the field of astrophysics.

Some of the astrophysical sites which can be evoked for X-ray line emissions are :

- Novae and supernovae; the discovery of gamma ray lines associated with these objects would make it possible to study explosive nucleosynthesis.

- Dense objects : neutron stars, and black holes which could give rise via processes of accretion to gamma ray line emissions. The determination of these lines and their possible gravitational shifts would provide information on the physical properties of these objects.

- The interstellar medium, particularly the regions rich in dust for which the detection of nuclear gamma-ray lines could, for example, provide information on the nature and size of grains.

Gamma-ray lines may also indicate the presence of low-energy (< 100 MeV/nucleon) cosmic radiation. This is of interest since such radiation is difficult to measure in the vicinity of the solar system owing to the presence of the solar

magnetic cavity.
The astrophysical interest of the study of these nuclear gamma-ray lines has been substantially developed in numerous articles, in particular by Clayton, Ramaty and Lingenfelter [3 to 7] .

## INSTRUMENTATION

These γ-ray lines may be looked for with the aid of two types of instruments; either sodium iodide crystals, which can have large dimensions, but whose spectral resolution is limited to some 6 or 7 % at 1 MeV, or germanium crystals, which are more difficult to obtain with large dimensions and require cooling down to temperatures of less than about 130°K [54]. The use of elaborate cryogenic systems aboard satellites adds greatly to the complexity of the experiments and presently limits their lifetime. However these detectors present the considerable advantage of having a resolution at 1 MeV lower by a factor of 20 to 30 than that of the NaI scintillators.

We do not intend to weigh the two techniques against each other but it is important to determine where each should be used. The first is especially indicated in the search for lines with very weak fluxes and a width of over several tens of keV. However, if the lines expected are very narrow, a considerable gain in the signal/noise ratio can be obtained using cooled germanium type detectors. In this case such detectors, despite the drawbacks described, are of great interest, since they facilitate the unambigous identification of spectral lines and thus the discovery of the elements at the origin of these emissions. Moreover they may provide insight into some of the physical parameters of the source through the study of the line profile and its possible redshift.

Many of the instruments likely to provide information on nuclear gamma-ray lines have been cited in a recent paper [8]. We shall recall here only the main experiments intended for fine gamma spectrometry (Table 1).

TABLE I - GERMANIUM DETECTORS

| | DIMENSIONS OF GE CRYSTALS | SHIELDING | FIELD OF VIEW FWHM | | |
|---|---|---|---|---|---|
| BALLOON EXPERIMENTS : | | | | | |
| JET PROPULSION LABORATORY (19)<br>50 keV - 10 MeV | ~ 4x45 $cm^3$ Ge(Li)<br>≃ 55 $cm^2$ | CsI(Na)<br>6.4 cm | 30° | | |
| CENTRE D'ETUDE SPATIALE DES RAYONNEMENTS - C.E.A./S.E.P.<br>100 keV - 8 MeV (57) | 140 $cm^3$ Ge(Li)<br>~ 20 $cm^2$ | NaI(Tl)<br>5 cm | 60° | | |
| BELL SANDIA LABORATORIES<br>100 keV - 5 MeV (38) | HP Ge<br>130 $cm^3$ | NaI(Tl) 200 kg<br>15.2 cm | 15° | | |
| RIKKYO UNIVERSITY TOKYO<br>0,2 - 6 MeV (56) | Ge(Li) 31$cm^3$<br>12 $cm^2$ | PLASTIC SCINTILLATOR 10 cm | 110° | | |
| GODDARD SPACE FLIGHT CENTER | HP Ge<br>3x100 $cm^3$ | NaI(Tl)<br>6,5 cm | 15° | | |
| SATELLITE EXPERIMENTS : | | | | | |
| LOCKHEED LABORATORY<br>1) 40 keV - 2,7 MeV (58) | 2x50 $cm^3$ Ge(Li)<br>2x15,2 $cm^2$ | TUNGSTEN 30g/$cm^2$<br>1,5 cm PLASTIC | 90° | LAUNCH OCT. 72<br>LIFE TIME 7 MONTHS | SATELLITE 550-650 km POLAR ORBIT SUN SYNCHRONOUS |
| 2) P 78 1 (9)<br>50 keV - 2,7 MeV | 2x85 $cm^3$<br>HP Ge | NaI(Tl)POLYSCIN<br>5 cm | 90° | LAUNCH FEBRUARY 79 | |
| GODDARD SPACE FLIGHT CENTER<br>ISEE-C* (11)<br>50 keV - 6,5 MeV | 35 $cm^3$<br>(12,5 $cm^2$)<br>HP Ge | DESIGNED FOR γ RAY BURST SPECTROSCOPY | ISOTROPIC | LAUNCH AUG. 78<br>230 EARTH RADII ABOUT LAGRANGIAN POINT | |
| JET PROPULSION LABORATORY (47)<br>HEAO-C 50 keV - 10 MeV | 4x100 $cm^3$<br>HP Ge<br>4x23 $cm^2$ | CsI(Na)<br>6,6 cm | 30° | SATELLITE ~ 370 km<br>i = 25° LAUNCH SEPTEMBER 79 | |

* ON ISEE-B THE EXPERIMENT OF CLINE AND HOVESTADT USES 6 CdTe DETECTORS OF 1 $cm^2$ - γ RAY BURST EXPERIMENT.

At the present time these experiments are mostly performed from balloons, using a detector or a mosaic of detectors, either of lithium-doped or intrinsic germanium with typical volumes ranging from 100 to 150 $cm^3$. The detectors must be surrounded by shielding materials. But in this type of experiment the use of passive or semi-active materials is definitely a disadvantage : the role of the anticoincidence shield is not only to reduce the background, but also to eliminate Compton scattering taking place within the main crystal, since the choice of these detectors is based on the search for lines.

In addition to these balloon experiments, four satellite experiments have been launched :

1) the experiment of Nakano et al.[9] which was launched in 1972 aboard a polar-orbit satellite with a detector protected by passive tungsten shielding. A new experiment by the same group was put into orbit in February 1979 [10],(Table 1). However, the polar orbit and the use of passive shielding in the first experiment resulted in high background, more in fact than in most balloon experiments.

2) the experiment of Cline and Hovestadt for detecting gamma bursts, launched aboard ISEE-B, consisting of six CdTe detectors of 1 $cm^2$.

3) the GSFC high-purity Ge detector aboard ISEE-C for searching for line emission from $\gamma$ ray bursts [11].

The present sensitivity of the experiments using germanium detectors and designed to search for gamma ray lines is around $10^{-3}$ photon $cm^{-2}s^{-1}$ for non-broadened lines; however in the case of Leventhal's balloon experiment [38] , the use of a large active anticoincidence makes it possible to reduce the sensitivity to some $10^{-4}$ photon $cm^{-2}s^{-1}$ for narrow lines.

Of course, nuclear gamma-ray spectrometry suffers from the same problems as gamma-ray astronomy in the same energy range; for instance in balloon experiments we again encounter problems involving the presence of a high atmospheric background and processes of neutron activation which give rise to intense interference lines. In Germanium some of these lines appear with the strongest intensities at 53, 67, 139 and 198 keV. Other lines associated with the activation of the Sodium and Cesium in the anticoincidence systems have also been detected. The search for the 511 keV line from extraterrestrial objects is especially delicate because this line is so intense in the atmosphere.

Despite these experimentation problems results are available on cosmic gamma-ray lines, obtained for the most part from balloon experiments, reviews on this subject have already been presented by Jacobson [55] and Chupp [50] . Moreover the Goddard Symposium on $\gamma$-ray Spectroscopy in Astrophysics [59] gives most of the results available today.

## SOLAR GAMMA RAYS

Although this symposium is devoted to non-solar gamma rays, it seems useful to recall that it was the observation of the sun that brought about the discovery of astrophysical nuclear gamma line radiation from the August 4 and 7 flares by the 7th Orbiting Solar Observatory (OSO 7) [12][13]. For the August 4 flare, the observed lines at 2.22, 0.51 , 4.44 and 6.13 MeV had fluxes of $(2.8 \pm 0.22) \times 10^{-1}$, $(6.3 \pm 2) \times 10^{-2}$, $(3.0 \pm 1.0) \times 10^{-2}$ and $(3.0 \pm 1.0) \times 10^{-2}$ photon $cm^{-2}s^{-1}$, respectively. The event lasted about $10^3$ seconds. Lines at 2.22 and 0.51 MeV were also detected during the decay phase of the August 7 flare, but only upper limits could be set on the 4.44 and 6.13 MeV lines.

These gamma line emissions are of great significance, since they are associated with nuclear reactions of accelerated particles in flares at the chromosphere or low corona level. For example, the strongest line, at 2.223 MeV, is due to the capture of neutrons on hydrogen ; these neutrons are themselves produced by nuclear reactions resulting from the interaction of particles accelerated at the moment of the flare with solar matter. The emission at 511 keV results from the annihilation of positrons. These positrons may be produced in flares mainly through short-lived radioactive positron emitters such as $^{11}C$, $^{13}N$, $^{14}O$, $^{15}O$ and the first nuclear level of $^{16}O$ at 6.052 MeV.

Gamma continuum and line emissions have been observed during more recent events [14],[15]. The results are grouped in Table 2 [15]. The large intensity of the 2.23 MeV line (1 photon $cm^{-2}s^{-1}$) observed by HEAO A during the July 11, 1978 event should be noted (Figure 1).

TABLE 2 - GAMMA RAY FLARES (FROM HUDSON (15))

| DATE | SOURCE OF DATE | KIND OF OBSERVATION | PEAK FLUXES PH$(CM^2 SEC)^{-1}$ > 1 MeV | PEAK FLUXES PH$(CM^2 SEC)^{-1}$ 2,2 MeV LINE |
|---|---|---|---|---|
| 1967 MAY 23 | ERS-18 | CONTINUUM | 3 | |
| 1972 AUG. 4 | OSO-7 | LINE + CONT. | 3 | .22 ± 0,2 |
| 1972 AUG. 7 | OSO-7 | LINES | - | .046 ± .013 |
| 1977 NOV. 22 | PROGNOZ-6 | CONT.+ LINE | 0,5 | 0,15 ± 0,05 |
| 1978 JULY 11 | HEAO-1 | LINE + CONT. | 16 | 1,1 ± 0,2 |

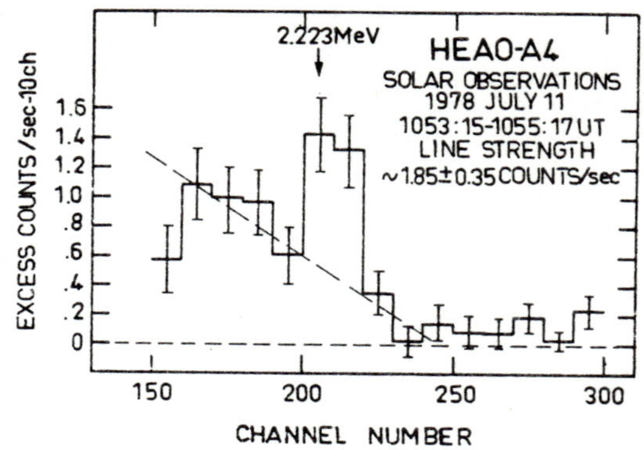


Fig. 1 : 2.2 MeV line observed by HEAO-A4 (From Hudson [15] ).

## CYCLOTRON LINES

Even if cyclotron lines are not typically in the field of γ ray astronomy, their astrophysical implications are so important that it is interesting to discuss them briefly. Their existence for Her X-1 seems now to have been clearly established by the excellent experiments of Trumper et al. [1,2] . The lines appear at 58 keV, they may also appear at 110 keV, but this is less clear (Figure 2).

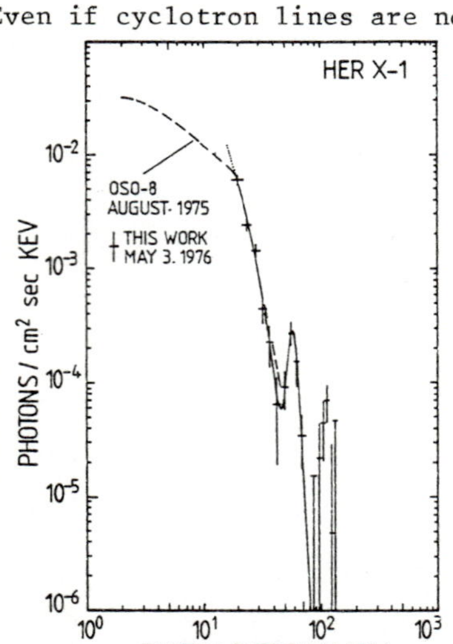


Fig. 2 : Deconvoluted spectrum of Her X-1 observed by Trümper et al. [2].

Given their great astrophysical importance, other experiments have also attempted to detect them. The main results were listed by Trümper at the Cospar Symposium in Innsbruck 1978 [16] . The most recent observations of HEAO-A [17][18] provide confirmation of these results and show evidence for variations by a factor of 2 on a scale of one day in the continuum and in the lines. Interpretations of the observations of these line features and associated theoretical considerations concerning the presence of a very intense magnetic field as the origin of the radiation are given in the paper by Trümper et al. [16] .

In this energy range, observations of the Crab Nebula by means of high spectral resolution experiments [19] have shown the presence of a line at 73 keV (Figure 3) whose intensity is $(3.8 \pm 0.9) \times 10^{-3}$ photon $cm^{-2}.sec^{-1}$ and whose width is less than 4.9 keV-FWHM at 73.3 ± 1 keV; this corresponds to a power of $2 \times 10^{35}$ erg $sec^{-1}$, assuming the distance of the Crab Nebula to be about 2 kpc from Earth[20]. If this line originates in the Crab Nebula, it might be associated with cyclotron emission. Preliminary results from a transmediterranean flight made in 1978 between Sicily and Spain by the French experiment CESAR do not seem to confirm the presence of such a line at this intensity; our $2\sigma$ upper limit is : $2.2 \times 10^{-3}$ photon $cm^{-2}s^{-1}$. Strickman et al. [21] have also provided a $2\sigma$ upper limit to this feature ($1.5 \times 10^{-3}$ photon $cm^{-2}sec^{-1}$), lower by a factor of 2.5 than the flux given by Ling et al. Perhaps, as in the case of Her X-1, such a line may exhibit considerable variability.

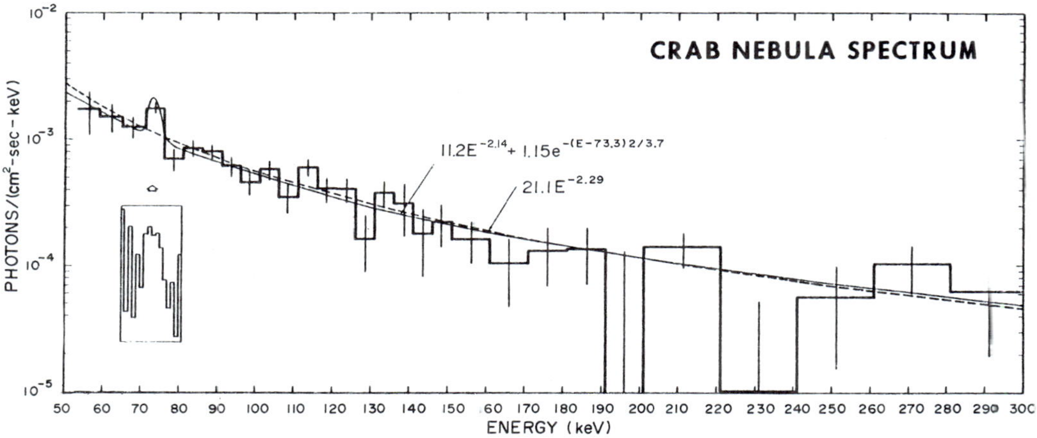


Fig. 3 : Low energy spectrum of the total emission of the Crab Nebula (10 June 1974) Solid and dashed curves indicate respectively power law fits to the data with and without a Gaussian line shape at 73 keV. The inset shows details of the fit in the vicinity of the possible line (From Ling et al. [20]).

## GALACTIC GAMMA RAY LINES

In this range the results are still very fragmentary and correspond essentially to the observation of the central regions of the Galaxy. Only the presence of radiation at 511 keV seems to be clear at the present time.

The pioneering work of Haymes is worthy of reexamination, since for a long time it gave the only evidence of the presence of nuclear gamma ray-emission. The main results are indicated in Table 3 [22]. These correspond to the observation of the Galactic Center as part of a project involving four balloon flights made between 1969 and 1974 [23][24] .

TABLE 3 - GALACTIC CENTER SPECTRAL LINES (FROM HAYMES ET AL. (24))

| | | | |
|---|---|---|---|
| MEASURED ENERGY (MEV) | 0.53 ± 0.01 | 1.2 - 2.0 | 4.6 ± 0.1 |
| FLUX ($10^{-4}$ PHOTONS/$CM^2$-SEC) | 8.0 ± 2 3 | 26 ± 6 | 9.5 ± 2.7 |
| INSTRUMENTAL RESOLUTION (MEV-FWHM) | 0.060 | 0.180 | 0.206 |
| LINE WIDTH (MEV-FWHM) | 0.07 ± 0.02 | -- | 0.7 |
| POSSIBLE ORIGIN | $\beta^+$ EMITTERS | $Mg^{24}$, $Ne^{20}$, $Si^{28}$ | $C^{12}$ |
| PREDICTED ENERGY (MEV) | 0.511 | 1.37, 1.63, 1.78 | 4.43 |

In the first flight in November 1970, from Parana, Argentina (Cutoff 11.5 GV), the detector was an NaI scintillator, 4 inches in diameter and two inches thick, within a thick NaI(Tl) anticoincidence shield collimator defining 24° FWHM. A line at 473 ± 30 keV with a flux of $(1.8 \pm 0.5) \times 10^{-3}$ photon $cm^{-2}sec^{-1}$ was reported. This spectral feature was confirmed by another balloon flight in November 1971 using the same apparatus. The same group launched a new instrument in April 1974 with a smaller angular resolution, 15° FWHM, compared to the above instrument with 24° FWHM. The results obtained (Table 3) are not in perfect agreement with the first results we have just examined. The possible origin of these differences has been discussed by Haymes et al. [24].

These results therefore bear evidence of the presence of lines in particular at 511 keV (Figure 4) and at 4.6 MeV, which may correspond to the deexcitation of Carbon 12 (first excited level) (Figure 5). We shall examine the problem of these two lines.

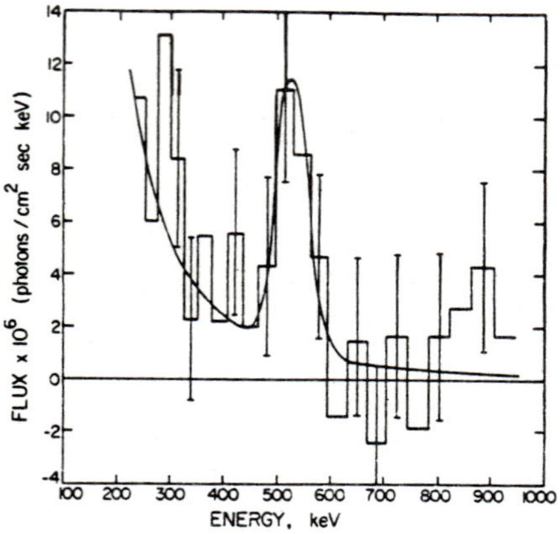


Fig. 4 : 0.5 MeV line observed from the Galactic Center at the 3.5 σ confidence level (From Haymes et al. [24 ]).

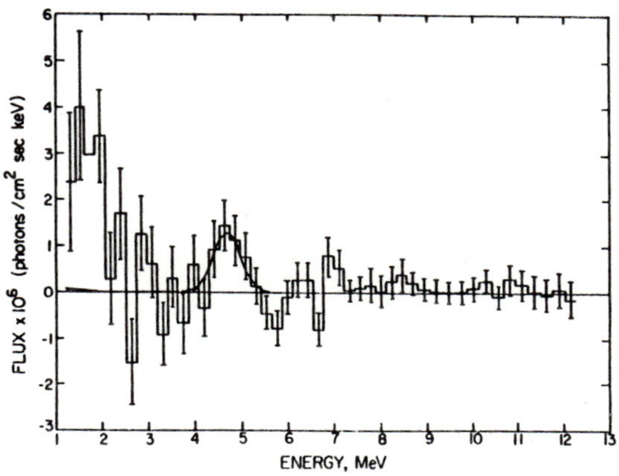


Fig. 5 : Evidence between 4 and 5 MeV of an excess at the 3.5 σ confidence level from the Galactic Center region. The sum of the overlapping spectral lines between 1.2 and 2 MeV is at the 4.1 σ level (From Haymes et al. [24 ] ).

a) <u>The 511 keV line</u>

The presence of this line is now clearly established from the observations of Leventhal et al. [25] [26], (Figure 6).

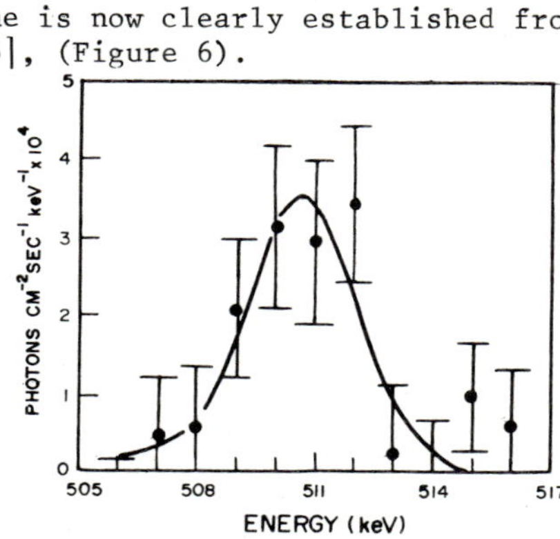


Fig. 6 : Energy spectra around 511 keV observed from the Galactic Center region by Leventhal et al. [26] .

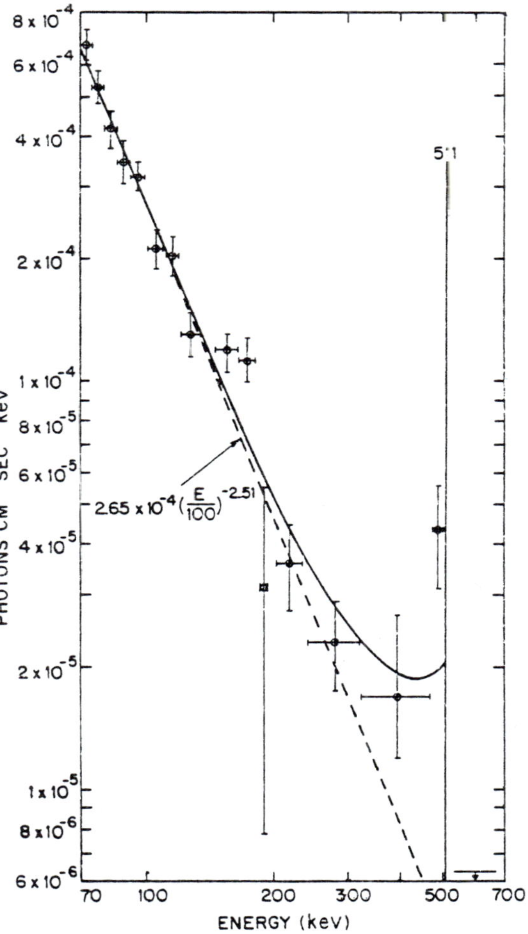


Fig. 7 : Differential photon spectrum at the top of the atmosphere from the Galactic Center region. The solid curve is a least squares fit to the data (From Leventhal et al. [26] ).

This observation was made in 1977 with a balloon-borne high-purity Ge detector with a 15° aperture centered on the Galactic Center. The observed spectrum is given in Figure 7.

The excess observed at 511 keV is significant at the 5.6 $\sigma$ level, corresponding to a corrected flux of $(1.22 \pm 0.22) \times 10^{-3}$ photon $cm^{-2}sec^{-1}$. This gives a line intensity of $1.4 \times 10^{37}$ erg $sec^{-1}$. It should be noted that measurements made with a French omnidirectional detector [60] point to an upper limit of $4 \times 10^{-3}$ photon $cm^{-2}sec^{-1}$ from the Galactic disc. This value is only compatible with Leventhal's results if the radiation does not represent a uniform contribution over the whole disc. A more recent French experiment [61] gives a flux at 511 keV $(4.19 \pm 1.56) \times 10^{-3}$ photon $cm^{-2}sec^{-1}$ from the Galactic Center region, the field of view being 50° FWHM.

A detailed study of the spectrum obtained by Leventhal et al. seems to reveal the presence of the three-photon positronium continuum. This contribution can be explained by positron annihilation in a low-density and low-temperature medium [27]. These results are a good example of the interest of a high spectral-resolution detector making of possible to understand the conditions under which radiation at 511 keV may occur. For example if such radiation took place in cold interstellar clouds its width would be about 5 keV [27]. On the other hand, the observed 3.2 keV upper limit on the width of the line may indicate that positron annihilation appears in an ionised medium [7] Thus Ramaty et al.[7] suggest that the annihilation may occur in the HII region at the Galactic Center, the warm ionized component of the interstellar medium or supernova remnants.

It seems difficult at the present time, given the high flux of the line to attribute it to the interaction of low-energy cosmic radiation with interstellar matter in the central region of the Galaxy. A recent estimate of 0.511 MeV line production by this process gives a flux of $7 \times 10^{-5}$ photon $cm^{-2}sec^{-1}rad^{-1}$ or about 2 % of the line intensity observed in a 15° cone around the Galactic Center [7] . Furthermore, if this line were only due to this process it would be accompanied by other detectable lines, e.g. 478 keV ($^{7}Li$), 1.37 MeV ($^{29}Mg$) and 4.43 MeV ($^{12}C$) [29] . Leventhal et al. [49] did not observe lines at 478 keV or 1.37 MeV.

It therefore seems necessary to look for other sources, such as explosive nucleosynthesis, in supernovae. Pulsars such as the Crab may emit high positron fluxes and the observation of the line at 511 keV may indicate the injection of positrons into the interstellar medium by a large number of pulsars.

More exotic suggestions have also been made, such as evaporation or explosion of primordial black holes [26].

b) The 4.43 MeV line.

This line was first noted by Haymes [24] . However its intensity is presently widely disputed. The early results of the HEAO-A4 experiment seemed to indicate the presence of this line in the results of the observation of the Galactic Center However, more recent results [30] indicate at 3σ a limit well below the reported 3σ detection by Haymes. This 3σ upper limit is 0.35 of the reported flux, if it is due to a narrow line, and 0.60 for a line at the reported width of 700 keV [30] (Figure 8), but the field of view of the two experiments is quite different.

Other experiments [31][32] were unable to observe the flux at 4.43 MeV, and the upper limits indicated are not consistent with the distribution of 4.4 MeV line intensity being uniform on $-50° < \ell < 22°$. All of the reported results to date in this line are given in Table 4.

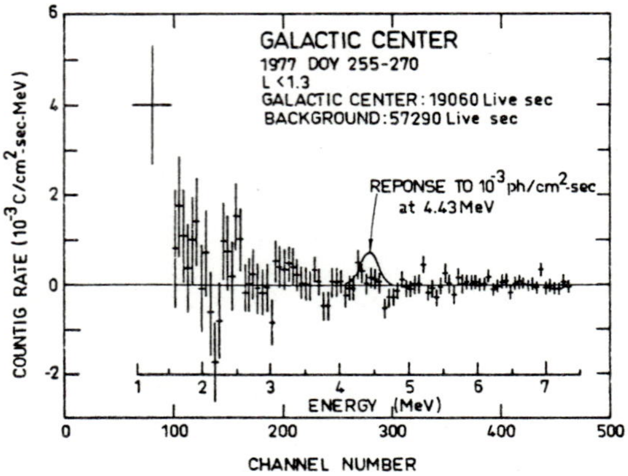


Fig. 8 : The measured photon spectrum from the Galactic Center with HEAO-A4 (From Matteson et al. [30]).

TABLE 4 4.43 MEV LINE

| | OBSERVED FLUX (PH CM$^{-2}$S$^{-1}$) | OBSERVED REGION | FLUX (PH CM$^{-2}$S$^{-1}$RAD$^{-1}$) | DETECTOR DESIGN AREA (CM$^2$) | APERTURE (FWHM) |
|---|---|---|---|---|---|
| RICE UNIVERSITY (24) | $(9.5 \pm 2.7) \times 10^{-4}$ | $-4° \leq \ell^{II} \leq +6°$ | $(5.4 \pm 1.5) \times 10^{-3}$ | 182 | 13° |
| APOLLO 16 (31) | 2 σ UPPER LIMIT $3.4 \times 10^{-3}$ | $-50° \leq \ell^{II} \leq 22°$ | 2 σ UPPER LIMIT* $2.7 \times 10^{-3}$ | 58 | OMNIDIRECTIONAL |
| CENTRE D'ETUDE SPATIALE DES RAYONNEMENTS (32) | 2 σ UPPER LIMIT $2 \times 10^{-3}$ | $-45° \leq \ell^{II} \leq +45°$ | 2 σ UPPER LIMIT* $1.27 \times 10^{-3}$ | 78.5 | 90° |
| HEAO-A4 (30) | 3 σ UPPER LIMIT $3.3 \times 10^{-4}$ (NARROW LINE) $5.7 \times 10^{-4}$ (BROAD LINE) | | | 120 | 43° |

* WITH THE HYPOTHESIS OF UNIFORM EMISSIVITY ALONG THE GALACTIC DISK.

Compared with the Rice group's results [24], the upper limits given by Apollo 16 [31] and the French CESR experiment [32] can be used to calculate :

$(\phi\ 4.4\ \mathrm{MeV})_{Apollo}/(\phi\ 4.4\ \mathrm{MeV})_{Rice} < 3.6$ and $(\phi\ 4.4\ \mathrm{MeV})_{CESR}/(\phi\ 4.4\ \mathrm{MeV})_{Rice} < 2$, respectively. This suggests that the 4.4 MeV line, if it exists, is limited to longitudes between ± 12° and comes from within 2 kpc of the Galactic nucleus. In a recent paper [33] these results were used to calculate the flux and spectrum of low-energy cosmic rays which may result from these observations. The total energy density of the low-energy cosmic rays $U_{CR}$ is estimated taking into account the results of Ramaty et al. [6] :

$$U_{CR} = \frac{4\pi\ d^2\ I_{4.4\ MeV}}{q_{4.4\ MeV}} \quad \frac{m_p}{M_{gas}}$$

with $m_p$ the mass of the proton and $M_{gas}$ the total mass of gas in the region. The 4.4 MeV emissivity $q_{4.4}$ is $1.5 \times 10^{-25}$ gamma $sec^{-1}.(H\ atm)^{-1}(eV.cm^{-3})^{-1}$ in the most favorable case where both the cosmic rays and the interstellar gas have a metal-enhanced composition. The energy density is calculated for two cases :

a) the line emission comes from a region with R < 300 pc, whose gas mass is $1.3 \times 10^8\ M_\odot$ : then $U_{CR} \sim 470$ eV $cm^{-3}$.

b) the line emission originates in a region with R < 2 kpc which may have a gas mass of $10^9\ M_\odot$ ; then $U_{CR} \sim 60$ eV $cm^{-3}$.

Audouze et al. have examined the stability of a gas subjected to the above-mentioned flux of low-energy cosmic radiation; they conclude that in case b) the disc cannot be stable; in case a) the energy density for low energy cosmic rays which is about $10^{-9}$ erg $cm^{-3}$ does not induce instabilities but this enormous density raises problems which the authors have analyzed [33] .

In another paper Bussard et al. [34] discuss the problem of the line intensity at 4.43 MeV in conjunction with the emission of a line at 6.8 keV by the 2p to 1s transition in fast Hydrogen-and Helium-like Iron ions, after electron capture to excited levels and after collisional excitation. They evaluate ratios of the X- and gamma-ray line intensities. They conclude also that it is difficult to produce the γ-ray flux given by Haymes et al. [24] without encountering problems with X-ray results. They suggest two possibilities to avoid these difficulties ; either the source region is so hot that the charge exchange and X-ray production are eliminated, or the region is opaque to X rays but not to gamma rays [34].

In conclusion, the fluxes obtained at 511 keV and 4.43 MeV are for now difficult to interpret.However the 4.4 MeV line seems now more controversial than the 0.5 MeV line. It should be noted in this respect that flux excesses in the region of 1 to 2 MeV were also noted by Haymes but seem unconfirmed by the HEAO-A4 experiment.

c) Other lines.

For the Galactic Center Leventhal [26] reports flux excesses detected between 4 and 5σ for lines at 170 keV, 1.611 MeV and 3.70 MeV. During a flight made on February 14, 1977 from Brazil the French CESAR experiment co-designed by the CESR and the CEA of Saclay gives at 170 and 1610 keV 2σ upper limits ($8.10^{-4}$ photon $cm^{-2}sec^{-1}$ and $1.1 \times 10^{-3}$ photon $cm^{-2}sec^{-1}$) which don't exclude the possibility of these lines.

Other results concerning lines have been reported : Walvaren et al. [51] observed a possible spectral feature at 1.16 MeV from the region ($\ell$ = 345°, b = 5°). The reported flux was $(3.4 \pm 1.5) \times 10^{-3}$ photon $cm^{-2}sec^{-1}$. Leventhal et al. [52] looked at this region and gave a 2σ upper limit of less than $1.8 \times 10^{-3}$ photon $cm^{-2}sec^{-1}$. Other upper limits have been reported by the same group from Nova Cygni 1975 and

Nova Serpentis [38] . Imhof et al. [53] with their satellite experiment have reported the possible presence of 2 other lines at 1121 and 1369 keV. These lines could be emitted from the galactic plane in the longitudinal region $\sim$ 255° to 328°.

All these possible lines lead one to realize the need for extreme caution in searching to establish their reality, so it is quite desirable to have many observations of the same region in order to confirm the results of one experimenter.

This situation is reminiscent of the situation which existed at the beginning of high-energy gamma-ray astronomy, when many flux excesses were localized in different regions of the sky with in fact very few objects having been observed by more than one experimenter. An examination of the ensemble of results has shown that in general the excesses seen by at least two or more experimenters were much more worthy of confidence, and that some of them were in agreement with the results of SAS 2 or COS B [35] .

Furthermore, the 511 keV line in particular seems to be very difficult to establish because at this energy the local production and the atmospheric flux are high. Moreover the intensity of this line depends strongly on the latitude, the residual presure and the zenith angle of the detector for balloon experiments. Nevertheless even if the presence of such a line of non-terrestrial origin is particularly difficult to assess, it seems now that it is the most credible line of galactic origin.

## TRANSIENT GAMMA RAY LINES

The appearance of nuclear gamma ray lines at different energies was noted by Jacobson et al. [36] during a Northern Hemisphere flight designed to observe the Crab Nebula. During this flight (10 June 1974) an intense transient event, lasting only 20 minutes, was observed. Both the intensities of the lines and their energies are indicated in Table 5 and Figure 9.

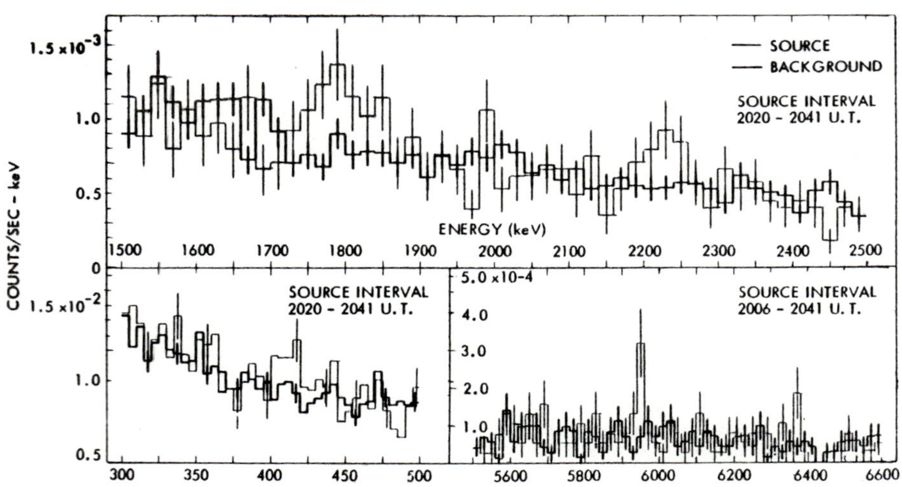


Fig. 9 : Observed lines during the 10 June 1974 transient event (From Jacobson et al. [36] ).

TABLE 5 : GAMMA RAY LINES OBSERVED DURING THE 10 JUNE 1974 TRANSIENT EVENT - (FROM JACOBSON ET AL. (36))

| LIVE ENERGY (keV) | COUNT RATE $(SEC)^{-1}$ | FLUX* $(CM^2 - SEC)^{-1}$ | FWHM (keV) | TIME INTERVAL SPANNED (U.T.) |
|---|---|---|---|---|
| 413.2 ± 1.8 | $(5.5 \pm 1.6) \times 10^{-2}$ | $(7.0 \pm 2.0) \times 10^{-3}$ | 15 | 2020 - 2041 |
| 1789.7 ± 6.0 | $(6.78 \pm 1.55) \times 10^{-2}$ | $(3.15 \pm 0.74) \times 10^{-2}$ | 95 | 2026 - 2041 |
| 2218.6 ± 6.3 | $(2.67 \pm 0.87) \times 10^{-2}$ | $(1.51 \pm 0.49) \times 10^{-2}$ | 70 | 2020 - 2041 |
| 5946.5 ± 3.7 | $(7.06 \pm 2.18) \times 10^{-3}$ | $(1.47 \pm 0.46) \times 10^{-2}$ | 25 | 2006 - 2041 |

*ASSUMES SOURCE ON THE DETECTOR AXIS

All of the verifications which are performed in order to confirm the effective presence of an intense flux of gamma photons arriving during this period of time were carried out, and it appears that the lines observed are quite significant from a statistical standpoint . Some of these have energies which correspond to clearly identified processes theoretically expected, while others correspond to no predicted energy. Despite this problem, models have been conceived, particularly that of Lingenfelter et al. [37] . These authors indicate the conditions under which a binary system could emit both lines with known energies, such as those at 511 keV, 2.23 MeV; 7.632 and 7.646 MeV $^{56}Fe$ lines, and redshifted lines when the radiation occurs in the vicinity of a neutron star with a strong gravitational shift. For example, the lines observed at 0.41, 1.8 MeV and 5.9 MeV would correspond to a z of 0.28 on the 511 keV, 2.23 MeV and 7.6 MeV $^{56}Fe$ lines, and they would imply a neutron star mass of about 1.4 to 1.8 $M_{\odot}$ and a radius of 10-13 km.

It is interesting to note in this respect that Leventhal, during a flight designed to observe the Crab Nebula, detected a line at 400 keV originating in this region [38] even though no excess in this region was observed two years earlier by Ling et al. [39] (Figure 10).

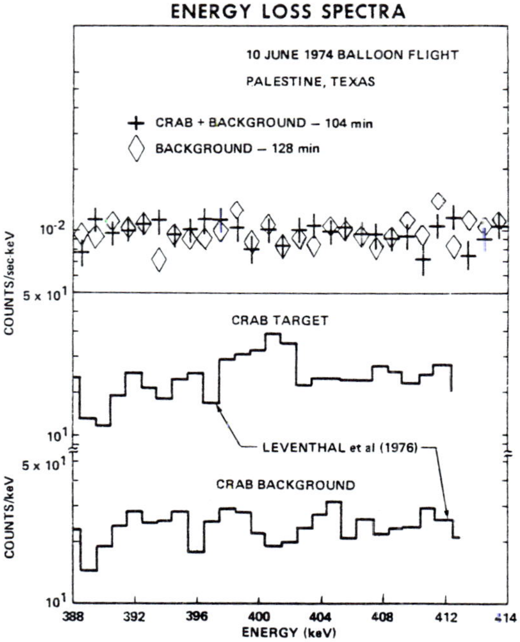


Fig. 10 : Comparison in the 400 keV region of the results obtained by Leventhal et al. [38] and Ling et al. during the observation of the Crab Nebula (From Ling et al. [39] ).

If these last two observations may be considered valid, it would be interesting to imagine that they correspond to a transient event which, given the field of view of the two experiments, could have the same spatial origin. In this respect it should be pointed out that in gamma-burst spectra no line has actually been identified [40] [41] . Nevertheless more recent results [42] seem to show a possible feature around 300-400 keV in some γ-ray bursts (9/18/78, 9/21/78).

## EXTRAGALACTIC GAMMA RAY LINES

Hall et al. [43] observed nuclear gamma lines at the 3.3σ confidence level at 4.4 and 1.6 MeV coming from the radio galaxy Cen A (Figure 11). No 511 MeV line was observed. These are currently the only results indicating the presence of gamma-ray lines in extragalactic objects. Although at medium energies the Seyfert NGC 4151 galaxy has been observed several times [44,45,46], no line has been seen in particular by Hayme's detectors, which are the most likely to detect nuclear gamma-ray lines.

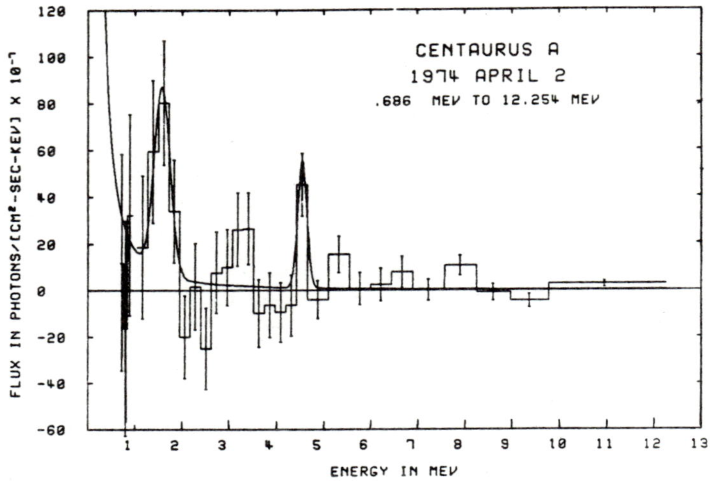


Fig. 11 : Centaurus A spectrum at energies above 0.686 MeV (From Hall et al. [43]).

## FUTURE EXPERIMENTS

Nuclear gamma-ray line astronomy, which began modestly 10 years ago has not yet been supported by a vast program of satellite investigation. The reported results were essentially obtained with balloon experiments ; in most cases they are not reliable, especially if we use the criteria given by Chupp [50] to estimate the confidence level of a reported line. It appears that except for the solar γ line and in my opinion the 0.5 MeV line from the Galactic Center, all other observed lines need to be confirmed. Even for the 0.5 MeV line, the galactic distribution remains to be studied in detail.

The most promising experiment, HEAO-C, will be launched at the end of 1979 and its sensitivity will be an order of magnitude better than most of the balloon experiments available at the present time, at least for non-broadened gamma-line studies [47] . This experiment, which includes 4 pure germanium detectors of about 100 $cm^3$ each (Figure 12) should advance this new astronomy.

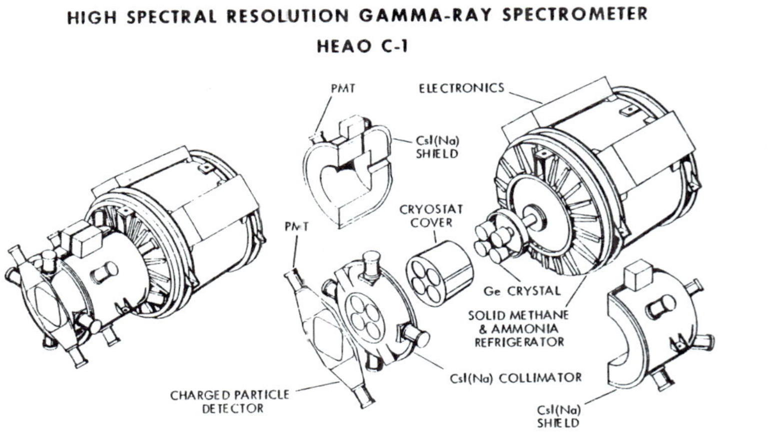


Fig. 12 : High spectral resolution gamma-ray spectrometer HEAO-C-1 (From Mahoney et al. [47] ).

Furthermore, the SMM experiment [48], a mosaic of sodium iodide detectors with sensitivity increased by a factor of about 10 compared with the OSO 7 experiment, should also contribute largely to the study of gamma lines in the field of solar physics.

It is undeniable that for the future the very important Gamma Ray Observatory will represent a completely decisive element for gamma-ray astronomy and especially nuclear γ-ray astronomy ; indeed the planned experiments include

- a mosaic of 12 pure germanium detectors, with a total volume of about 2100 $cm^3$, providing an instrument sensitivity of $10^{-5}$ photon $cm^{-2}sec^{-1}$. Such an instrument could then measure nuclear gamma line fluxes corresponding to the most recent theoretical estimates, such as those given by Ramaty et al. [7] .
- a large NaI array which will have a greater sensitivity for broadened lines.

Beyound the present possibilities of cooled detectors, more work must be done in the field of new detectors. One example is CdTe or $HgI_2$ detector. While these do not, of course, have the resolution of the germanium detectors, they can operate without any cryogenic systems and are thus much more simple, but the main problem is to increase their dimensions substantially.

REFERENCES

1 J. Trümper, W. Pietsch, C. Reppin, B. Sacco, E. Kendziorra, and R. Staubert Annals New York Academy of Sciences 302, 538 (1977)
2 J. Trümper, W. Pietsch, C. Reppin, and W. Voges and R. Staubert and E. Kendziorra Ap. J., 219, L105, (1978)
3 D.D. Clayton, Preprint to "International Symposium and Workshop on Gamma-Ray Astrophysics", NASA Goddard (1973)
4 D.D. Clayton, in γ ray Astrophysics, NASA SP 339 p. 263 (1973)
5 R.E. Lingenfelter and R. Ramaty, Physics Today, March (1978)
6 R. Ramaty, B. Kozlovsky and R.E. Lingenfelter, NASA T.M. 79654 September(1978)
7 R. Ramaty and R.E. Lingenfelter, Nature, 278, March (1979)
8 G. Vedrenne, in New Instrumentation for Space Astronomy XXI, 151-168 (1978)
9 G.H. Nakano, W.L. Imhof, R.G. Johnson, IEEE Trans. Nucl. Sc. NS-21, 159, (1974)
10 G.H. Nakano and W.L. Imhof, in Gamma Ray Spectroscopy in Astrophysics, NASA T.M. 79619, 529 (1978)

11 B.J. Teegarden, G. Porreca, D. Stilwell, U.D. Desai and T.L. Cline and Hovestadt, in Gamma Ray Spectroscopy in Astrophysics,NASA T.M.79619, 516 (1978)
12 E.L. Chupp, D.J. Forrest, P.R. Higbie, A.N. Suri, C. Tsai and P.P. Dunphy Nature, 241, 333 (1973)
13 E.L. Chupp, D.J. Forrest, A.N. Suri, Proceedings of the IAU-COSPAR Symposium D.Reidel Publishing Company, 68 (1974)
14 G. Chambon, K. Hurley, M. Niel, R. Talon, and G. Vedrenne and O.B. Likine, A.V. Kouznetsov, and I.V. Estouline, in Gamma Ray Spectroscopy in Astrophysics, NASA T.M. 79619, 70 (1978)
15 H.S. Hudson, HEAO-1 data presented at "Workshop on Particle Acceleration", La Jolla, California (1979)
16 J. Trümper, in IAU/COSPAR Symposium on X-Ray Astronomy, Innsbruck, Pergamon Press - Space Research V. III (1978)
17 J.L. Matteson, D.E. Gruber, R.E. Rothschild and J.A. Hoffman, in IAU/COSPAR Symposium on X-Ray Astronomy, Innsbruck, Pergamon Press - Space Research Vol.III (1978)
18 J.L. Matteson and D.E. Gruber and J.A. Hoffman, in Gamma Ray Spectroscopy in Astrophysics, NASA T.M. 79619, 386 (1978)
19 A.S. Jacobson, R.J. Bishop, G.W. Culp, L. Jung, W.A. Mahoney and J.B. Willett N.I.M., 127, 115 (1975)
20 J.C. Ling, W.A. Mahoney, J.B. Willett and A.S. Jacobson, Ap. J. 231, 896(1979)
21 M.S. Strickman, W.N. Johnson and J.D. Kurfess, Ap. J. Lett. to be published
22 C.A. Meegan, in Gamma Ray Spectroscopy in Astrophysics, NASA T.M. 79619, 223 (1978)
23 W.N. Johnson and R.C. Haymes, Ap. J., 184, 103 (1973)
24 R.C. Haymes, G.D. Walvaren, C.A. Meegan, R.D. Hall, F.T. Djuth, and D.H. Shelton, Ap. J. 201, 593 (1975)
25 M. Leventhal and C.J. Mac Callum and P.D. Stang, Ap. J. 225 : L11 (1978)
26 M. Leventhal and C.J. Mac Callum, 9th Texas Symposium Munich Germany, Dec.(1978)
27 R.W. Bussard and R. Ramaty and R.J. Drachman, Ap. J. in press (1979)
28 R. Ramaty, F.W. Stecker, and D. Misra, J.G.R., 75, 1141 (1970)
29 T.A. Rygg and G.J. Fishman, Proc. 13th Conf. Cosmic. Ray, 1, 472, Denver (1973)
30 J.L. Matteson, P.L. Nolan and L.E. Peterson, in IAU/COSPAR Symposium on X-Ray Astronomy, Innsbruck, Pergamon Press - Space Research V.III (1978)
31 D. Gilman, A.E. Metzger, and R.H. Parker and J.I. Trombka, Ap. J. 229, 753-761 (1979)
32 P. Mandrou, Thèse de Doctorat d'Etat n° 872, Université Paul Sabatier - Toulouse (1979)
33 J. Audouze, J. Lequeux, J.L. Masnou, J.L. Puget, accepted for publication Ast. and Ap. (1979)
34 R.W. Bussard, and R. Ramaty and K. Omidvar, Ap. J. 220 : 353-362 (1978)
35 G. Serra, Thèse de Doctorat d'Etat n° 871, Université Paul Sabatier - Toulouse (1979)
36 A.S. Jacobson, J.C. Ling, W.A. Mahoney, J.B. Willett, in Gamma Ray Spectroscopy in Astrophysics, NASA T.M. 79619 , 228 (1978)
37 R.E. Lingenfelter, J.C. Higdon and R. Ramaty, in Gamma Ray Spectroscopy in Astrophysics, NASA T.M. 79619, 252 (1978)
38 M. Leventhal and C. Mac Callum and A. Watts, Ap. J. 216, 491-502 (1977)
39 J.C. Ling, W.A. Mahoney, J.B. Willett and A.S. Jacobson, Nature,270, 36 (1977)
40 A.E. Metzger and R.H. Parker and D. Gilman, L.E. Peterson and J.I. Trombka , Ap. J. 194, L19-L25 (1974)
41 W.L. Imhof, G.H. Nakano, R.G. Johnson, J.R. Kilner, and J.B. Reagan and R.W. Klebesadel and I.B. Strong, Ap. J. 198, 717-725 (1975)
42 E.P. Mazets, S.V. Golenetsky, V.N. Ilinsky, V.N. Panov, R.L. Aptekar, Yu. A. Gurian, I.A. Sokolov, Z.A. Sokolova, T.V. Kharitonova, Ioffe Physics Technical Institute Leningrad, USSR(1979)
43 R.D. Hall, C.A. Meegan, G.D. Walraven, F.T. Djuth, and R.C. Haymes, Ap. J. 210, 631-641 (1976)

44 V. Schonfelder, Nature 274, 344 (1978)
45 G. di Cocco, G. Boella, F. Perotti, R. Stiglitz, G. Villa, R.E. Baker, R.C. Butler, A.J. Dean, S.J. Martin and D. Ramsden, Nature 270, 319 (1977)
46 C.A. Meegan and R.C. Haymes Submitted to Ap. J. (1979)
47 W.A. Mahoney, J.C. Ling, J.B. Willett and A.S. Jacobson, in Gamma Ray Spectroscopy in Astrophysics, NASA T.M. 79619, 541 (1978)
48 D.J. Forrest, in Gamma Ray Spectroscopy in Astrophysics, NASA T.M.79619,558(1978)
49 M. Leventhal, C.J. Mac Callum, P.D. Stang, in Gamma Ray Spectroscopy in Astrophysics, NASA T.M. 79619, 169 (1978)
50 E.L. Chupp, in Gamma Ray Spectroscopy in Astrophysics, NASA T.M. 79619, 42 (1978)
51 G.D. Walvaren and R.C. Haymes, Nature 264, 42 (1976)
52 M. Leventhal, C.J. Mac Callum, and P.D. Stang, Nature 275, 200 (1978)
53 W.L. Imhof and G.H. Nakano, Ap. J. 214, 38 (1977)
54 G.H. Nakano, W.L. Imhof and J.B. Reagan, Space Science Instrumentation 2,219 (1976)
55 A.S. Jacobson, American Physical Society, April (1977)
56 M. Yoshimori, H. Watanabe, K. Okudaaira, Y. Hirasima and H. Murakami, to be published in the Australian Journal of Physics (1979)
57 F. Albernhe, D. Boclet, J. Claisse, Ph. Durouchoux, P. Frabel, J. Marques, da Costa, E. Olivier, P. Pagnier, R. Rocchia, G. Vedrenne, Proc. 12th Eslab Symposium, Frascati, Italy (1977)
58 G.H. Nakano, W.L. Imhof, R.G. Johnson, IEEE Trans. Nucl. Sc., NS 21, 159 (1974)
59 Symposium Gamma Ray Spectroscopy in Astrophysics , NASA Technical Memorandum 79619 August 1978, T.L. Cline and R. Ramaty ed.
60 F. Albernhe, G. Vedrenne and I.M. Martin, accepted for publication J.G.R.(1979)
61 Ph. Durouchoux, D. Boclet, R. Rocchia, F. Albernhe, J.F. Le Borgne, G. Vedrenne, J. Marques da Costa, This volume(1979)

SESSION 3

# Cosmic Gamma-ray Lines and Bursts

*Chairman:* **S. HAYAKAWA**

# COSMIC GAMMA-RAY LINES: THEORY

R. E. Lingenfelter* and R. Ramaty**

*Physics Department, University of California, San Diego, La Jolla, CA, U.S.A.
**Laboratory for High Energy Astrophysics, Goddard Space Flight Center, Greenbelt, MD, U.S.A.

## ABSTRACT

The various processes that lead to gamma-ray line emission and the possible astrophysical sources of such emission are reviewed. In particular we consider the processes of nuclear excitation, radiative capture, positron annihilation and cyclotron radiation, which may produce gamma-ray line emission from such diverse sources as the interstellar medium, novae, supernovae, pulsars, accreting compact objects, the galactic nucleus and the nuclei of active galaxies. The significance of the relative intensities, widths and frequency shifts of the lines are also discussed. Particular emphasis is placed on understanding those gamma-ray lines that have already been observed from astrophysical sources.

## INTRODUCTION

Gamma ray line astronomy is an exciting new field and one with enormous promise, for gamma-ray lines can reveal directly basic nuclear processes in astrophysics that have so far been studied only indirectly.

Here we briefly review the theoretical basis of gamma-ray line astronomy. In particular we consider the physical processes leading to gamma-ray line emission, the principal gamma-ray line energies associated with each, the likely astrophysical sources of line emission and their expected entensities. We will also discuss what can be learned about the sources from measurements of the intensities, widths and shifts of the gamma-ray lines, and at the same time we will try to understand those cosmic gamma-ray lines which have already been observed. The line observations are reviewed by Vedrenne [1] in this volume and also in recent general reviews[2,3,4] of gamma ray line astronomy, including solar flare line emission.

## LINE EMISSION PROCESSES

There are several basic processes which lead to gamma ray line emission: nuclear de-excitation, radiative capture, positron annihilation and cyclotron radiation.

Excited nuclear levels which decay by photon emission can be populated in a variety of ways: inelastic collisions and spallation reactions by energetic particles, fusion reactions in the processes of nucleosynthesis and decay of radionuclei produced in all of these ways. The principal de-excitation lines expected from energetic-particle interactions are those at 4.438 MeV and at 6.129 MeV primarily from $^{12}C(p,p')^{12}C^{*\,4.439}$, $^{16}O(p,p\,\alpha)^{12}C^{*\,4.439}$, and $^{16}O(p,p')^{16}O^{*\,6.131}$. These and many other lines from energetic-particle interactions have recently been reviewed [5]. Observation of such lines above the electron bremsstrahlung background requires an average electron energy $\overline{E}_e < 1$ MeV and an average ion energy $\overline{E}_{ion} > 1$ MeV/nucleon, or $T_e < 10^{10}K < T_{ion}$. The principle de-excitation lines expected from explosive nucleosynthesis are those at 0.847 MeV and 1.238 MeV from decay of synthesized $^{56}Ni$ to $^{56}Co$ to $^{56}Fe^{*\,0.847}$ and $^{56}Fe^{*\,2.085}$. These and other lines from nucleosynthesis, for example the 1.809 MeV line from $^{26}Al$ decay, have been previously reviewed [2,6,7]

The principle gamma-ray line from radiative capture of neutrons made in inelastic nuclear interactions is expected to be that at 2.223 MeV from $^{1}H(n,\gamma)^{2}H$ although emission at 7.632 and 7.646 MeV from $^{56}Fe(n,\gamma)^{57}Fe$ may also be important, depending on the relative abundance of iron compared to hydrogen. Observation of these lines requires an average density of $\gtrsim 10^{16} H/cm^3$ in order that a significant fraction of the neutrons are captured before they decay. Gamma-ray line emission at 0.511 MeV results from the annihilation of positrons which can be produced by a variety of processes. These include the decay of radionuclei and $\pi+$ mesons, de-excitation of nuclear levels, e.g. $^{16}O^{*\,6.052}$, which decay by $e^+$-$e^-$ pair emission, and pair production by photons in matter and in strong magnetic fields. The yield of 0.511 MeV photons per positron annihilation depends [8] on the relative rates of direct annihilation, which yields 2 photons of 0.511 MeV per positron, and of annihilation from bound states of positronium which yields 0.5 photon of 0.511 MeV per positron. Only positronium in the singlet spin state annihilates into 2 photons of 0.511 MeV while those in the triplet state, which is three times more probable, annihilate into three photons with a continuum spectrum below 0.511 MeV. These relative rates of annihilation have been studied [9,10] as functions of temperature, density and degree of ionization.

Cyclotron line emission at gamma-ray energies can be produced [11,12,13] in very strong magnetic fields ($B > 10^{12}$ gauss) by electron transitions between quantized levels of the electron momentum perpendicular to the magnetic field. The primary line is $E = [(1+2B/B^*)^{\frac{1}{2}}-1]\, M_e C^2 \approx (B/B^*) M_e C^2$ if $B << B_*^{*}$ where $B^* = M_e{}^2 C^3/eh = 4.413 \times 10^{13}$G. The observed width of the line is strongly dependent on both the electron temperature and the angle between the observer and the direction of the magnetic field [14].

## SOURCES OF ASTROPHYSICAL LINES

We now consider a variety of astrophysical sites where these line emission processes might be expected to be active: the interstellar medium, novae, supernovae, neutron stars and black holes. Although gamma-ray line emission must have been produced by such reactions as $^{1}H(n,\gamma)^{2}H(p,\gamma)^{3}He$ within the first $10^4$ s of the Big Bang, the universe at that time was so opaque to gamma rays that they were quickly degraded to ambient photons. Moreover, calculations of nucleosynthesis in the Big Bang suggest that no significant number of gamma-ray emitting radionuclei could have survived for over $10^7$ years until the universe finally became transparent. Thus, quite aside from the enormous cosmological redshift such lines would suffer, no detectable primordial gamma-ray line emission is expected.

## INTERSTELLAR MEDIUM

Diffuse gamma-ray line emission in the interstellar medium is expected both from low-energy cosmic-ray interactions with the gas and dust and from long-lived radionuclei produced in the continuing process of stellar nucleosynthesis.

The low-energy cosmic ray intensity in the interstellar medium is not known because solar modulation excludes the bulk of the low-energy particles from the inner solar system. But a rough upper limit to the gamma-ray line emission expected from these cosmic rays has been calculated [5], assuming a local interstellar density of 1 eV/cm$^3$ and substantial gradients along the line of sight towards the galactic center in both the cosmic-ray energy density and heavy-element abundances. If the cosmic-ray intensity were increased any further, X-ray line emission at ∿ 6.8 keV due to charge exchange of fast Fe ions with interstellar matter would exceed the observational upper limit on this line, and if the heavy-element abundances along the line of sight were increased, the absorption of X-rays would be larger than observed. Such low-energy cosmic-ray interactions in the interstellar medium, can produce a rich gamma-ray spectrum (see Figure 1) which includes many nuclear deexcitation lines. Of special interest are the very narrow lines such as that at 6.129 MeV from $^{16}$O, resulting from deexcitation of nuclei in interstellar grains [15]. The line broadening, which in gases is caused by the recoil velocities of the excited nuclei, is greatly reduced in solids where the excited nuclei or their radioactive progenitors can come to rest before deexcitation. The width of such lines would thus result primarily from differential galactic rotation, ranging up to about 200 km/sec along a line of sight and Doppler-shifting the lines over as much as several keV. The detection of these gamma-ray lines from low-energy cosmic-ray interactions in the interstellar medium with forthcoming instrumentation could not only measure the unknown interstellar intensity of the low-energy cosmic rays, but also provide unique information on the distribution, motion, composition and size of interstellar dust grains.

The strongest resolvable lines in the diffuse galactic background, however, may be from explosive nucleosynthesis [16,17,6] at 1.809 MeV from $^{26}$Al decay to $^{26}$Mg, and at 1.332 and 1.173 MeV from $^{60}$Fe decay, with mean lives of 1.1 x 10$^6$ and 4.3 x 10$^5$ yrs respectively. The bulk of these nuclei produced in supernova explosions should be fairly well mixed in the interstellar medium before their decay. They should produce a diffuse source of very narrow line emission concentrated along the galactic plane. Assuming, from explosive-nucleosynthesis calculations, that about 0.1% of the present $^{26}$Mg results from $^{26}$Al decay, and about 1% of the $^{60}$Ni results from $^{60}$Fe decay and that the rate of nucleosynthesis has decreased with time so that the rate averaged over the age of the Galaxy is roughly ten times the present rate, then the intensities of these three lines should all be about the same, 5 x 10$^{-5}$ photon cm$^{-2}$s$^{-1}$rad$^{-1}$ in the general direction of the galactic center. The discovery [18] of anomalous $^{26}$Mg in the Allende meteorite lends support to the possibility of significant $^{26}$Al production by explosive nucleosynthesis in supernovae. The corresponding differential intensity of the $^{26}$Al and $^{60}$Fe lines is about 2 x 10$^{-2}$ photon cm$^{-2}$s$^{-1}$rad$^{-1}$MeV$^{-1}$ roughly an order of magnitude larger than the diffuse background intensity at these energies (Figure 1). Observation of these lines would provide the first direct evidence for the explosive nucleosynthetic origin of the heavy elements, demonstrate that nucleosynthesis is a continuing process in the Galaxy and measure its present rate.

Currently the best observed [19,20] galactic gamma-ray line is at 0.511 MeV from the direction of the galactic center. The observed [19] line at 510.7 ± 0.5 keV has an intensity of 1.2 x 10$^{-3}$ photons cm$^{-2}$s$^{-1}$ and a full width at half maximum (FWHM) of less than 3.2 keV. This observation was made in 1977 with a balloon-borne high-purity Ge detector of opening angle 15°, centered on the galactic center. The line energy clearly identifies it as positron annihilation radiation; there is also some evidence for the three-photon continuum from triplet positronium annihilation, implying ∿90% annihilation via positronium. This fraction is consistent with positron annihilation in a low-density (< $10^{15}$ $cm^{-3}$) and low-temerature (< $10^6$K) medium [9], and implies about 0.6 photon in the 0.511 MeV line per positron.

The 3.2 keV upper limit on the width of the 0.511 MeV line suggests [10] that the bulk of the positrons annihilate in an ionized medium ($n_e/n_H \gtrsim 0.05$), since in cold interstellar clouds, where the bulk of the positronium is formed before the positrons thermalize, the line width (∿ 5 keV) would exceed the observed upper limit. Possible sites of annihilation could be supernova remnants, HII regions at the galactic center and the warm ionized component of the interstellar medium. Observations with higher spatial resolution are needed to determine whether the emission comes from a broad region of the galactic disk or just from a discrete source at the galactic center.

If the observed 0.511 MeV line is diffuse emission from the galactic disk, it cannot result from low-energy cosmic-ray interactions which at most could produce about 1.6 x $10^{-5}$ photon $cm^{-2}s^{-1}$ at 0.511 MeV (Figure 1) within $15^o$ field of view - only about 1% of the observed intensity. High-energy (≳ 1 GeV) cosmic-ray interactions in the interstellar medium producing $\pi^+$ mesons which decay to positrons, also cannot be the source of the annihilating positrons. Since $\pi^o$ mesons which decay by high-energy gamma-ray emission are also produced in these interactions, the observed [21] flux (∿3 x $10^{-5}$ photon $cm^{-2}s^{-1}$) of high-energy (>100 MeV) gamma-rays from a $15^o$ cone centered on the galactic center implies that not more than about 2% of the observed 0.511 MeV line can be from the annihilation of positrons from $\pi^+$ decay.

The most likely source of these positrons seems to be the relatively long-lived positron-emitting radioisotopes produced by nucleosynthesis during the explosive ejection of matter in supernova explosions [22,23,24]. The principal positron-producing decay chains are shown in Table 1, together with the relevant mean lives, accompanying deexcitation line energies, and numbers of positrons and photons per disintegration. The supernova yields relative to the $^{56}$Ni yield are based on solar-system abundances, and the assumption that all the $^{56}$Fe, $^{57}$Fe and $^{44}$Ca in the Galaxy, and 1%, 0.5% and 0.1% of the galactic $^{60}$Ni, $^{22}$Ne, and $^{26}$Mg, respectively, are produced via the decay chains shown in this table. The absolute yield of $^{56}$Fe assumes a total of 1.7x$10^9$ supernovae [6]. This corresponds to a rate of occurrence of supernovae in the Galaxy that was much higher in the past, so that the average rate over the age of the Galaxy is several times the present rate.

Although nucleosynthesis takes place in dense regions which are opaque to gamma rays, the expanding envelope can become transparent if there is sufficient delay between the synthesis of the isotopes and the emission of the photons. Moreover, even if the lifetime of $^{56}$Co is not long enough for the supernova to become transparent to its deexcitation lines, a significant fraction of the positrons from $^{56}$Co may escape from the supernova [23] and annihilate in regions which are transparent to the gamma rays. Since the slowing-down and annihilation times of the positrons can be longer than the interval between supernova explosions this could lead to an essentially steady galactic emission at 0.511 MeV.

It has been estimated [23] that about 0.1 of the positrons from $^{56}$Co decay could escape from the supernova. This escape fraction would make $^{56}$Co the dominant positron source for processes of nucleosyntheis. The production of ∿2 x $10^{43}$ positrons $s^{-1}$ in the Galaxy, as indicated by the 0.511 MeV line observation [19], would imply a present rate of nucleosynthesis of ∿$10^{45}$ iron nuclei $s^{-1}$ This is about 10% of the average rate of nucleosynthesis over the age of the Galaxy, and is consistent with current models [25] of galactic evolution. The fraction of escaping positrons from supernovae is very uncertain [23,26], however, and requires further theoretical study. If the escape fraction for $^{56}Co$ were much lower than 0.1, $^{22}Na$ decay could become the dominant positron source, although this yield [27] is very uncertain as well. Novae may also produce a $^{22}Na$ yield [28] of about $10^{48}$ nuclei per nova explosion, but the observed 0.511 MeV line emission would require about 600 novae per year which is an order of magnitude higher than current estimates.

One test of the explosive nucleosynthetic origin of the positrons would be the detection of the associated deexcitation gamma ray lines, especially the 1.809 MeV line from $^{26}Al$ decay, discussed above.

TABLE 1 Gamma Ray Producing Decay Chains From Supernovae and Novae

| Decay Chain | Mean Life (yr) | Nuclei per Supernova | Photon Energy (Mev) | Photons or Positrons per disintegration |
|---|---|---|---|---|
| $^{56}Ni \rightarrow ^{56}Co \rightarrow ^{56}Fe$ | 0.31 | $3x10^{54}$ | 0.847 | 1 |
| | | | 1.238 | 0.70 |
| | | | 2.598 | 0.17 |
| | | | 1.771 | 0.16 |
| | | | 1.038 | 0.13 |
| | | | $e^+$ | 0.2 |
| $^{57}Co \rightarrow ^{57}Fe$ | 1.1 | $7x10^{52}$ | 0.122 | 0.88 |
| | | | 0.014 | 0.88 |
| | | | 0.136 | 0.12 |
| $^{22}Na \rightarrow ^{22}Ne$ | 3.8 | $2x10^{52}$ | 1.275 | 1 |
| | | ($10^{48}$/nova) | $e^+$ | 0.9 |
| $^{44}Ti \rightarrow ^{44}Sc \rightarrow ^{44}Ca$ | 68. | $5x10^{51}$ | 1.156 | 1 |
| | | | 0.078 | 1 |
| | | | 0.068 | 1 |
| | | | $e^+$ | 0.94 |
| $^{60}Fe \rightarrow ^{60}Co ^{60}Ni$ | $4.3x10^{5}$ | $5x10^{50}$ | 1.332 | 1 |
| | | | 1.173 | 1 |
| | | | 0.059 | 1 |
| $^{26}Al \rightarrow ^{26}Mg$ | $1.1x10^{6}$ | $4x10^{50}$ | 1.809 | 1 |
| | | | 1.130 | 0.04 |
| | | | $e^+$ | 0.85 |

NOVAE

Accretion from a large, cool companion onto a white dwarf leading to thermonuclear runaway in the CNO cycle has been suggested as the cause of novae [e.g.28]. If so, novae could also be sources of gamma-ray line emission from the decay of synthesized radionuclei. The easiest line to detect is at 1.275 MeV from $^{22}Na$ decay. For the $^{22}Na$ yield per nova [28] as given in Table 1, the flux of the 1.275 MeV line from a nova at 1 kpc, in photon $cm^{-2}s^{-1}$, is $\sim 8x10^{-5} exp(-t/3.8$ years). The limit of $8x10^{-4}$ photon $cm^{-2}s^{-1}$ set by Leventhal et al. [29] on Nova Cygni (distance = 1.5 kpc) 9 months after the explosion is larger by more than an order of magnitude than the expected flux from this nova.

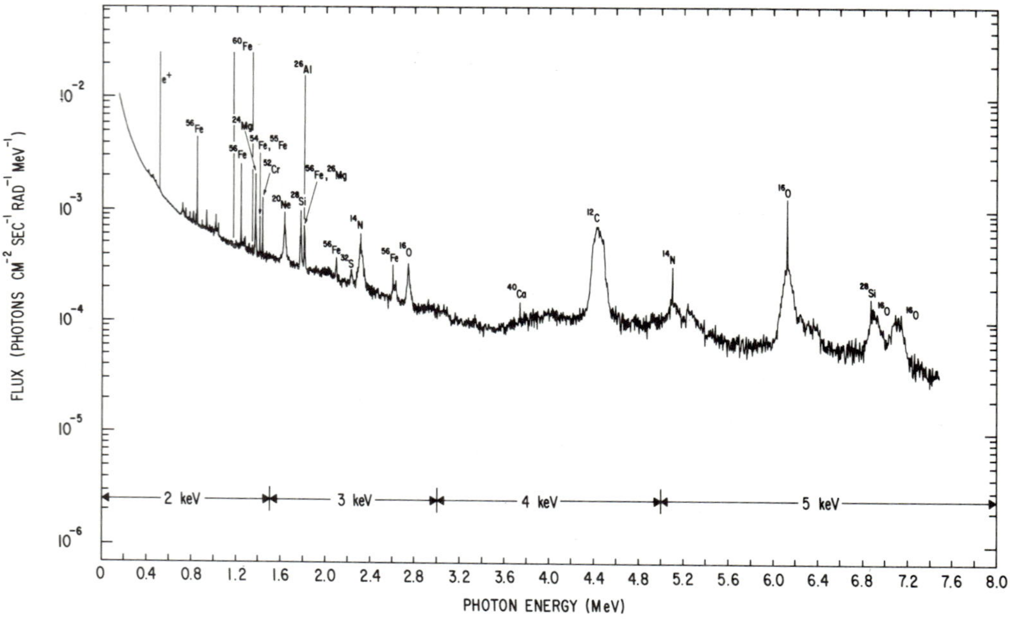


Fig. 1. Calculated [5] gamma-ray spectrum from low-energy cosmic-ray interactions with gas and dust in the direction of the galactic center.

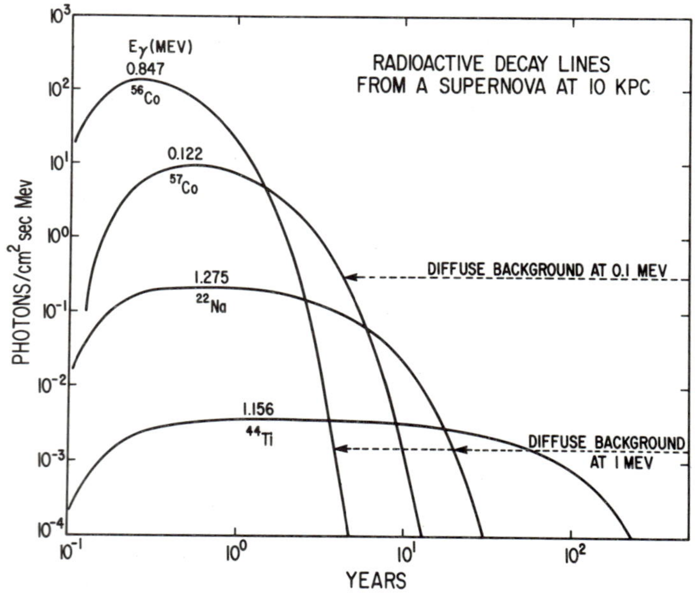


Fig. 2. Intensities [6,22] of gamma-ray lines from the decay of nucleosynthetic radionuclei in the expanding nebula of a young supernova 10 kpc away, shown as a function of time after the explosion. The yields of radionuclei are given in Table 1 and the line widths of 8,56,77 & 85 keV for the 0.122,0.847,1.156 & 1.275 MeV lines are based on an expansion velocity of $10^4$ km s$^{-1}$.

Nevertheless, for an estimated nova rate of ~40 per year, a few closeby novae should occur during a two year gamma-ray astronomy mission. With a detector sensitivity of $\sim 2\times10^{-5}$ photon $cm^{-2}s^{-1}$ for a line of FWHM ~10 keV at 1 MeV, an individual nova could be detected from a maximal distance of about 2 kpc. But because the time between nova explosions is much shorter than the $^{22}Na$ mean life, a detector with a broad field of view observing the galactic plane in the direction of the galactic center would probably observe an essentially diffuse emission at 1.275 MeV of $\sim 10^{-4}$ photon $cm^{-2}s^{-1}rad^{-1}$.

## SUPERNOVAE

Supernovae may be the most intense discrete sources of gamma-ray lines through nuclesynthesis of relatively long-lived radionuclei that decay with gamma-ray emission after the explosive nebula has become transparent to such radiation. The principle radionuclei and the associated gamma-ray lines expected [6,16,22] from these sources are listed in Table 1.

As can be seen the strongest lines expected from a supernova are those from $^{56}Co$ decay, if the exploding nebula becomes transparent to gamma radiation in a time at least comparable to the $^{56}Co$ decay mean life. This condition is dependent on both the mass and the expansion velocity of the ejected material, but a supernova nebula of $\sim 1M_{\odot}$ expanding with a velocity of $\sim 10^4$km $s^{-1}$, should start to become transparent to gamma rays after about 0.3 yrs, which is just sufficient to allow significant gamma ray line emission from $^{56}Co$ decay to escape. For such conditions the expected [6,22] intensity at the Earth of the various decay lines from a supernova at a distance of 10 kpc is shown in Figure 2, as a function of time after the explosion.

For differing periods of time, all of these lines should be detectable above the diffuse background. The predicted luminosities of the $^{56}Co$ decay lines are so large that for nearly a year after a supernova explosion on the farthest edge of the galaxy, the line intensity seen at the Earth should be greater than that observed in the strongest lines during the August 1972 solar flares [30]. But because the $^{56}Co$ decay lines exceed the diffuse background for less than three years, these lines should be observable from any galactic supernova less than 10% of the time, if the rate of occurrence of galactic supernovae is about 1 to every 25 years. The line most likely to be observed therefore is that at 1.156 MeV from $^{44}Sc$ decay, which should have an intensity from some galactic supernova roughly comparable to that of the diffuse background all of the time. The 1.275 MeV line from $^{22}Na$ and the 0.122 MeV from $^{57}Co$ should also be observable ~50% and ~10% of the time, respectively.

If, however, the bulk of galactic nucleosynthesis occurs only in supernovae of the most massive stars which are much less frequent, then the probability of observing gamma ray line emission would be greatly reduced even though the yield of radionuclei per event might be a factor of 30 greater 25 . Moreover, the strongest lines would not be those from $^{56}Co$ decay because, for a nebula roughly 30 times more massive, an expansion velocity of $10^4$ km/sec would not make it transparent to gamma radiation for at least a couple of years, by which time the gamma ray emissivity from $^{56}Co$ decay would have fallen below that from $^{57}Co$, making it the strongest line.

There is also a small chance of observing gamma-ray lines from supernovae in nearby galaxies if most supernovae contribute significantly to galactic nucleosynthesis with yields such as those in Table 1. Supernovae in the Large Magellanic Cloud could provide observable line emission at 1.275 MeV from $^{22}Na$ decay roughly 3% of the time and 0.847 MeV from $^{56}Co$ decay about 1% of the time. The probability of observing the 0.847-MeV line from a supernova in M31 is also about 1% and the chance of observing it from a supernova in some other galaxy within 3 Mpc is about 4%. But if only the less-frequent, massive supernovae are the major sites of nucleosynthesis, then the chance of seeing gamma-ray line emission from radioactive decay in supernovae in any galaxy other than ours drops to about 0.1%. In either case the supernova in NGC5253 in May 1972 could have produced detectable line emission at 0.847 MeV from $^{56}Co$ decay with an intensity of $\sim 6 \times 10^{-5}$ photon $cm^{-2}s^{-1}$, roughly equal to the background. Unfortunately no detectors of sufficient sensitivity were being flown at the time.

A search for these gamma-ray lines can help determine how large the yields of such nuclei are from explosive nucleosynthesis, and what portion of the supernovae do make significant contributions to galactic nucleosynthesis.

The possible detection of the $^{44}Ti$ decay line at about 1.16 MeV reported [31] from a point in the galactic plane ($\ell=345^o$) has not been confirmed by recent, more sensitive, observations [32].

## NEUTRON STARS AND BLACK HOLES

Gamma-ray line emission from neutron stars is expected from a variety of processes: cyclotron radiation, positron annihilation, neutron capture, and nuclear interactions in accreting matter. Line emission from each of these processes may already have been detected.

Cyclotron line emission appears to have been observed [33,34,35] from the binary pulsar Her X-1, at about 58 keV with somewhat variable intensity on the order of $10^{-3}$ photons $cm^{-2}s^{-1}$. If this line is cyclotron radiation in a strong magnetic field of the neutron star, the line energy would imply a field strength of $\sim 5 \times 10^{12}$G or less depending on the gravitational redshift. This is consistent with fields that might be expected near the surface of a pulsar. But since the X-ray continuum seems to imply a temperature of 20 keV or more, the reported line width (FWHM) of $\sim$12 keV severely constrains [14] the angle between the direction of observation and the magnetic field to be very close to $90^o$.

A line at about 73 keV has been reported [36] from the direction of the Crab pulsar, but was not seen in a more recent observation [37]. Its very narrow width, FWHM $\sim$5 keV, also makes it extremely difficult to reconcile with present models of cyclotron radiation.

Significant electron-positron pair production could occur [38] if such strong magnetic fields ($\sim 10^{12}$G) exists around yound neutron stars and positron annihilation near the neutron-star surface could lead to detectable gravitationally-redshifted 0.511-MeV line emission. Such line emission may have been observed [29] at 0.400 MeV with an intensity of $(2.2 \pm 0.7) \times 10^{-3}$ photon $cm^{-2}s^{-1}$ from the direction of the Crab Nebula, using a balloon-borne, high_resolution Ge detector. If this is positron annihilation radiation redshifted by Z of $\sim$0.28 from the Crab pulsar the implied mass of the neutron star is $\sim$1.4 to $\sim$1.8 $M_o$, depending on the equation of state. The reported line width of $\sim$5 keV also places an upper limit on the magnetic field $\lesssim 10^{12}$G since the annihilation-line width [39] should be $\gtrsim M_e C^2 \ln\sqrt{2}\,(B/B^*)$. Two years earlier a similar detector [40] did not observe this feature when the Crab Nebula was in the center of the field of view, but a burst at energies of 0.40 to 0.42 MeV was observed from the same general direction. This burst also showed up in line emission at higher energies and we will discuss the event in more detail below.

Accretion of matter onto neutron stars and black holes has frequently been suggested as a mechanism for both steady state and transient, continuum X and gamma-ray emission. Such processes, where ion energies may exceed several MeV, could also lead to significant gamma-ray line emission from nuclear interactions of the infalling matter with that of a neutron-star surface [41] or from quasithermonuclear excitation of matter, either in a standing shock above the surface of a neutron star or in an accretion disk around a neutron star or a black hole [42]. Line emission stronger than the associated electron-bremsstrahlung background could result as long as the effective ion and electron temperatures are such that $T_{ion} > 1$ Mev $> T_e$. This will occur if thermalization of the infalling matter is incomplete, because the gravitational energy released in the accreting matter is originally almost entirely in the nuclei, rather than being equipartitioned with the electrons. Dynamical models [e.g. 43,44] of such accretion suggest that thermal equilibrium between the ions and electrons is not achieved and that the effective ion temperatures may be an order of magnitude or more greater than the electron temperatures.

The strongest expected deexcitation lines are those from $^{12}C^*$ and $^{16}O^*$ at 4.438 and 6.129 MeV. They should be thermally broadened and shifted by as much as an MeV. The gravitational broadening, unlike the thermal, is not symmetrical but essentially reflects the radial dependence of the square of the accreting ion density around the neutron star or black hole. Positrons and neutrons are also generated in such processes so that there should be additional line emission from positron annihilation at 0.511 MeV, and for accretion around neutron stars there might be significant neutron capture at or near the star surface where the density exceeds $10^{16}$ H cm$^{-3}$, leading to line emission at 2.223 MeV from capture on hydrogen and at 7.632 and 7.646 MeV from capture on iron. These radiative-capture lines could be gravitationally red-shifted by an MeV or more, but, if the radiation is produced in a thin surface layer with a temperature $< 10^7$K, both the gravitational and thermal broadening could be much smaller. Neutron capture on iron also leads to strong X-ray line emission at 14.4 keV, comparable in intensity to that of the combined 7.632 and 7.646 MeV line.

High resolution measurements of both continuous and transient X-and gamma-ray source spectra above ~0.01 MeV may thus reveal significant nuclear-line-emission, and determination of the broadening and redshift of such lines could provide unique information on the accretion process. Although no survey at gamma-ray energies has been made of possible accretion sources, there are two measurements of transient events which may be from unidentified accretion sources--that of the 27 April 1972 gamma-ray burst with a low-resolution detector [45] on Apollo 16 and that of 10 June 1974 with a balloon-borne high-resolution Ge detector [46].

The energy spectrum of the 27 April 1972 burst shows enhanced emission at about 4 MeV, and it is tempting to attribute this enhancement to slightly redshifted and broadened line emission at 4.438 MeV from $^{12}C^*$ deexcitation. But the measured intensity is not high enough to permit resolution of such a line with any statistical significance.

The transient gamma-ray line emission observed [46] on 10 June 1974 lasted about twenty minutes and was characterized by strong emission in four relatively narrow energy bands or lines with no detectable continuum. The four observed lines at energies of 0.40-0.42 MeV, 1.74-1.86 Mev, 2.18-2.26 MeV and 5.94-5.96 MeV had average intensities of $7 \times 10^{-3}$, $3.2 \times 10^{-2}$, $1.5 \times 10^{-2}$, and $1.5 \times 10^{-2}$ photon cm$^{-2}$s$^{-1}$, during the event. These lines showed statistically significant increases in emission compared both to the intensity in the adjacent continuum and to that in the same energy bands before and after this event. A simultaneous increase was observed in the shield detectors with an intensity consistent with that observed just in the lines. This transient source appears to lie within about 20° of $\ell = 195°$. No simple nuclear schemes can account for all four observed lines, since the lines at about 0.41 and 5.95 MeV cannot be identified with any direct emission lines which are not coupled with stronger companion lines which were not observed. All of these lines however can be identified [47] with lines from the most intense nuclear emission processes expected in nature, positron annihilation and neutron capture on hydrogen and iron, if a self-consistent range of gravitational redshift is assumed. In particular the 5.935-5.960 MeV line in the transient event can be identified with the two strongest lines from neutron capture on the iron crust of a neutron star having a surface gravitational redshift of $z \simeq 0.285$. Because it has the maximum binding energy per nucleon $^{56}Fe$ is expected to be the dominant constitutent of neutron star crusts and the relative intensity of this line compared to the hydrogen capture line requires enormous enrichment in iron compared to solar abundances. This line alone strongly suggests a neutron star source for the transient emission on 10 June 1974.

The 0.40-0.42 MeV and the 1.74-1.86 MeV lines can similarly be identified as the positron annihilation and neutron capture in a hydrogen atmosphere of density $n_H > 10^{16}$ H cm$^{-3}$ from the surface z of 0.285 up to an altitude corresponding to a z of about 0.24. Such an atmosphere is presumably a temporary feature formed by infall of gas from an accretion disk. This minimum hydrogen density is required in order that a significant fraction of the neutrons are captured before they decay. The width of the ~0.4 MeV line also places a limit on the average temperature of this atmosphere of $<2 \times 10^6$ K. The unshifted ($z \simeq 0 \pm .02$) line at 2.22 MeV requires a comparable high density region well above the neutron star surface and can be understood as neutron capture in an accretion disk at a distance corresponding to $z < .02$ and in the atmosphere of a binary companion.

A surface redshift of ~ 0.285 implies a neutron star mass of about 1.4 to 1.8 $M_\odot$ and a radius of 10 to 13 km, depending on the assumed equation of state [48]. The observed 25 keV full width of the ~5.95 MeV line reduced by the instrumental broadening of about 5 keV and the $^{56}Fe$ doublet separation of 14.4 keV gives an emitted width of about 10 keV. If the line is produced uniformly over the surface of the neutron star, this width gives a lower limit for the rotation period $P > \frac{E_\gamma}{\Delta E_\gamma} \frac{4\pi r \sin\theta}{c} = \frac{\sin\theta}{4}$ sec, where $\theta$ is the angle between the rotation axis and the direction of observation. Observation of the ~5.95 MeV and ~1.8 MeV lines from the surface of the neutron star also places an upper limit on the surface magnetic field of $10^{12}$ G, since pair production by these gamma-rays in the magnetic field would greatly attenuate emission from the surface if the field were much stronger.

Thus it was suggested [47] that the neutrons and positrons which were captured and annihilated in the vicinity of the neutron star and its binary companion were all produced by nuclear reactions occurring when gas, flowing episodically from the companion, was gravitationally accelerated to an energy of > 1 MeV and collided with gas in the outer part of the accretion disk around the neutron star. This is shown schematically in Figure 3. There are, however, unresolved problems with such a model, since for any likely source distance (>pc) the total $^2H$ and $^{57}Fe$ produced by neutron capture per unit area of the neutron star surface would be many times the gamma ray attenuation length, requiring stripping, connection or other processes which could bring or maintain $^1H$ and $^{56}Fe$ at the star surface.

The observation [29] of a possible redshifted positron-annihilation line at ~0.400 MeV from the general direction of the Crab Nebula on 10-11 May 1976 may also be variable emission from the same source as the 10 June 1974 event, since the two observations had overlapping fields of view and the identification of the lines in both observations implied essentially the same neutron star surface redshift. We should also note that if the 10-11 May 1976 & 10 June 1974 line emission is for the same source, then the limit on the rotation period P (sec) > 0.25 sin Θ for the 10 June 1974 source effectively rules out the Crab pulsar with a period of 0.033 s as the source.

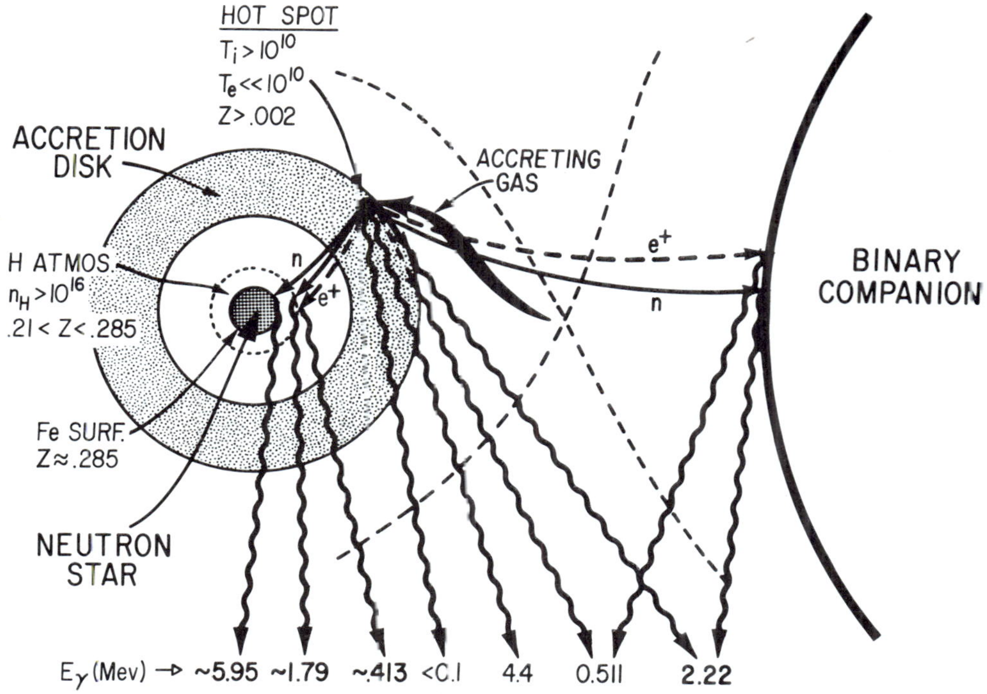


Figure 3. Schematic model [47] for the 10 June 1974 transient event.

Finally it has been suggested [47] that spherically accreting blackholes may be very strong gamma-ray line sources, emitting a large fraction of their total luminosity in the form of gamma-ray lines. High efficiencies (>10%) for gamma-ray line emission can result from spherical accretion around a black hole if $t_{ff} < t_{eq} < t_{rad}$, where the three times are the free fall time of accreting matter to the event horizon, the energy-equipartition time between accreting ions and electrons, and the radiation time required for the electrons to radiate away the released gravitational energy of the accreting matter.

The first inequality, $t_{ff} < t_{eq}$, implies that only a fraction, $\sim t_{ff}/t_{eq}$, of the total gravitational energy released in accretion is transferred from the ions to the electrons, so that the bulk of this energy is carried over the event horizon still in the ions. Taking the free-fall time to the event horizon from some distance < 2 x $10^3$ GM/$c^2$ at which the ion temperature could start to exceed $10^{10}$K but the electron temperature is still <$10^{10}$K, it can be shown [47] that $t_{ff} \lesssim t_{eq}$ requires a column density of matter above the event horizon N < 6 x $10^{26}$ H cm$^{-2}$. Since the attenuation length for 4.4 MeV gamma rays in hydrogen is $\sim 10^{25}$ H cm$^{-2}$, the upper limit on the column density above the event horizon places no added constraint on the thickness of the emission region.

The second inequality, $t_{eq} < t_{rad}$, implies that most of the energy transferred to the electrons goes into thermal energy, heating them up, since they can radiate away only a fraction, $\eta \approx t_{eq}/t_{rad}$, of that energy. Assuming that bremsstrahlung emission is the principle radiative loss process for the electrons, the fraction of energy radiated may be rewritten $\eta \approx \dot{w}_{brem}/\dot{w}_{ion}$. where $\dot{w}_{brem}$ is the rate of energy loss of the electrons by bremsstrahlung and $\dot{w}_{ion}$ is the rate of energy transfer of the ions to the electrons. This ratio $\dot{w}_{brem}/\dot{w}_{ion}$ will [47] be less than unity if $T_e < 10^{10}K < T_{ion}$. As discussed above, this condition is already required in order to observe nuclear gamma-ray lines such as that at 4.4 MeV above a bremsstrahlung continuum. Such accretion around a massive blackhole may account for broad line emission at about 4.4 MeV and about 1.6 MeV reported [49] from the radiogalaxy Centaurus A, but further observations are required to confirm these lines.

ACKNOWLEDGEMENTS

The research of R.E.Lingenfelter has been supported in part by National Science Foundation Grant AST 78-10289 and travel to the Symposium was funded by the Smithsonian Foreign Currency Program.

REFERENCES

1. G. Vedrenne, this volume.
2. R.E. Lingenfelter and R. Ramaty, Phys. Today, 31:3, 40 (1978).
3. E.L. Chupp, in Gamma Ray Spectroscopy in Astrophysics, T.L. Cline and R. Ramaty, ed., NASA, 1978, 42.
4. R. Ramaty and R.E. Lingenfelter, Nature, 278, 127 (1979).
5. R. Ramaty, B. Kozlovsky and R.E. Lingenfelter, Astrophys. J. Supp. 40, (1979).
6. D.D. Clayton, Gamma Ray Astrophysics, NASA, 1973, 263.
7. W.D. Arnett, Gamma Ray Spectroscopy in Astrophysics, T.L. Cline and R. Ramaty eds., NASA, 1978, 310.
8. F.W. Stecker, Astrophys. Space Sci., 3, 579 (1969).
9. C.J. Crannell, G. Joyce, R. Ramaty, and C. Werntz, Astrophys. J., 210, 582 (1976).
10. R.W. Bussard, R. Ramaty and R.J. Drachman, Astrophys. J. 228, 928 (1979).
11. Yu. N. Gnedin and R.A. Sunyaev, Astron. Astrophys., 36, 379 (1974).
12. J.K. Daugherty and J. Ventura, Astron. Astrophys., 61, 723 (1977).
13. J.K. Daugherty and J. Ventura, Phys. Rev. D, 18, 1053 (1978).
14. R.W. Bussard, Astrophys. J., in press (1979).
15. R.E. Lingenfelter and R. Ramaty, Astrophys. J., 211, L19 (1977).
16. R. Ramaty and R.E. Lingenfelter, Astrophys. J., 213, L5 (1977).
17. W.D. Arnett, Ann. N.Y. Acad. Sci., 302, 90 (1977).
18. T. Lee, D.A. Papanastassiou and G.J. Wasserberg, Astrophys. J., 211, L107 (1977).
19. M. Leventhal, C.J. MacCallum and P.D. Stang, Astrophys. J., 225, L11 (1976).
20. R.C. Haymes, G.D. Walraven, C.A. Meegan, R.D. Hall, F.T. Djuth, and D.H. Shelton, Astrophys. J., 201, 593 (1975).
21. C.E. Fichtel, R.C. Hartman, D.A. Kniffen, D.J. Thompson, G.F. Bignami, H.B. Ogelman, M.E. Ozel, and T. Tumer, Astrophys. J., 198, 163 (1975).
22. D.D. Clayton, S.A. Colgate, and D.J. Fishman, Astrophys. J., 155, 75 (1969).
23. S.A. Colgate, Astrophys. Space Sci., 8, 457 (1970).
24. D.D. Clayton, Nature Phys. Sci., 244, 137 (1973).
25. W.D. Arnett, Astrophys. J., 219, 1008 (1978).
26. W.D. Arnett, Astrophys. J., 230, L37 (1979).
27. D.D. Clayton, Astrophys. J., 198, 151 (1975).
28. J.W. Truran, S.G. Starrfield and W M. Sparks, Gamma Ray Spectroscopy in Astrophysics, T.L. Cline and R. Ramaty, eds., NASA, 1978, 315.
29. M. Leventhal, J.C. MacCallum, and A.C. Watts, Astrophys. J., 216, 491 (1977).

30. E.L. Chupp, D.J. Forrest, P.R. Higbie, A.N. Suri, C. Tsai, and P.P. Dunphy, Nature, 241, 333 (1973).
31. G.D. Walraven and R.C. Haymes, Nature, 264, 42 (1976).
32. M. Leventhal, C.J. MacCallum and P.D. Stang, Nature, 275, 200 (1978).
33. J. Trümper, W. Pietsch, C. Reppin, W. Voges, R. Staubert and E. Kevdziorra, Astrophys. J., 219, L105 (1978).
34. J.L. Matteson, D.E. Gruber, R.E. Rothschild and J.A. Hoffman, in X-Ray Astronomy, W.A. Baity and L.E. Peterson eds., Pergamon, Oxford, 1979, 493.
35. J. Trümper, in X-Ray Astronomy, W.A. Baity and L.E. Peterson eds., Pergamon, Oxford, 1979.
36. J.C. Ling, W.A. Mahoney, J.B. Willett and A.S. Jacobson, Astrophys. J., 231, in press (1979).
37. M.S. Strickman, W.N. Johnson and J.D. Kurfess, Astrophys. J., 230, L15 (1979).
38. P.A. Sturrock, Astrophys. J., 164, 529 (1971).
39. J.K. Daugherty and R.W. Bussard, Astrophys. J., in press (1979).
40. J.C. Ling, W.A. Mahoney, J.B. Willett and A.S. Jacobson, Nature, 270, 36 (1970).
41. R. Ramaty, G. Borner and J.M. Cohen, Astrophys. J., 181, 891 (1973).
42. J.C. Higdon and R.E. Lingenfelter, Astrophys. J., 215, L53 (1977).
43. S.L. Shapiro and E.E. Salpeter, Astrophys. J., 198, 671 (1975).
44. S.L. Shapiro, A.P. Lightman and D.M. Eardley, Astrophys. J., 204, 187 (1976).
45. A.E. Metzger, R.H. Parker, D. Gilman, L.E. Peterson and J.I. Trombka, Astrophys. J. 194, L19 (1974).
46. A.S. Jacobson, J.C. Ling, W.A. Mahoney, and J.B. Willett in Gamma Ray Spectroscopy in Astrophysics, T.L. Cline and R. Ramaty eds., NASA, 1978, 228.
47. R.E. Lingenfelter, J.C. Higdon and R. Ramaty, in Gamma Ray Spectroscopy in Astrophysics, T.L. Cline and R. Ramaty, eds., NASA, 1978, 252.
48. B. Börner and J.M. Cohn, Astrophys.J., 185, 959 (1973).
49. R.C. Hall, C.A. Meegan, G.D. Walraven, F.T. Djuth, D.H. Shelton and R.C. Haymes, Astrophys. J., 210, 631 (1976).

# 50 keV - 3 MeV HIGH RESOLUTION OBSERVATION OF THE GALACTIC CENTER REGION

Ph. Durouchoux*, D. Boclet*, R. Rocchia*, F. Albernhe**, J. F. Leborgne**, G. Vedrenne** and J. Marques da Costa***

*Section d'Astrophysique, Centre d'Etudes Nucléaires de Saclay, France
**Centre d'Etudes Spatiales des Rayonnements, Toulouse, France
***Instituto de Pesquisas Espacias, Sao Paulo, Brazil

## ABSTRACT

In this paper, we present our results obtained with a high-resolution gamma-ray spectrometer which has been flown successfully two times in Brazil in 1977. The main objective was the observation of the galactic-center (GC) region, which has been performed during more than 8 hours.

Numerous gamma-ray lines issued from Ge(n, $\gamma$) reactions, and other elements present in the gondola are detected. Furthermore, a positive detection of the 511 keV positron-annihilation line coming from the direction of the galactic center is detected in both flights. The measured flux in the line is : $(4.2 \pm 1.6)10^{-3}$ photon cm$^{-2}$ s$^{-1}$ and our data seem to fit an extended emissive region.

## INSTRUMENTATION

The instrument consists of a large Ge(Li) crystal (140 cm$^3$) cooled at liquid nitrogen temperature and actively shielded by a sodium iodide crystal.

A complete description of the instrumentation has been already published [1].

## THE FLIGHTS

Two balloon flights have been successfully carried out in Brazil (Feb. 77) on board a stabilized gondola.

The background total counting rate was about 10 counts s$^{-1}$ in the 0.1-8 MeV energy range at floating altitude (2.75 g cm$^{-2}$ residual atmosphere, 12 GV rigidity cut-off).

Ground spectral resolution of about 3.5 keV (F.W.H.M.) at 1 MeV unfortunately increased during the ascent of the balloon, and then remained constant at about 8 keV during both flights. Due to the location of the southern-hemisphere launching site, the GC was seen at the zenith. During the first flight (flight 7, Feb. 14, 1977) the axis of the detector was first pointed for 2 hours 20° off the zenith direction, and then, looking at the vertical during the transit of the GC in the meridian plane of the detector. For the second flight (flight 9, Feb. 19, 1977) the axis of the detector was set up looking toward the zenith during all the flight.

## RESULTS AND INTERPRETATION

Counting rate. For flight 9, the GC was on the axis of the detector at 11.20 U.T. In fig. 1, the counting rate vs time is shown for the 30-73 keV energy range, with 30 mn bins. Fig. 2 and 3 show the counting rate vs time around the energy of the 511 keV annihilation line after pressure-absorption and dead-time corrections.

A bump is present in both observations, when the GC is on the axis of the detector.

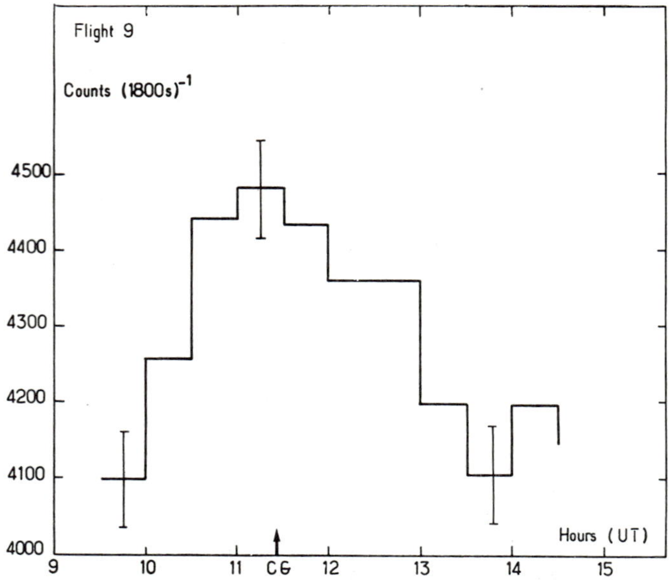


Fig.1 30-73 keV counting rate vs time $\Delta t$=30 mn

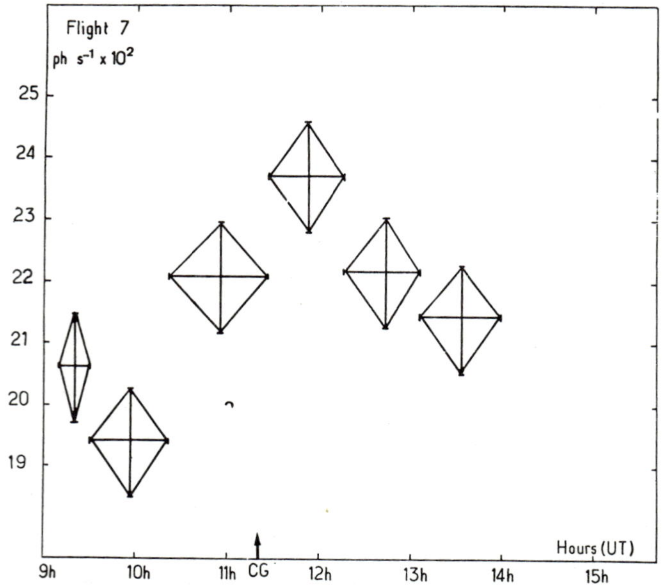


Fig.2 511 keV counting rate vs time $\Delta E$=10.5 keV (residual pressure 2.75 mb)

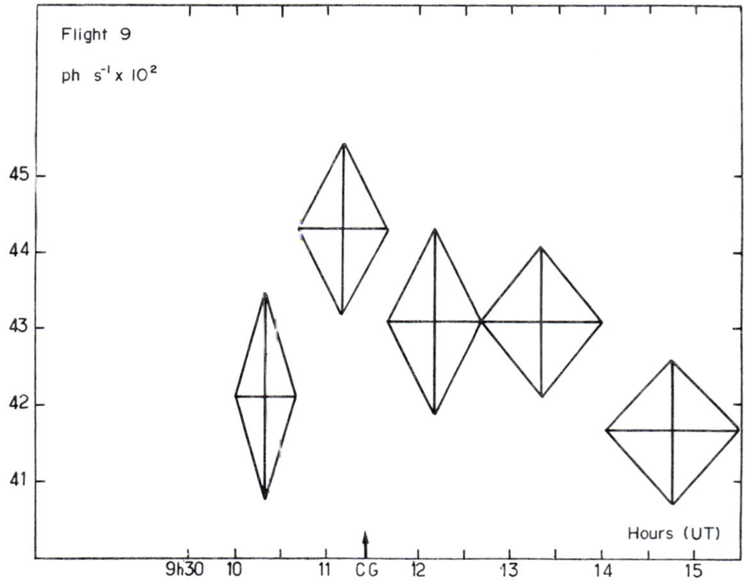


Fig.3 511 keV counting rate vs time ΔE=30 keV (residual pressure 2.25 mb)

In order to detect any artefact, we plotted in fig. 4 the counting rate vs time for the two sides of the line, i.e. 495-501 keV and 517.5-523.5 keV ranges.

No excess with statistical significance was detected.

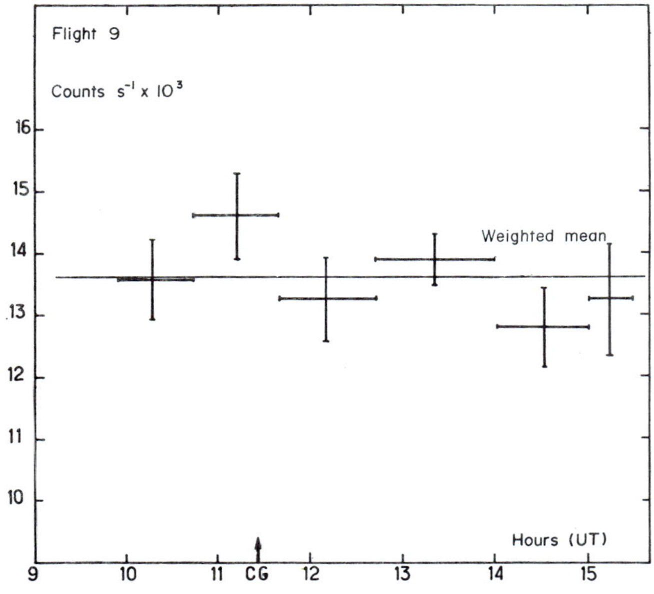


Fig.4 Counting rate vs time(495-501 keV ; 517.5-523.5 keV)

Spectrum. After including detector area, efficiency, dead time and atmospheric absorption factors, the excess near 511 keV energy represents a mean flux for both flights of :

$$(4.2 \pm 1.6)\ 10^{-3}\ \text{photon cm}^{-2}\ \text{s}^{-1}$$

The differential photon spectrum from the GC direction for the energy range near 511 keV is shown in fig.5.

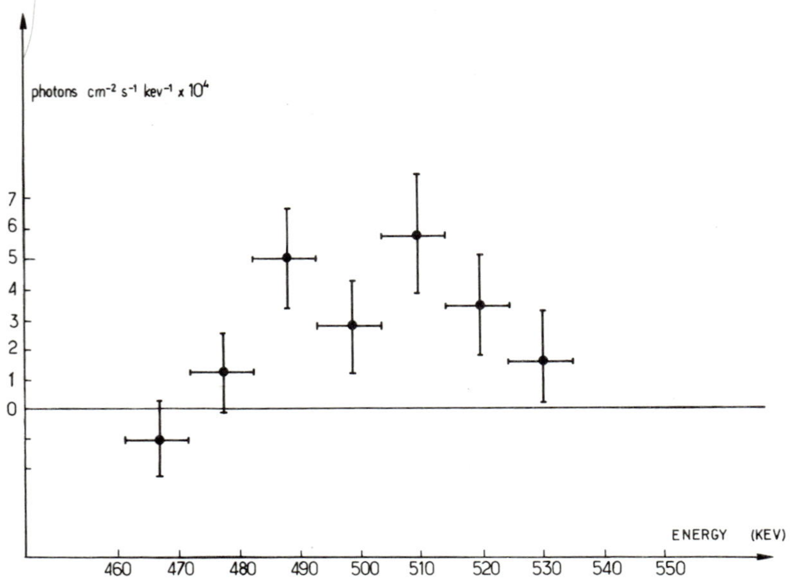


Fig.5 Differential photon spectrum from the G.C. direction for the energy range near 511 keV

In the light of these results, one can make the following comments :

. our results exhibit a positive detection of annihilation line from the GC region but with a low level of statistical significance (2 7 $\sigma$) compared with the 5.5$\sigma$ result of Leventhal.

. 511 keV is within the energy statistical and instrumental errors.

. our flux is not really in conflict with Leventhal's flux $(1.22 \pm 0.22)10^{-3}$ photon cm$^{-2}$ s$^{-1}$ if we consider that our field of view was about 4 times higher than that of Leventhal and that the emissive region can be large (a disk) rather than a point source.

. the increase of the 511 keV counting rate when the target is in the f.o.v. of the detector for both flights is a strong indication of flux coming from the GC.

. the constant counting rate in the channels corresponding to the sides of the 511 keV line during the transit of the G.C. in the meridian plan of the detector enforces the confidence of our detection.

Astrophysical implication. The detected line centered on 511 keV involves no gravitational redshift : that is to say that if the annihilation is related to neutron stars or other compact objects, it must not take place at the surface of such stars.

A broad feature between 1 and 2.3 MeV seems present in our data, as well as other possible lines, but only a fraction of our data has been processed to date and the remainder is presently being analyzed.

Future experiments with greater sensitivities and higher spatial resolution are strongly needed to provide more precise results in this exciting field of astronomy and the Gamma Ray Spectroscopy Experiment on board the GRO satellite planned to carry high resolution gamma-ray line astronomy forward in the 1980's is promising.

## REFERENCES

1. D. Boclet, J. Claisse, Ph. Durouchoux, E. Olivier, P. Pagnier, R. Rocchia, F. Albernhe, M. Frabel, G. Vedrenne, J. Marques da Costa, Proceedings of the 12th ESLAB Symposium Frascati, ESA SP-124, 293 (1977).

2. M. Leventhal, C.J. MacCallum, Proceedings of the 9th Texas Symposium on Relativistic Astrophysics, Munich, Germany, Dec. 1978.

# OBSERVATIONAL PROPERTIES OF GAMMA RAY BURSTS

K. Hurley

*Centre d'Etude Spatiale des Rayonnements, Toulouse, France*

ABSTRACT

The present status of gamma-ray burst astronomy is reviewed, with emphasis on the progress made recently by dedicated gamma burst experiments. Results on both the statistical properties of gamma-ray bursts (log N-log S and spatial distributions) and on their individual properties (time histories and energy spectra) are discussed.

## I - INTRODUCTION

The first gamma-ray burst was recorded in 1967 [1]. Today, 12 years and some 80 gamma bursts later, gamma-ray burst astronomy has emerged as a unique field, distinct from the classical gamma-ray astronomy of constant sources. This distinction may be justified on several grounds. First, and most important, although we do not yet know the origin of gamma-ray bursts, we do know where they do not originate : no gamma-burst has been associated with a steady gamma or hard X ray source [48,51] . Second, the temporal variations of "steady" gamma-ray sources occur on time scales far removed from those characterizing gamma-bursts. Third, gamma burst astronomy offers the possibility of localizing burst sources to minutes and even tens of seconds of arc, again far removed from the localization capability of classical gamma-ray astronomy. Finally, the relatively simple gamma burst instrumentation bears little resemblance to the detectors of gamma-ray astronomy.

Gamma-burst astronomy is currently well into its second generation. The first generation was that of the discovery and study of bursts with instruments, over relatively short baselines ($\lesssim$ 1 light second), not designed specifically for this purpose. The pioneering work of the Los Alamos and Goddard groups [1,4,7,12,15] paved the way for the second generation, that of dedicated gamma-burst instrumentation over lengthening baselines ($\sim$ hundreds of light seconds); this generation was inaugurated with the launch in January 1976 of the Helios 2 satellite, which carried a dedicated gamma-burst detector into a heliocentric orbit [48,51] . Today, the Pioneer Venus Orbiter and the Venera 11 and 12 probes complete the triangulation network over interplanetary distances.

The main purpose of this paper is to review the observations of gamma-ray bursts which have been made during the second generation. Previous review papers [12, 23,29,30] have treated the first generation observations. No attempt will be made to do justice to the large body of theoretical work which has appeared since 1973; reviews on this subject have also appeared [16,18,24] .

## II - OBSERVATIONS

Table I lists the gamma-ray bursts which have been observed by at least two separate detectors, since 1967; Figure 1 shows the number detected per year, and the satellites which participated in the observations (a summary of satellite instrumentation appears in Table II).

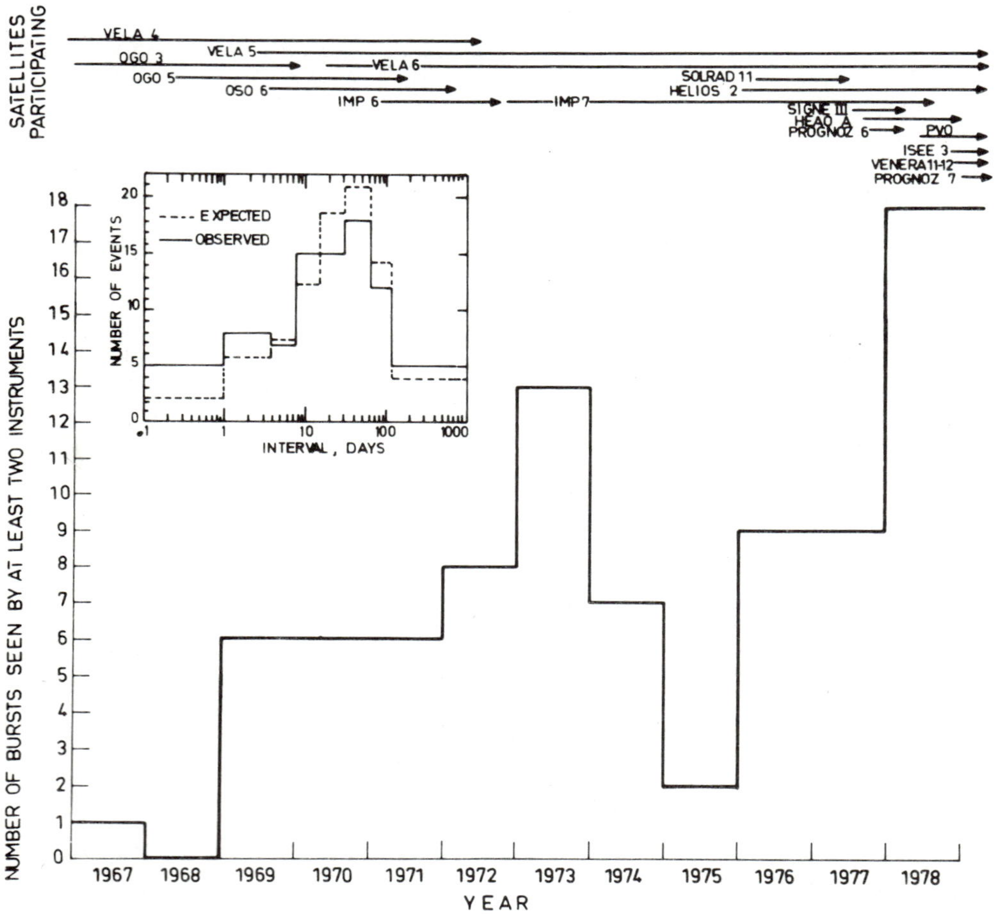


Figure 1 : The number of confirmed gamma-ray bursts detected per year, as a function of the year; the lifetimes of the various satellites participating are indicated above. Inset : the arrival time statistics of the gamma ray bursts (solid line), and the curve which would be expected based on Poisson statistics (dashed line).

An inset in Figure 1 shows the arrival time statistics of gamma-ray bursts, based on all the events except the first (good sensitivity and coverage were not achieved until the launch of Vela 5A and 5B in 1969). This curve is compared to the distribution expected on the basis of Poisson statistics; while a departure from the expected interval statistics appears at short arrival times, the agreement between the two distributions is such that no strong case can yet be made for the existence of non-random arrival times (i.e., repeating sources) on this basis alone ($\chi^2$ = 7.41 for 6 degrees of freedom, or a confidence of 28 %). At the same time, however, it should be noted that on the basis of rather imprecise localizations, the 15 March 1971 and 12 March 1972 events might be related, as might the events of 21 and 25 July, 1973 [29] .

## III - SPATIAL DISTRIBUTION

Figure 2 shows the distribution of the arrival directions of 23 gamma-ray bursts in galactic coordinates. Here we have positions for 16 bursts occurring up to 1973 [29] , for the events of 28 January 1976 [48] and 16 August 1976 [45] and 5 positions which have not yet appeared in the literature. This map is intended to give only the most general characteristics of the spatial distribution, since the localizations often involve two alternate positions several degrees or more square, or large or peculiarly shaped error boxes. An inset in Figure 2 shows the number of bursts above a given latitude as a function of galactic latitude, as well as the number expected for an isotropic distribution (where alternate positions exist, their latitudes have been averaged for this plot). It is now generally believed that the distribution is at least roughly isotropic, although some clustering about the galactic plane may be present [58] .

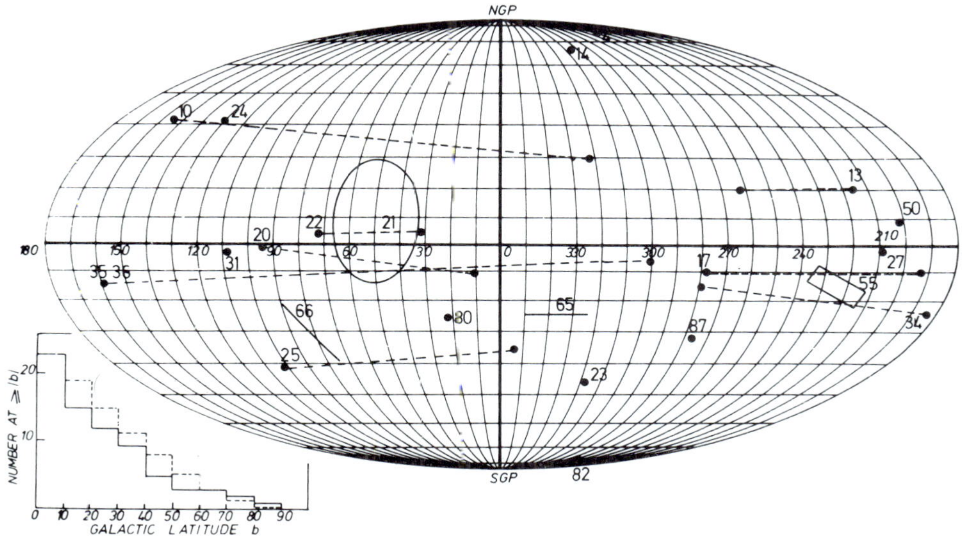


Figure 2 : A plot in galactic coordinates of 23 burst arrival directions; numbers refer to Table I. Points connected by dashed lines are alternate positions, and error box sizes vary from several arc-minutes square to many degrees square. Inset: the number of directions above a given galactic latitude as a function of galactic latitude (solid line), and the number expected for an isotropic distribution (dashed line). Latitudes of alternate positions have been averaged for this plot.

## IV - BURST TIME HISTORIES

The simple structured burst of 24 November 1978 (Figure 3) illustrates the basic phenomenon : here we have a single-peaked burst, lasting about 1 second, preceeded 1 second earlier by a short precursor. Precursors are often associated with gamma bursts; multiple peaks, as seen in Figure 4, are also common. The latter figure shows the complex event of 4 November 1978, which has at least 8 distinct peaks, and exhibits time structure down to about 30 ms.

Where time histories have been measured at both low and high energies (say, below and above 100 keV) a tendency has sometimes been noted for the low energy component to last longer (another way of stating this is to say that the spectra tend to soften outside the main event - see Section V) or to exhibit less fine time structure [2,4,13,37 ] . It has been pointed out, however, that this is not always the case, and may in fact be a threshold or sensitivity effect [42 ] .

It the energy range measured is disregarded, it may be said that the shortest events observed (Nos. 6 and 25, Ref. 7 and No. 87, Refs. 59 and 62) display a single spike lasting only about 0.1 second, while the longest, No. 24, which lasted 80 s [2],and No. 4, which lasted 110 s [14,17,37], are quite complex.

Although periodicity is not a feature of typical gamma-ray bursts, it has been noted in one case: the unusual event of 5 March 1979 [57,60,62] . For reasons which will be given later, however, it appears that this event may not fall into the category of "typical" gamma-ray bursts.

Searches for rapid fluctuations have been carried out on many events. One search for structure down to 5 μs for event No. 27 revealed microbursts of about 60 ms width [13] ; 10 ms structure was reported for event No. 54[40]. An unconfirmed event observed by balloon had 5 ms structure [28] . Burst observations made by the French Signe experiments aboard the Prognoz 6, Prognoz 7, and Venera 11 and 12 spacecraft, which have large detectors and time resolution down to 1/512 s, have not yet revealed significant structure below about 16 ms. The counterpart to peaks, namely valleys, has also been observed. Just as the narrow peaks may either be superimposed on a relatively high burst counting rate or may occur independently, the valleys may represent either a partial decrease in the counting rate, or a total cutoff, at which the emission goes to zero. Examples may be found in the high time resolution plots of the 4 November and 19 November, 1978 events [54] .

Fine time structure is certainly significant; but what it signifies is not at all clear. It is often stated that it indicates a characteristic source dimension d, through the relation $d \lesssim v\,\Delta t$, where $\Delta t$ is the time scale of the structure, and v is a characteristic velocity (the speed of light or the Alfven velocity come to mind). It must be recalled, however, that at least two classes of sources exist for which this is not true : sources in relativistic expansion, and sources involving beams of photons sweeping past the observer. In addition, the term "time structure" is used by different authors to signify either the rise or fall time (either of which may be a zero to maximum or e-folding time), or the pulse width, which is especially hard to determine for a narrow pulse superimposed on a high burst counting rate.

This discussion of time histories may be expanded to ask what is known about emission from burst sources at times long before and after the main event. The 14 May 1972 burst arrival direction was observed by the UCSD OSO-7 Cosmic X-Ray Telescope for about 12 hours before and after the event, and upper limits to the emission were set at about 1/400 of the peak burst flux [2] .

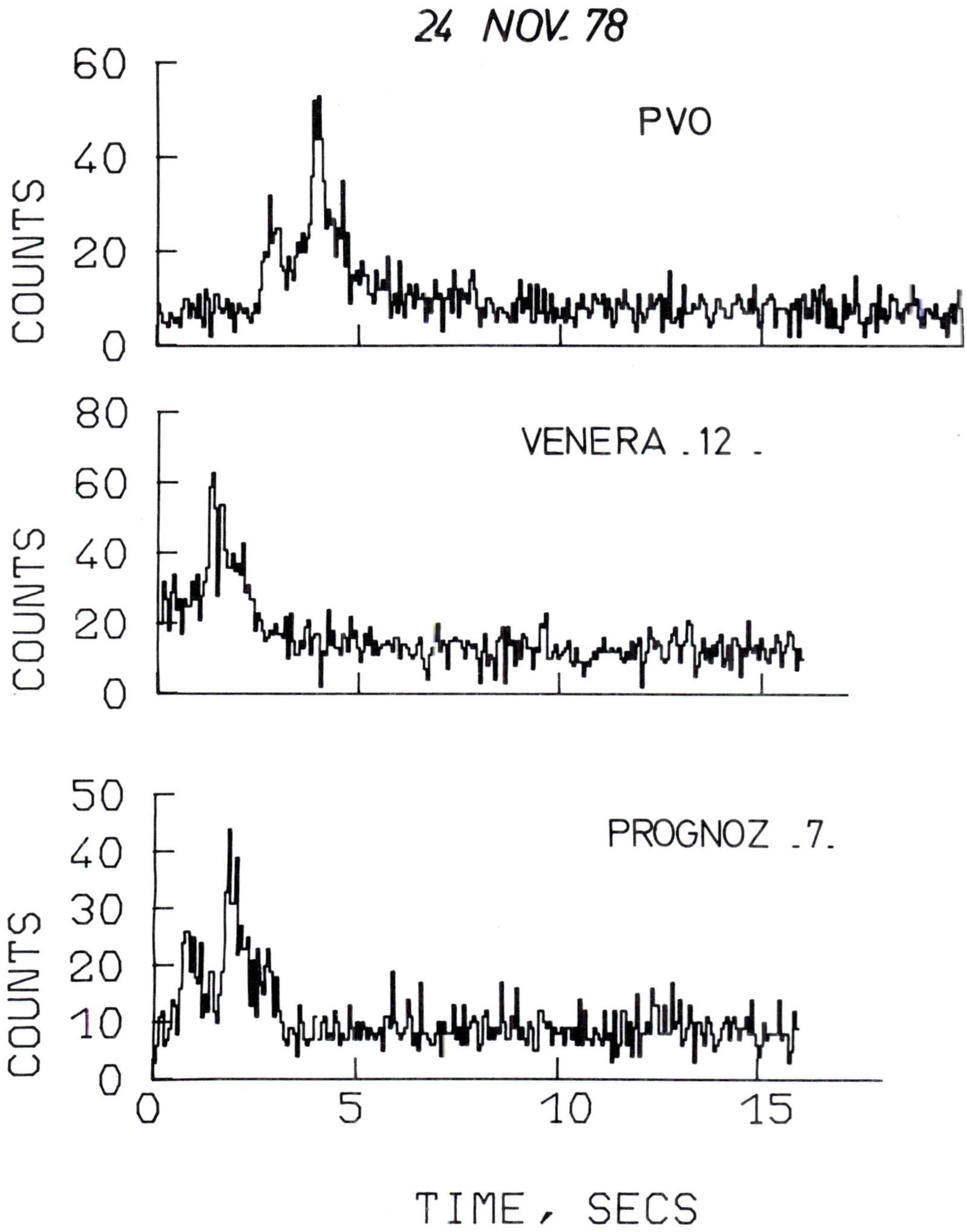


Figure 3 : Time histories of the 24 November 1978 event as seen by Pioneer Venus Orbiter, Venera 11, and Venera 12.

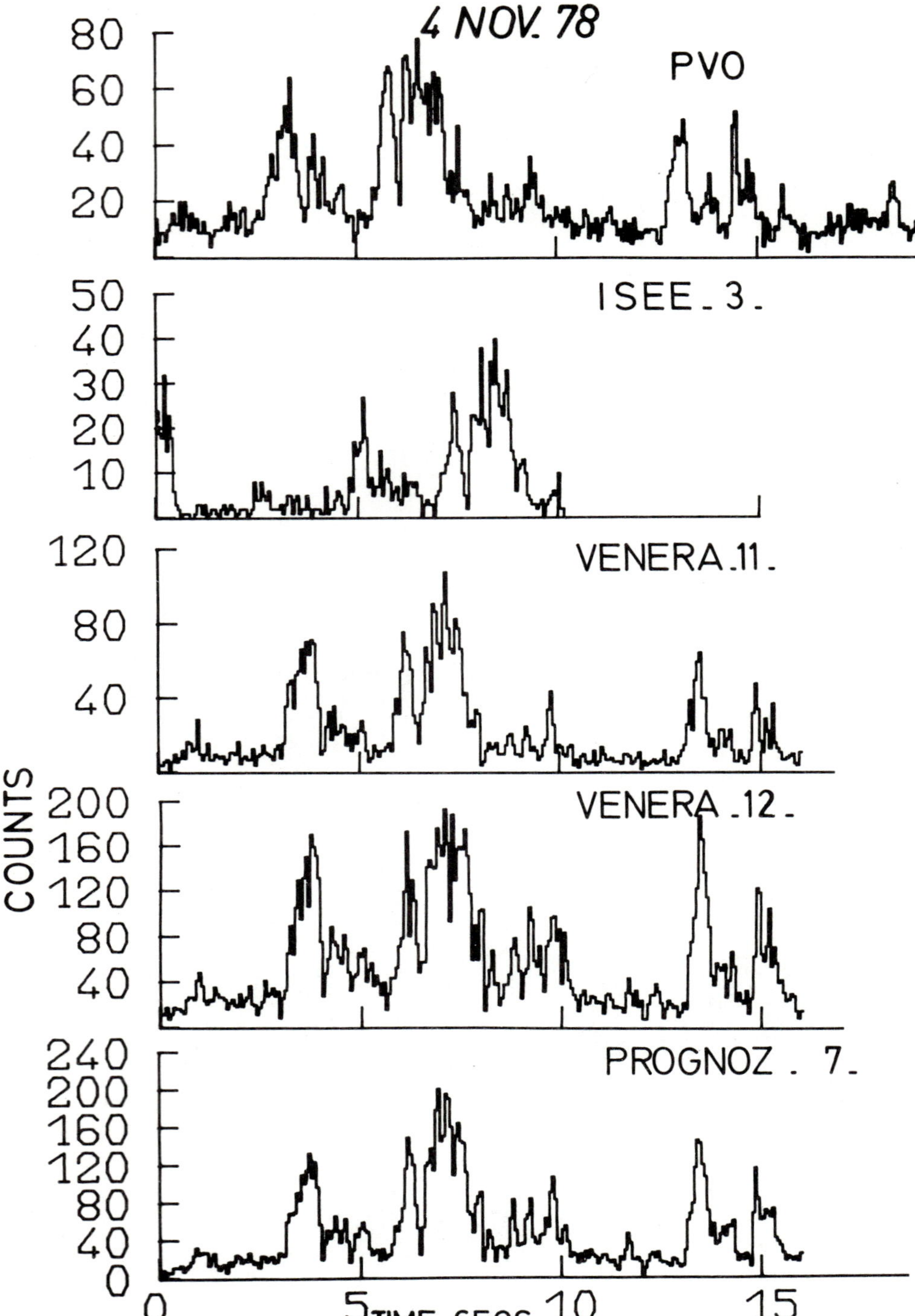


Figure 4 : Time histories of the 4 November 1978 event as seen by ISEE-C, Pioneer Venus Orbiter, Venera 11, Venera 12, and Prognoz 7.

Table 1 indicates the spectral data available for various bursts from different instruments. Almost all the current, dedicated gamma-burst experiments have spectral analysis, some on time scales down to 0.25 s. Figure 5 summarizes the spectral determinations which have appeared. The most complete remains that done using an Apollo-16 observation of the 27 April 1972 event [11], which extends from 2 keV to 7.9 keV and from 67 keV to 5.1 MeV. The OSO-7 and IMP-6 observations extend from 7 to 1500 keV [2] . The IMP-7 observations [20] of 9 events from 100 keV to 1100 keV indicate a remarkable consistency from one event to another, suggesting the "standard" spectrum shown (arbitrary normalization). The OGO-5 and OSO-6 measurements [41,42] of six gamma-bursts in the hard X-ray range may be used to define a range of spectral shapes from $E^{-1}$ to $E^{-2.5}$ (shown with arbitrary normalization) in the 20-130 keV range, and appear to indicate that gamma-ray bursts are not related to the softer X-ray burst phenomenon.

Evidence for spectral variability in a given event has frequently been presented [37,42]. More intense parts of a burst appear to have harder spectra in some cases [2,4,13], but softer in at least one case [11]. For one event which exhibited spectral variability (No. 27), it was demonstrated that the event-averaged spectrum still resembled closely the "standard" spectrum of Figure 5 [20]. The existence of a standard spectrum may indicate that bursts are generated by a process which reaches some sort of limiting conditions. It may also be used as a criterion to search for candidate events in data; some single-spacecraft events (not listed in Table I) have been found in this way, and are quite likely to be valid gamma-ray bursts [30]. The current generation of quasi-omnidirectional burst detectors will undoubtedly shed more light on these aspects of spectral variability, but before detailed spectral determinations can be done, it is necessary to determine the arrival directions in order to apply the appropriate corrections.

The detection of line emission in burst spectra would be of major importance in determining the emission mechanism. To date, no evidence for line emission has been found in three experiments [4,11,13], but a fourth [50] indicates that a line in the 300-400 keV range may be present in the spectra of events 72 and 73. This same experiment has found evidence for a line around 430 keV in event 87 [57]. The results of the Goddard Germanium Spectrometer experiment aboard ISEE-C, which observed several gamma bursts with high energy resolution in 1978, are anxiously awaited.

We may expand the study of gamma burst spectra to ask what is known about emission in other parts of the electromagnetic spectrum. No radio [19,33,45] , optical [8] or ultra high energy gamma emission [34] has been found in coincidence with gamma ray bursts, which leads to the overall upper limit spectrum shown in an inset in Figure 5.

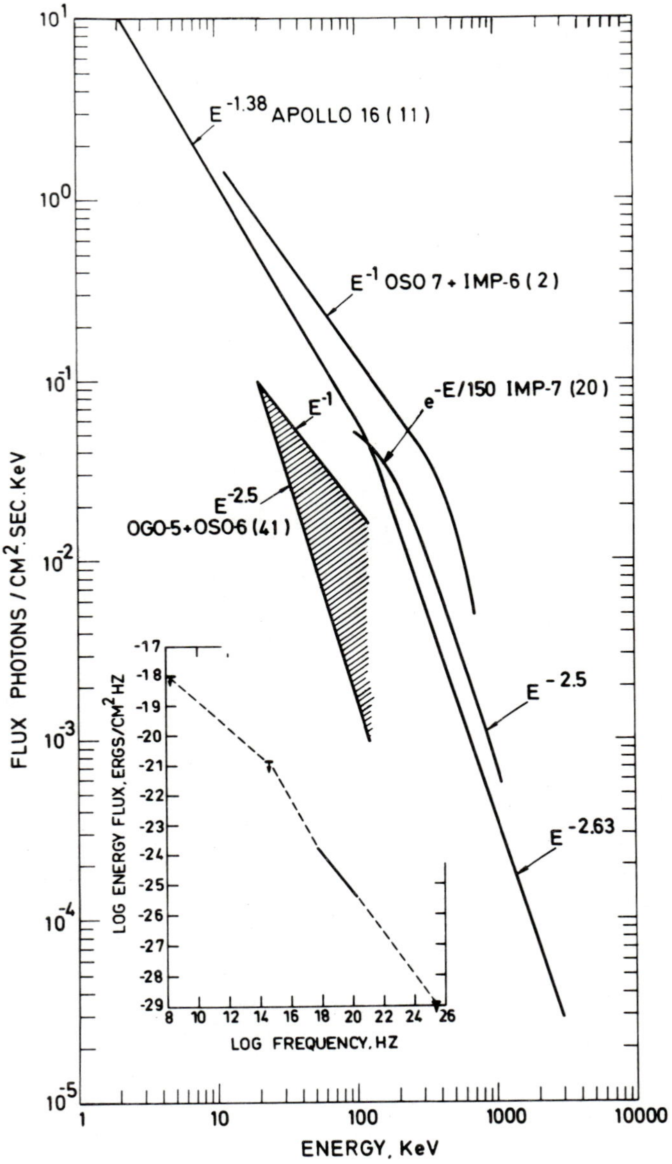


Figure 5 : Spectra of gamma-ray bursts from 2 keV to 2 MeV. The exp(-E/150) spectrum and the $E^{-1}$ to $E^{-2.5}$ spectrum are shown with arbitrary normalization. Inset: Three upper limits to the electromagnetic radiation associated with gamma ray bursts in other energy ranges ; 151 MHz radio emission, 2 eV optical emission, and $10^5$ MeV upper limits are shown. The solid line represents a typical X-and gamma-ray spectrum ; the dashed line does not represent measurements.

## VI - BURST INTENSITIES AND LOG N-LOG S DISTRIBUTIONS

We come now to what is probably the most controversial subject in gamma-burst astronomy. It was recognized very early that precise source localizations would not be accomplished soon (they are, in fact, only now being done - see Section VII) and that general information about possible classes of sources could be obtained more quickly from the intensity (S) - number (N) distribution, similar to the procedure used in radio source counts. Thus an $S^{-1.5}$ relation indicates an isotropic source distribution, an $S^{-1}$ relation is characteristic of a disc distribution, and so on, if the analysis is limited to very simple (in fact, oversimplified) models. A recent analysis of various spatial distribution models has just appeared [53]. But whereas the log N-log S distributions constructed for radio sources are based on steadily emitting sources, and the data are corrected for such relatively well known effects as completeness of the survey, angular resolution, source confusion, and noise, the distributions of burst sources involve the observation of fast transient signals, whose intensities may be difficult to estimate. In addition, many satellite detection systems have trigger criteria which must be satisfied before data from a candidate event is stored in the memories, and this constitutes a selection of the events to be studied. Since some of the events will not be selected because they do not meet the trigger criterion, the number of bursts N is subject to underestimation, and the intensity of the event may be underestimated because not all of it has been observed.

It was not at all surprising, then, to find that the early satellite surveys [9, 20,26,30,31] showed log N-log S distributions which, after following an $S^{-1.5}$ power law down to some given intensity (typically $\sim 3 \times 10^{-5}$ ergs/cm$^2$), then turned over and remained flat down to the smallest intensities observable with these instruments ($\sim 10^{-6}$ ergs/cm$^2$) (Figure 6). This was rapidly realized to be an instrumental effect.

Radio source surveys are frequently based on hundreds or even thousands of sources, but with only about 8 bursts/year detectable using satellite instrumentation, it was clear that large area, sensitive detectors were needed to study the problem, and that these were best placed on balloon platforms. Here, in principle, one is free from trigger criteria, since all the data may be telemetered to the ground in real time and searched later ; but the selection of bursts still poses numerous problems. First, some selection criterion must be established, and this is controversial [39], and second, it must be demonstrated that candidate events detected are not magnetospheric [38] or cosmic-ray induced [32,47] in origin. On the other hand, detector sensitivities down to about $5 \times 10^{-8}$ ergs/cm$^2$ can be achieved, and several events/day are expected at this level on the basis of some models.

An alternate technique has been employed by the Leningrad group [50]; this involves detector arrays on multiple satellites, such as the Venera 11 and 12 spacecraft. These instruments have relatively large areas and modest (several degree) localization capability, using comparisons of count rates in different detectors. The advantage here is that one can construct fairly rapidly not only the log N-log S relation, but also the burst strength as a function of galactic latitude, in order to test proposed classes of sources. This can also be done using burst arrival directions from triangulated bursts, of course, but triangulation is a more time consuming procedure.

Figure 6 shows the log N-log S distributions derived by various groups using different techniques. The interpretation of these results is not completely clear. For example, it is often stated that a departure from the $S^{-1.5}$ law constitutes a proof of galactic origin. It has been shown, however, that a metagalactic origin

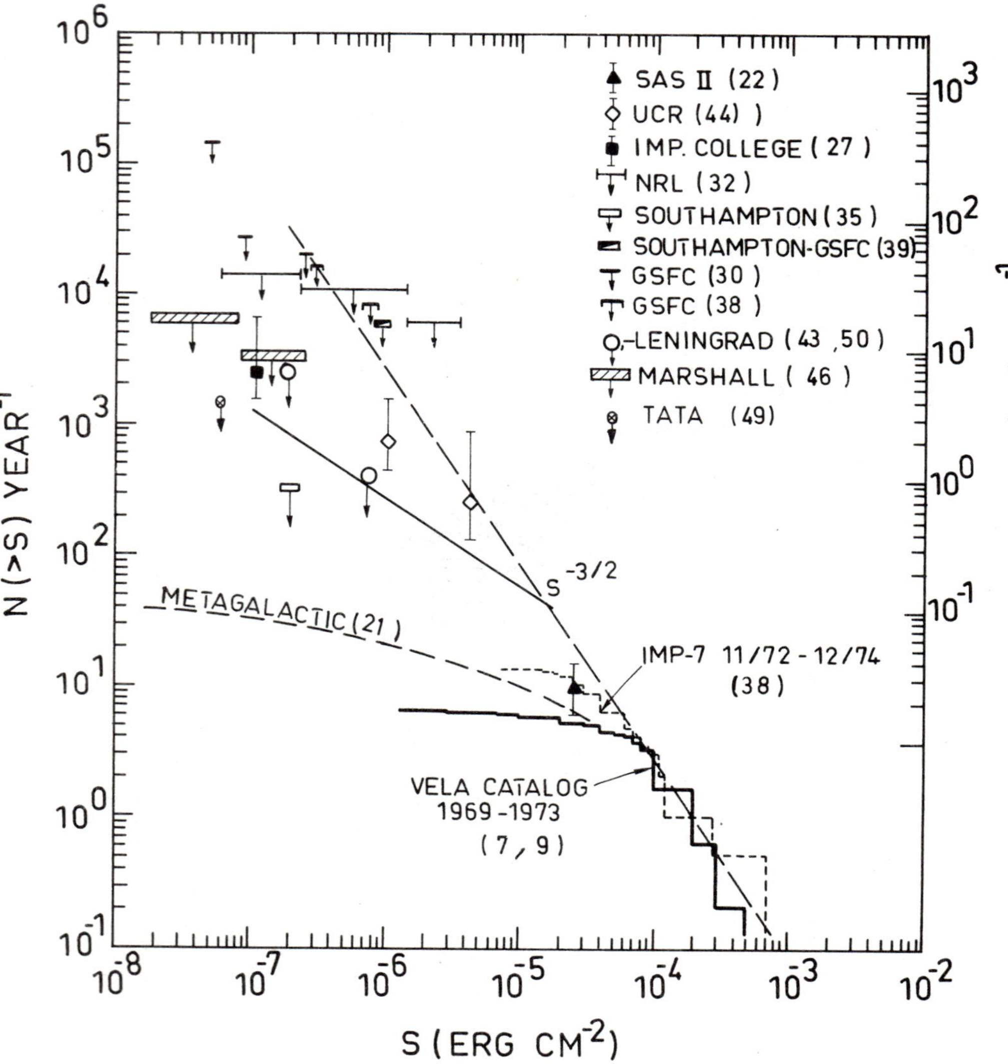


Figure 6 : The log N-log S relation as determined by different groups. The $S^{-1.5}$ curve is shown for reference (dashed line), as is the log N-log S curve for metagalactic origin models.

could also give an intensity distribution which turns over at low intensities, much like a galactic distribution[21] . There may well be good reasons for preferring to believe that gamma ray bursts are galactic in origin, but the intensity distribution alone is insufficient evidence at present. If, however, we exclude extragalactic models a priori, then the log N-log S curve may be interpreted as giving evidence that thin disc models may be ruled out, making Population I objects unlikely candidates [53] .

## VII - LOCALIZATION

One of the ultimate goals of gamma-ray burst astronomy is to provide accurate source locations which will encourage optical searches for candidate objects. "Accurate", in this case, means error boxes which are a maximum of about 5'X5', and preferably 1'x1' or less. At present, the only way of achieving these accuracies is by the triangulation method : the arrival times of a burst are recorded at two satellites, and this defines an annulus of arrival on the celestial sphere, whose width depends on the errors in the arrival time difference and spacecraft positions. A third satellite defines a second annulus which intersects the first at two positions, in general. A fourth satellite, if it is out of the plane of the first three, can be used to remove the ambiguity.

It is perhaps worth reviewing briefly the limitations of this method. The width of a given annulus is determined by three errors. First, there is the error in the knowledge of the spacecraft position. This rarely exceeds 10 light-ms, and is the smallest error in most practical cases. Second, there will be an error in the calibration of the spacecraft clock with respect to Universal Time : uncertainties of 10-50 ms can occur here. Finally, there is an uncertainty in the arrival time difference between the two spacecraft due to the cross correlation between the two burst time histories and its associated statistical errors. A reasonable estimate here is 20-100 ms, the exact value depending on the presence or absence of fine time structure in the event. An important question which is often asked is how one can in fact estimate the magnitude of the first two errors. There are three methods currently in use. First, multi-satellite observations of solar flare X-ray bursts which may be associated with well localized optical flares provide a rough check of the time bases. In practice, this method is good only to perhaps several hundred milliseconds, due to the fact that solar X ray bursts do not display enough fast time structure to reduce the cross-correlation errors. Second, many experiments have a "trigger by command" system, which allows one to trigger the memory readout by ground command. Knowing the light travel time to the spacecraft allows one to compare the time read into the memory from the spacecraft with Universal Time on the ground. The accuracy of this method is of the order of 100 ms or less, depending upon the unknown time delays involved in sending and receiving commands. Finally, the most accurate check involves redundant solutions of burst arrival directions. Three spacecraft can be used to generate two annuli on the celestial sphere, which will, in general, intersect somewhere even if errors are present in the triangulation. But the annulus generated by the addition of a fourth spacecraft will not, in general, intersect either of the first two intersections if errors are present. Thus the addition of a fourth spacecraft serves as a check against the other three.

Taking 100 ms as a maximum total error, it can be shown [30] that spacecraft separations of $2x10^5$ km (first generation) give error boxes several degrees on a side, while separations of about 1 A.U. (second generation) reach the required 1' accuracy. An intermediate case is that in which the first and second spacecraft are separated by a distance of $\sim$ 1 A.U., while the second and third are only $\sim 2x10^5$km apart [48,51] which results in error boxes in the shape of long, thin arcs. In these error boxes, one finds no steady or transient X ray sources, no gamma ray sources, no radio or infrared sources, no quasars, Seyfert galaxies, globular clusters, flare stars, pulsars, white dwarfs or nearby stars.

## VIII - GAMMA RAY BURST SOURCE CANDIDATES

Perhaps a more positive way to view the progress which has been made recently is to point out that some theories of gamma-bursts may indeed be ruled out by the observations. Supernovae [5] do not appear to be valid candidate objects. The relativistic dust grain hypothesis [6] may be excluded by the fact that gamma bursts have been observed over very long baselines. Flare stars [3,25] and white dwarfs [10,36] also seem to be unlikely candidates. A tendency seems to be emerging for burst intensities to be inversely correlated with galactic latitude [51] ; if confirmed, this would also be evidence against disc population models.

If we assume that burst sources radiate isotropically, then a $10^{-4}$ erg/cm$^2$ burst represents an intrinsic energy $E_i = 10^{34} R^2_{pc}$, where $R_{pc}$ is the distance in parsecs to the source. Thus a local (100 pc) origin implies energies around $10^{38}$ erg, galactic origin (10 kpc) implies $10^{42}$ erg, an extragalactic origin (30 Mpc) requires $10^{48}$ erg, and metagalactic origin (1000 Mpc) requires $10^{52}$ erg. There is only one event at present for which an estimate of the intrinsic energy may be made : the 5 March 1979 event, which has been localized to 1'x1' [55,61]. The error box contains the supernova remnant N49, a soft X ray emitter in the Large Magellanic Cloud. If that is indeed the origin of the event, then $E_i \sim 10^{43}$ erg. It would be premature to generalize on the basis of this event, however, because it is quite atypical. First, it displays a rise time of less than 0.25 ms, followed by a quasi-exponential decay, and pulsations - a striking departure from all other time histories; second, its spectrum is softer than the typical gamma burst spectrum [62].

## IX - FUTURE WORK

Many of the events detected in late 1978 (Table I) are now being triangulated ; several arc minute square error boxes will be obtained for them in the weeks and months to come. This will remain a typical error box size for some years to come. The third generation of gamma ray burst astronomy will begin some time after 1983, when the two Solar Polar Mission spacecraft will take gamma-burst detectors far out of the ecliptic plane, to provide localizations on the order of 10's of arc seconds. At the same time, an array of large area detectors aboard the Gamma Ray Observatory will investigate the log N-log S relation down to intensities around $10^{-8}$ erg/cm$^2$.

## ACKNOWLEDGEMENTS

I am grateful to all the groups who furnished and discussed their results before publication. This work was supported by CNES Contract 77-CNES-212.

REFERENCES

1. Klebesadel, R.W., Strong, I.B., and Olson, R.A., Ap.J.Letters, 182, L85 (1973)
2. Wheaton, W.A., Ulmer, M.P., Baity, W.A., Datlowe, D.W., Elcan, M.J., Peterson, L.E., Klebesadel, R.W., Strong, I.B., Cline, T.L., and Desai, U.D., Ap.J. Letters, 185, L57 (1973)
3. Hudson, H., and Tsikoudi, V., Nature Physical Science, 245, 88 (1973)
4. Cline, T.L., Desai, U.D., Klebesadel, R.W., and Strong, I.B., Ap.J.Letters, 185, L1 (1973)
5. Colgate, S.A., Ap.J., 187, 333 (1974)
6. Grindlay, J.E., and Fazio, G.G., Ap.J.Letters, 187, L93 (1974)
7. Strong, I.B., Klebesadel, R.W., and Olson, R.A., Ap.J.Letters, 188, 1 (1974)
8. Grindlay, J.E., Wright, E.L., and Mc Crosky, R.E., Ap.J.Letters, 192, L 113 (1974)
9. Strong, I.B., and Klebesadel, R.W., Nature, 251, 396 (1974)
10. Chanmugam, G., Ap.J.Letters, 193, L75 (1974)
11. Metzger, A.E., Parker, R.H., Gilman, D., Peterson, L.E., and Trombka, J.I., Ap.J.Letters, 194, L19 (1974)
12. Strong, I.B., in ESRO SP-106, The Context and status of Gamma-Ray Astronomy, 9th ESLAB Symposium, Frascati, Italy, 1974
13. Nakano, G.H., Imhof, W.L., and Reagan, J.B., in ESRO SP-106, The Context and Status of Gamma-Ray Astronomy, 9th ESLAB Symposium, Frascati, Italy, 1974
14. Share, G.H., Meekins, J.F., and Kreplin, R.W., in ESRO SP-106, The Context and Status of Gamma-Ray Astronomy, 9th ESLAB Symposium, Frascati, Italy, 1974
15. Cline, T.L., and Desai, U.D., in ESRO SP-106, The Context and Status of Gamma-Ray Astronomy, 9th ESLAB Symposium, Frascati, Italy, 1974
16. Pacini, F., in ESRO SP-106, The Context and Status of Gamma-Ray Astronomy, 9th ESLAB Symposium, Frascati, Italy, 1974
17. Palumbo, G.G.C., Pizzichini, G., and Vespignani, G.R., Ap.J.Letters, 189, L9 (1974)
18. Ruderman, M., Annals of the New York Academy of Sciences, 262,164 (1975)
19. Baird, G.A., Delaney, T.J., Lawless, B.G., Griffiths, D.J., Shakeshaft, J.R., Drever, R.W.P., Meikle, W.P.S., Jelley, J.V., Charman, W.N., and Spencer, R.E., Ap.J.Letters, 196, L11 (1975)
20. Cline, T.L., and Desai, U.D., Ap.J.Letters, 196, L43 (1975)
21. Usov, V.V., and Chibisov, G.V., Astron. Zh., 52, 192 (1975)
22. Ogelman, H., Fichtel, C.E., Kniffen, D.A., Nature, 255, 208 (1975)
23. Strong, I.B., in Neutron Stars, Black Holes and Binary X-ray Sources, D. Reidel Publishing Co., Dordrecht-Holland, 1975
24. Prilutskii, O.F., Rozental, G.V., and Usov, V.V., Soviet Physics-Uspekhi, 18, 548 (1975)
25. Karitskaya, E.A., Astron. Zh., 52, 189 (1975)
26. Fabian, A.C., Nature, 256, 347 (1975)
27. Bewick, A., Coe, M.J., Mills, J.S., Quenby, J.J., Nature, 258, 686 (1975)
28. Herzo, D., Dayton, B., Zych, A., and White, R.S., Ap.J.Letters, 203, L115 (1976)
29. Klebesadel, R.W., and Strong, I.B., Ap.and Space Science, 42, 3 (1976)
30. Cline, T.L., and Desai, U.D., Ap.and Space Science, 42, 17 (1976)
31. Share, G.H., Ap.and Space Science, 42, 29 (1976)
32. Johnson, W.N., Kurfess, J.D., and Bleach, R.D., Ap.and Space Science, 42, 35 (1976)
33. Baird, G.A., Meikle, W.P.S., Jelley, J.V., Palumbo, G.G.C., and Partridge, R.B., Ap.and Space Science, 42, 69 (1976)
34. O Brien, S., and Porter, N.A., Ap.and Space Science, 42, 73 (1976)
35. Carter, J., Dean, A.J., Manchanda, R.K., and Ramsden, D., Nature, 262, 370 (1976)
36. Mullan, D.J., Ap.J., 208, 199 (1976)
37. Kane, S.R., and Anderson, K.A., Ap.J., 210, 875 (1976)
38. Cline T.L., Desai, U.D., Schmidt, W.K.H., and Teegarden, B.J., Nature, 266, 694 (1977)

39. Cline, T.L., and Schmidt, W.K.H., Nature, 266, 749 (1977)
40. Laros, J.G., Evans, W.D., Klebesadel, R.W., Olson, R.A., and Spalding, R.E., Nature, 267, 131 (1977)
41. Share, G.H., and Kane, S.R., in ESA SP-124, Recent Advances in Gamma-ray Astronomy, proceedings of 12th ESLAB Symposium, Frascati, Italy, 1977
42. Kane, S.R., and Share, G.H., Ap.J., 217, 549 (1977)
43. Mazets, E.P., Golenetski S.V., Aptekar, R.L., Il Inskij, V.N., and Panov, V.N., Pis'ma Astron. Zh., 4, 188 (1978)
44. White, R.S., Ryan, J.M., Wilson, R.B., Zych, A.D., and Dayton, W.B., Nature, 271, 635 (1978)
45. Sommer, M., and Muller, D., Ap.J.Letters, 222, L17 (1978)
46. Fishman, G.J., Meegan, C.A., Watts, J.W., and Derrickson, J.H., Ap.J.Letters, 223, L13 (1978)
47. Hurley K., Astron. and Ap., 69, 313 (1978)
48. Cline, T.L., Desai, U.D., Pizzichini, G., Spizzichino, A., Trainor, J., Klebesadel, R., Ricketts, M., and Helmken, H., Ap.J.Letters, 229, L47 (1979)
49. Agrawal, P.C., Damle, S.V., Gokhale, G.S., Naranan, S., and Sreekantan, B.V., in(Cospar) X-Ray Astronomy, Pergamon Press Oxford and New York, 1979
50. Mazets, E.P., Golenetskii, S.V., Il'Inskii, V.N., Panov, V.N., Aptekar, R.L., Gur'Yan Yu. A., Sokolov, I.A., Sokolova, Z. Ya, and Kharitonova, T.V., preprint, 599, Ioffe Institute, Leningrad, USSR
51. Cline, T.L., Desai, U.D., Pizzichini, G., Spizzichino, A., Trainor, J.H., Klebesadel, R.W., and Helmken, H., Accepted for publication, Ap.J.Lett., 1979
52. Evans, W.D., Glore, J.P., Klebesadel, R.W., Laros, J.G., Tech, E.R., and Spalding, R.E., Accepted for publication, Science, 1979
53. Fishman, G.J., preprint, submitted to Ap. J., 1979
54. Chambon, G., Hurley, K., Niel,M., Vedrenne, G., Zenchenko, V.M., Kuznetsov, A. and Estulin, I.V., This volume (1979)

55. Evans, D., Klebesadel, R., Laros, J., Cline, T. L., Desai, U., Teegarden, B., Spizzichini, G., Hurley, K., Niel, M., Vedrenne, G., Estulin, I.V., and Mazets, E.P., I.A.U. Circular No. 3356 (1979)
56. Mazets, E.P., Golenetskii, S.V., Il'Inskii, V.N., Panov, V.N., Aptekar, R.L., Gur'Yan Yu. A., Sokolov, I.A., Sokolova, Z. Ya., and Kharitonova, T.V., preprint 610, Ioffe Physico-Technical Institute, Leningrad, USSR, Sumbitted to Pis'ma Astronomitcheskii Zhurnal (1979)
57. Mazets, E.P., Golenetskii, S.V., Il'Inskii, V.N., Panov, V.N., Aptekar, R.L. Gur'Yan Yu. A., Sokolov, I.A., Sokolova, Z. Ya., and Kharitonova, T.V., preprint 617, A.F. Ioffe Physico-Technical Institute, Leningrad, USSR (1979)
58. Mazets, E.P., Golenetskii, S.V., Il'Inskii, V.N., Panov, V.N., Aptekar, R.L., Gur'Yan Yu. A., Sokolov, I.A., Sokolova, Z. Ya., and Kharitonova, T.V., preprint 618. A.F. Ioffe Physico-Technical Institute, Leningrad, USSR (1979)
59. Evans, W.D., Klebesadel, R.W., Laros, J.G., Cline, T.L., Desai, U.D., Teegarden, B., Pizzichini, G., Margon, B., Hurley, K., Niel, M., Vedrenne, G., and Estulin, I.V., paper OG 5-6, 16th International Cosmic Ray Conference, August 6-18, Kyoto, Japan (1979)
60. Barat, C., Chambon, G., Hurley, K., Niel, M., Vedrenne, G., Estuline, I.V., Kurt, V.G., and Zenchenko, V.M., Submitted to Astron. and Ap. (1979)
61. Evans, W.D., Klebesadel, R.W., Laros, J.G., Cline, T.L., Desai, U.D., Hurley,K., Niel, M., Vedrenne, G., Estuline, I.V., Kuznetsov, A.V., and Zenchenko, V.M., Submitted to Ap.J.Lett.(1979)
62. Cline, T.L., Desai, U.D., Pizzichini, G., Teegarden, B.J., Evans, W.D., Klebesadel, R.W., Laros, J.G., Hurley, K., Niel, M., Vedrenne, G., Estuline, I.V., Kuznetsov, A.V., and Zenchenko, V.M., submitted to Ap.J.Lett. (1979)

| No. | Date (Day,Month,year,seconds) | | | | | Satellites observing |
|---|---|---|---|---|---|---|
| 1 | 2 | July | 1967 | 51568 | | VELA(2) |
| 2 | 3 | July | 1969 | 26233 | | VELA(2) |
| 3 | 19 | July | 1969 | 13606 | | OGO-5 VELA |
| 4 | 7 | October | 1969 | 26790 | X | VELA(2) OSO-6* OGO-5* |
| 5 | 8 | October | 1969 | 59444 | | VELA(2) |
| 6 | 17 | October | 1969 | 11927 | | VELA(2) |
| 7 | 17 | October | 1969 | 78113 | | VELA(2) OGO-3 |
| 8 | 25 | January | 1970 | 18090 | | OSO-6* OGO-5* |
| 9 | 10 | July | 1970 | 19066 | | VELA(2) OGO-5 |
| 10 | 22 | August | 1970 | 60571 | X | VELA(3) OGO-5 |
| 11 | 1 | October | 1970 | 56532 | | OSO-6* OGO-5* VELA |
| 12 | 1 | December | 1970 | 72059 | | VELA(2) OSO-6* OGO-5* |
| 13 | 30 | December | 1970 | 25337 | X | VELA(3) |
| 14 | 2 | January | 1971 | 69056 | X | VELA(3) SAS-1 |
| 15 | 27 | February | 1971 | 62857 | | VELA(2) OSO-6* |
| 16 | 15 | March | 1971 | 40827 | | VELA(4) IMP-6* OGO-5 |
| 17 | 18 | March | 1971 | 55685 | X | VELA(3) IMP-6* SAS-1 OGO-5* OSO-6* |
| 18 | 21 | April | 1971 | 11919 | | VELA(2) SAS-1 |
| 19 | 30 | June | 1971 | 63059 | | VELA(2) IMP-6* SAS-1 OGO-5* |
| 20 | 17 | January | 1972 | 63556 | | VELA(3) IMP-6* SAS-1 COSMOS-461 |
| 21 | 12 | March | 1972 | 57194 | X | VELA(4) |
| 22 | 28 | March | 1972 | 49587 | X | VELA(2) IMP-6* |
| 23 | 27 | April | 1972 | 39512 | X | APOLLO-16* VELA(2) |
| 24 | 14 | May | 1972 | 13591 | X | VELA(3) IMP-6* OSO-7* TD-1 |
| 25 | 1 | November | 1972 | 68206 | X | VELA(3) |
| 26 | 13 | November | 1972 | 55758 | | IMP-7* VELA(RI) |
| 27 | 18 | December | 1972 | 73659 | X | VELA(2) IMP-7* 1972-076B* |
| 28 | 25 | January | 1973 | 54900 | | VELA(2) IMP-7* |
| 29 | 2 | March | 1973 | 84475 | | VELA(3) IMP-7* SAS-2 |
| 30 | 16 | April | 1973 | 45550 | | IMP-7* SAS-2 VELA |
| 31 | 7 | May | 1973 | 29071 | X | VELA(3) |
| 32 | 6 | June | 1973 | 25648 | | IMP-7* SAS-2 VELA(2) |
| 33 | 6 | June | 1973 | 67630 | | SAS-2 IMP-7* VELA(2) |
| 34 | 10 | June | 1973 | 75582 | X | VELA(3) IMP-7* |
| 35 | 21 | July | 1973 | 32113 | X | VELA(3) IMP-7* 1972-076B |
| 36 | 25 | July | 1973 | 61674 | X | VELA(3) IMP-7* |
| 37 | 18 | September | 1973 | 18634 | | VELA(2) TD-1 |
| 38 | 26 | September | 1973 | 57890 | | IMP-7* VELA(RI) |
| 39 | 17 | December | 1973 | 29348 | | IMP-7* VELA |
| 40 | 23 | December | 1973 | 2387 | | IMP-7* VELA |
| 41 | 21 | January | 1974 | 65122 | | IMP-7* VELA |
| 42 | 2 | April | 1974 | 66038 | | IMP-7* VELA(RI) |
| 43 | 23 | July | 1974 | 47488 | | IMP-7* VELA |
| 44 | 24 | July | 1974 | 11784 | | IMP-7* VELA(RI) |
| 45 | 23 | August | 1974 | 52544 | | IMP-7* VELA(RI) |
| 46 | 29 | September | 1974 | 67905 | | IMP-7* VELA |
| 47 | 14 | December | 1974 | 36893 | | IMP-7* VELA(RI) |
| 48 | 2 | November | 1975 | 43518 | | VELA(2) OSO-8 |
| 49 | 3 | December | 1975 | 3 | | VELA(3) ANS |
| 50 | 28 | January | 1976 | 26969 | X | VELA(2) HELIOS-2 IMP-7* ARIEL |
| 51 | 22 | March | 1976 | 55528 | X | VELA(3) HELIOS-2 IMP-7* SAS-3 |
| 52 | 7 | April | 1976 | 10464 | X | HELIOS-2 VELA IMP-7* OSO-8 |
| 53 | 19 | April | 1976 | 26771 | X | VELA HELIOS-2 IMP-7* OSO-8 |
| 54 | 12 | June | 1976 | 68618 | | VELA(3) SOLRAD |
| 55 | 16 | August | 1976 | 58532 | X | SOLRAD(2) VELA MPI* IMP-7* SAS-3 |
| 56 | 3 | September | 1976 | 53400 | | SOLRAD(2) VELA |

| | | | | | |
|---|---|---|---|---|---|
| 57 | 9 December | 1976 | 3780 | | HELIOS-2 SOLRAD VELA(2) |
| 58 | 20 December | 1976 | 63512 | | HELIOS-2 SAS-3 SOLRAD(2) VELA(3) |
| 59 | 7 January | 1977 | 79531 | | HELIOS-2 SOLRAD VELA(2) |
| 60 | 31 January | 1977 | 37764 | | SOLRAD VELA |
| 61 | 10 March | 1977 | 38392 | X | HELIOS-2 VELA OSO-8 SAS-3 |
| 62 | 10 April | 1977 | 48447 | | SOLRAD VELA(2) |
| 63 | 1 May | 1977 | 76560 | | HELIOS-2 SOLRAD VELA(2) |
| 64 | 8 July | 1977 | 46231 | X | HELIOS-2 VELA |
| 65 | 20 October | 1977 | 28496 | X | HELIOS-2 VELA HEAO-A(2)* PROGNOZ-6* |
| 66 | 29 October | 1977 | 42081 | X | HELIOS-2 VELA HEAO-A* PROGNOZ-6(RI) |
| 67 | 10 November | 1977 | 61964 | X | HELIOS-2 HEAO-A(2)* PROGNOZ-6* SIGNE-3 VELA |
| 68 | 8 May | 1978 | 72877 | | HELIOS-2 VELA |
| 69 | 19 May | 1978 | 26509 | | HELIOS-2 VELA |
| 70 | 21 May | 1978 | 78827 | | HELIOS-2 PVO* VELA |
| 71 | 14 September | 1978 | 60233 | | PVO ISEE-3 |
| 72 | 18 September | 1978 | 71364 | X | VENERA 11(2) ISEE-3 |
| 73 | 21 September | 1978 | 14155 | X | VENERA-11(2)* ISEE-3 PVO* VELA |
| 74 | 6 October | 1978 | 51917 | X | VENERA-11* VENERA-12(2)* |
| 75 | 12 October | 1978 | 62029 | X | VENERA-11* VENERA-12(2)* |
| 76 | 13 October | 1978 | 53676 | X | VENERA-11* VENERA-12(2)* |
| 77 | 19 October | 1978 | 66370 | X | VENERA-11* VENERA-12(2)* ISEE-3 PVO* |
| 78 | 25 October | 1978 | 86035 | X | VENERA-11* VENERA-12(2)* |
| 79 | 26 October | 1978 | 29033 | X | VENERA-11* VENERA-12(2)* |
| 80 | 4 November | 1978 | 58667 | X | PROGNOZ-7* VENERA-11(2)* VENERA-12(2)* ISEE-3(2)* VELA(4) PVO* |
| 81 | 15 November | 1978 | 75978 | X | VENERA-11(2)* VENERA-12(2)* ISEE-3 PVO* |
| 82 | 19 November | 1978 | 34112 | X | VENERA-11(2)* VENERA-12* PROGNOZ-7* ISEE-3(2) VELA(4) PVO* HELIOS-2 |
| 83 | 21 November | 1978 | 5735 | X | PROGNOZ-7* VENERA-11* VENERA-12* PVO* |
| 84 | 24 November | 1978 | 14127 | X | PROGNOZ-7* VENERA-11* VENERA-12* VELA PVO* HELIOS-2 |
| 85 | 13 December | 1978 | 86015 | X | PROGNOZ-7* VELA PVO* HELIOS-2 |
| 86 | 13 January | 1979 | 27360 | X | PROGNOZ-7* ISEE-3 VELA PVO* |
| 87 | 5 March | 1979 | 57124 | X | PROGNOZ-7* VENERA-11* VENERA-12* ISEE-3 PVO* HELIOS-2 |

* Indicates spectral information
X Indicates burst can be localized
TIME IS ARRIVAL TIME AT A SATELLITE
(N) Indicates event was seen by N satellites or instruments
(RI) Indicates rate increase

TABLE I : Gamma-Ray Bursts Detected by 2 or More Satellites.

TABLE II - Gamma Ray Burst Detectors -

| SATELLITE | DATES | ORBIT | DETECTORS* | ENERGY RANGE | REFERENCE |
|---|---|---|---|---|---|
| VELA 5A,5B | 5/69- | GEO. CIRC. $1.2 \times 10^5$ km | 6-10 $cm^3$ CsI each | 0.2-1 MeV | 1 |
| VELA 6A,6B | 4/70- | GEO. CIRC. $1.2 \times 10^5$ km | 6-10 $cm^3$ CsI each | 0.3-1.5 MeV | 1 |
| OGO-5 | 3/68-6/71 | GEO. $1.5 \times 10^5$ by $2.3 \times 10^5$ km | 9.5 $cm^2$ NaI | 10-200 keV | 37 |
| OSO-6 | 8/69-1/72 | GEO. CIRC. 500 km alt. | 3.3 $cm^3$ NaI | 23-800 keV | 42 |
| IMP-6 | 3/71-9/72 | GEO. $2 \times 10^5$ by $10^4$ km | 25.6 $cm^2$ CsI | 50-880 keV | 4 |
| IMP-7 | 9/72-9/78 | GEO. $2.8 \times 10^5$ by $1.4 \times 10^5$ km | 25.6 $cm^2$ CsI | 50-880 keV | 20 |
| HELIOS-2 | 1/76- | HELIO. 0.29x1.0 A.U. | 21.5 $cm^3$ CsI | > 100 keV | 48,51 |
| SOLRAD 11A, 11B | 4/76-12/76 | GEO. CIRC. $1.3 \times 10^5$ km | 2-43.4 $cm^3$ CsI each | .2-2.0 MeV | 40 |
| | 4/76-6/77 | | | | |
| SIGNE III | 6/77-3/78 | GEO. CIRC. 500 km alt. | 950 $cm^2$ CsI anti. | > 60 keV | |
| HEAO-A | 8/77-12/78 | GEO. CIRC. 450 km alt. | 2400 $cm^2$ CsI anti., | .1-1.6 MeV | |
| | | | LARGE AREA PROP. CTRS. | | |
| PROGNOZ-6 | 9/77-3/78 | GEO. $2 \times 10^5$ by 500 km alt. | 63 $cm^2$ NaI | .1-2.5 MeV | |
| | | | 750 $cm^2$ CsI anti. | > 100 keV | |
| ISEE-C | 8/78- | HELIO. 1.0 A.U. at $L_1$ | 22 $cm^2$ NaI | .125-1.25 MeV | |
| | | | HP Ge SPECTROMETER | .05-6.5 MeV | |
| PIONEER VENUS ORBITER | 5/78- | VENUSCENTRIC | 2-36 $cm^3$ NaI | .1-2.0 MeV | 26 |
| VENERA 11, 12 | 9/78- | HELIOCENTRIC | 2-63 $cm^2$ NaI each, | .1-2.5 MeV | 54 |
| | | | 6-50 $cm^2$ NaI each | 50-150 keV | 50 |
| PROGNOZ-7 | 11/78- | GEO. $2 \times 10^5$ by 500 km alt. | 63 $cm^2$ NaI | .1-2.5 MeV | |
| | | | 750 $cm^2$ CsI anti. | > 100 keV | |
| PROGNOZ | 1981 | GEO. $2 \times 10^5$ by 500 km alt. | 2-176 $cm^2$ NaI, | .1-2.5 MeV | |
| | | | 1-63 $cm^2$ NaI | | |
| VENERA 13, 14 | 1981 | HELIOCENTRIC | 2-63 $cm^2$ NaI each | .05-1.0 MeV | |
| SOLAR POLAR | 1983 | HELIOCENTRIC | 2-40 $cm^2$ CsI each | 20-200 keV | |
| GAMMA RAY OBSERVATORY | 1986 | GEO. CIRC. 400 km alt. | 6-2027 $cm^2$ scintillators | 60-600 keV | |

* Where height and diameter of a detector are approximately equal, the volume is given, rather then the surface area.

# GAMMA RAY BURST OBSERVATIONS BY THE 3 SATELLITE SIGNE NETWORK

G. Chambon*, K. Hurley*, M. Niel*, G. Vedrenne*,
V. M. Zenchenko**, A. V. Kuznetsov** and I. V. Estulin**

*Centre d'Etude Spatiale des Rayonnements, Toulouse, France
**Institute for Space Research (IKI), Moscow, U.S.S.R.

ABSTRACT

The Signe network, consisting of French experiments aboard the Soviet Venus probes Venera 11 and 12, and aboard the Soviet eccentric earth orbit Prognoz 7 satellite, has detected 16 gamma-ray bursts in the period September 1978- March 1979. The experiments and the Venera and Prognoz missions are described, and time profiles for two November events are presented. Future plans for a similar network to be launched in 1981 are discussed.

## I - INTRODUCTION

Various aspects of the Franco-Soviet collaboration for the study of gamma ray bursts have been presented at previous COSPAR Symposia [1,2] and elsewhere [3] . This collaboration involves French built gamma-ray burst detectors carried aboard Soviet spacecraft; the first dedicated gamma burst detector in this program was launched aboard the Prognoz-6 satellite in September 1977. This paper deals specifically with a set of three experiments launched in late 1978 aboard two Venus probes (Venera 11 and 12) and an eccentric earth orbiting satellite (Prognoz-7). These represent an important step in gamma ray burst astronomy for several reasons. First, the detectors used are the largest dedicated gamma-burst detectors yet put into orbit. Second, these detectors, having been designed and built in the same laboratory, have identical dimensions and time resolutions; this is especially important for accurate comparisons of time histories. Finally, these three experiments form, by themselves, a complete triangulation network.

## II - VENERA AND PROGNOZ MISSIONS

In the Venera mission, two spacecraft were launched in September, 1978, 5 days apart, and encountered Venus in late December. After the Venus encounter, both spacecraft were gravitationally scattered, and went into slightly different helic-centric orbits. On the outbound trajectory, a typical spacecraft separation was 0.1 A.U.. The Venera spacecraft were spin stabilized during most of the trajectory, but 3 axis stabilized near Venus encounter.

The Prognoz-7 satellite was launched in November 1978; the apogee is about $2x10^5$km., and lies outside the magnetosphere. The orbital period is about 4 days, and the satellite is spin stabilized with the spin axis oriented towards the sun.

### III - DETECTORS

The burst detector used is a 9 cm. diameter by 3.7 cm thick NaI(Tl) crystal, surrounded by a plastic anticoincidence jacket; the NaI and the plastic are viewed by two photomultipliers. The upper surface area of the NaI is 63.6 $cm^2$. Each Venera spacecraft has two of these detectors on board, facing in opposite directions so as to cover $4\pi$ sr. The total experiment weight is about 20 kg. The Prognoz satellite has one burst detector facing the sun, and a smaller, collimated detector on the side of the satellite; this detector has a gamma-ray burst detection mode [2] .

High time resolution data are stored into memories when a burst is detected. Among the data which are stored are 16 second time histories with 1/64 s resolution, and 2 s time histories with 1/512 s resolution for strong events. 6 channel pulse height spectra covering the energy range 80-1000 keV with 0.25 s resolution are also stored.

### IV - BURSTS DETECTED

From Table 1 of the previous paper [4] it can be seen that these three experiments detected 16 gamma-ray bursts in the period September 1978-March 1979. 11 of these were detected by experiments aboard other satellites.

Figure 1 shows three time histories of the 19 November 1978 event. Some six peaks are evident in these data, the narrowest being about 48 ms. wide. The 4 November 1978 event (Figure 2) is even more remarkable in this respect. Many spikes in the time histories may be correlated down to 16-32 ms; in additions, the emission clearly falls to zero between the second an third main peaks, for about 16 ms. This event has the finest time structure of any yet observed with our three instruments. Despite the fact that a 1/512 s memory is present, which has triggered on strong events, no significant structure has yet been found on time scales below about 16 ms.

### V - FUTURE WORK

Using these data alone, and assuming a maximum total error of 100 ms [4] , error boxes of maximum dimensions $\sim 1^\circ \times 1'$ may be derived. In most cases, the ambiguity inherent in the three satellite triangulation method can be removed by comparing count rates between the pairs of detectors aboard each satellite. Post-Venus encounter error boxes will be smaller than this, since the spacecraft separation increases during this period. These positional determinations can be used to unfold the observed spectra and derive the true incident spectra; this will be done over the next few months.

In most cases, much smaller error boxes can be generated by the addition of a fourth or fifth satellite, such as Helios-2 or Pioneer Venus Orbiter. The dimensions of these error boxes will be on the order of 1'x1', and the positions will, in addition, be redundantly determined, eliminating any chance of error [4] . This will permit accurate and efficient optical searches to be made. These error boxes will be derived over the next several months.

The Signe program will be continued in 1981 with another three satellite network. This will include two Venus probes with instrumentation identical to that described above, and a Prognoz satellite. The Prognoz will have two 176 $cm^2$ NaI burst detectors, as well as a smaller 63.6 $cm^2$ detector.

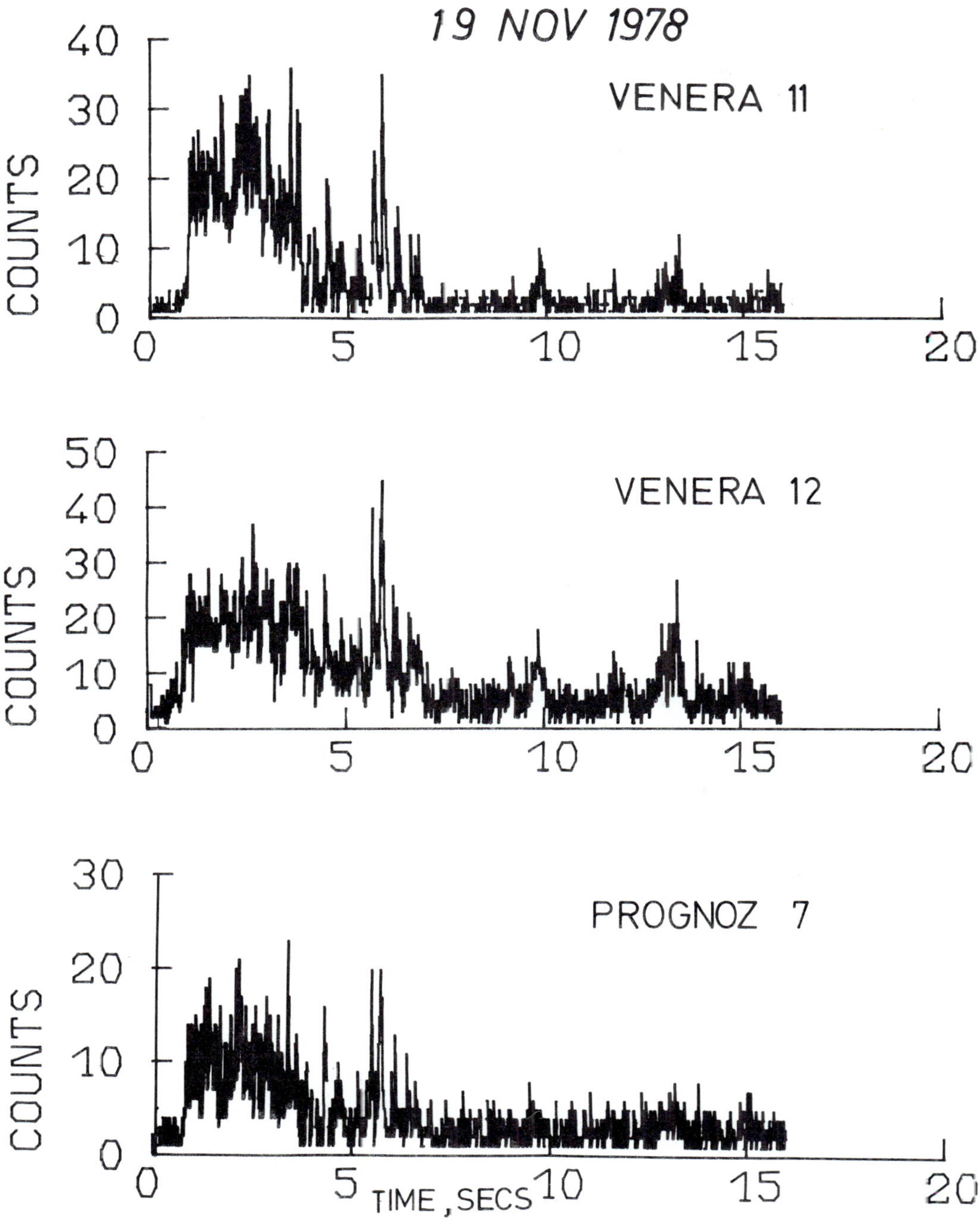


Figure 1 : The event of 19 November 1978 as seen by the Venera and Prognoz experiments. The time resolution is 16 ms.

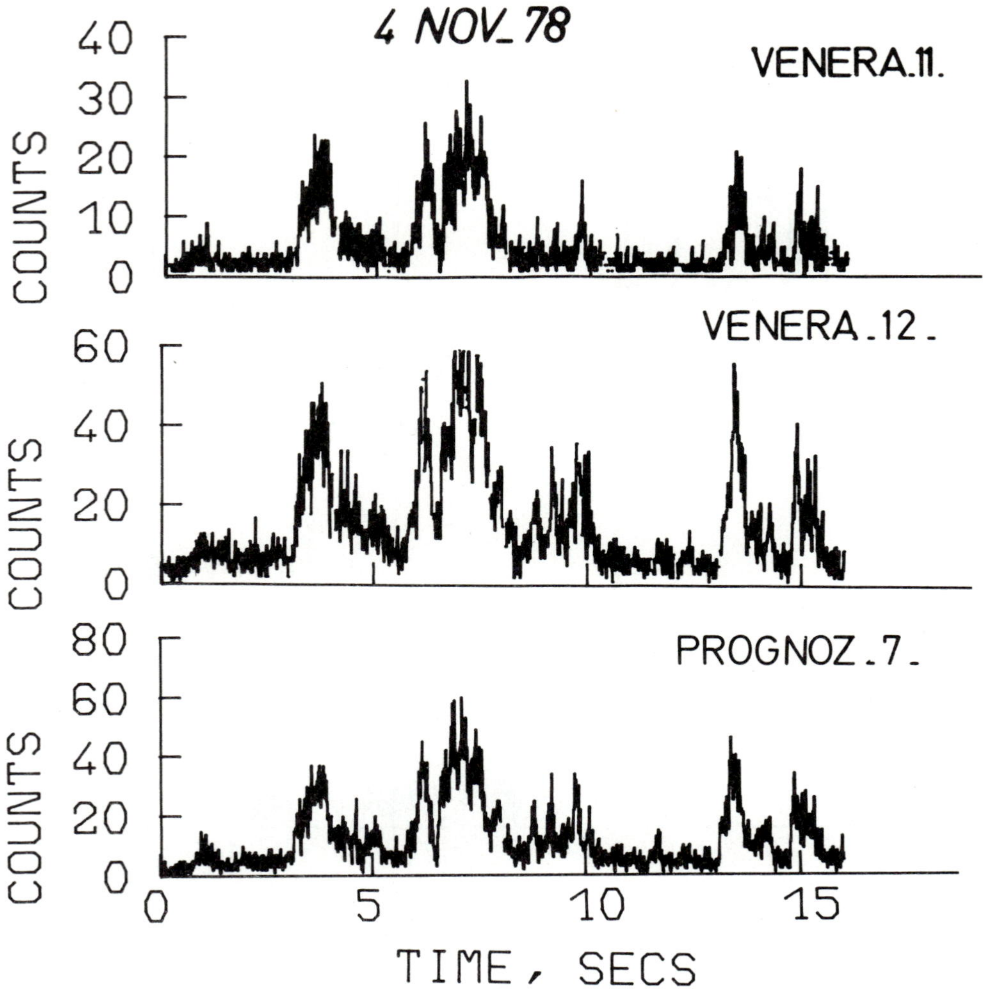


Figure 2 : The event of 4 November 1978 as seen by the Venera and Prognoz experiments. Time resolution is 16 ms.

REFERENCES


1. Niel, M., Hurley, K., Vedrenne, G., Estuline, I.V., and Melioransky, A.S. Ap. and Space Science, 42, 99 (1976)
2. Chambon, G., Hurley, K., Niel, M., Vedrenne, G., Estuline, I.V., and Melioransky, A.S., in (COSPAR) X-Ray Astronomy (1978)
3. Estulin, I.V., Kuznetsov, A.V., Vedrenne, G., Niel, M., and Hurley, K., Pis'ma Astron. Zh., 4, 1979
4. Hurley, K., (This volume).


ACKNOWLEDGEMENTS

This work was supported by CNES Contract 77-CNES-212.

SESSION 4

# Diffuse Galactic Gamma-ray Sources

*Chairman:* **G. F. BIGNAMI**

# OBSERVATIONS OF DIFFUSE GALACTIC GAMMA RAYS

G. A. Simpson

*NASA/Goddard Space Flight Center, Greenbelt, Maryland 20771, U.S.A.*

ABSTRACT

Observations of 35 MeV to several GeV diffuse gamma radiation from the galaxy are available from the SAS-2 and COS-B instruments. This paper gives a brief review of the experimental problem, a discussion of the high-latitude (local) galactic component, and a study of the more distant low-latitude emission from the galactic plane. Finally, the emerging observations in other energy ranges are mentioned.

INTRODUCTION

This paper is a review of the observations of galactic diffuse gamma radiation. Diffuse $\gamma$-ray observations in general are of interest because they bear upon three fields of astrophysics: galactic structure, cosmic rays, and cosmology. The cosmological implications are the subject of another review, so I will be concentrating on the connections with galactic structure and cosmic rays.

The bulk of the diffuse galactic gamma radiation which we observe is created in the collisions of cosmic rays with interstellar matter. 1/ Galactic structure is revealed in that the large-scale picture of the plane in the gamma radiation shows features similar to those seen in the distribution of interstellar matter. Cosmic-ray gradients and concentrations, localized and broad scale, are becoming discernible--linked, of course, to progress in estimates of the interstellar matter densities.

Observations of the diffuse galactic emission are not easily made: One requires very low background; and in studying the galactic plane high angular resolution is needed to distinguish the contribution of point sources from truly diffuse emission. Further, the flux of particles is low, $\sim 10^{-5}$ that of cosmic rays. To date, the main experimental results have been obtained with spacecraft in the energy region from 35 MeV to a few GeV. For these reasons, this paper will mainly be a discussion of SAS-2 and COS-B data.

---

1/ But see also Hayakawa, these proceedings.

EXPERIMENTAL TECHNIQUE

Gamma rays pass freely through a few gm/cm$^2$ of matter, leaving no track. However, they may interact catastrophically in the field of a heavy nucleus, converting to an electron-positron pair. In fact, this process is the dominant energy-loss mechanism at all energies above 50 MeV.[1] The pair retains information on the direction and energy of the incident $\gamma$-ray, and also provides a unique spatial signature which allows the event to be clearly distinguished from instrumental background. This is seen in Fig. 1, which shows the SAS-2[2] and COS-B[3] instruments.

The detectors consist of a series of thin plates of high Z material in which $\gamma$-rays have a high probability of converting to an electron positron pair. Interleaving the plates are x-y digital spark modules, which record the position of the ionized gas left along the tracks of the electron and the positron. The energy of the particles, and thus the $\gamma$-ray energy, can be inferred from the rate at which they scatter in traversing the plates (SAS-2) or by stopping them in a thick scintillator (COS-B). The direction of the $\gamma$-ray is the mean of the initial directions of the pair.

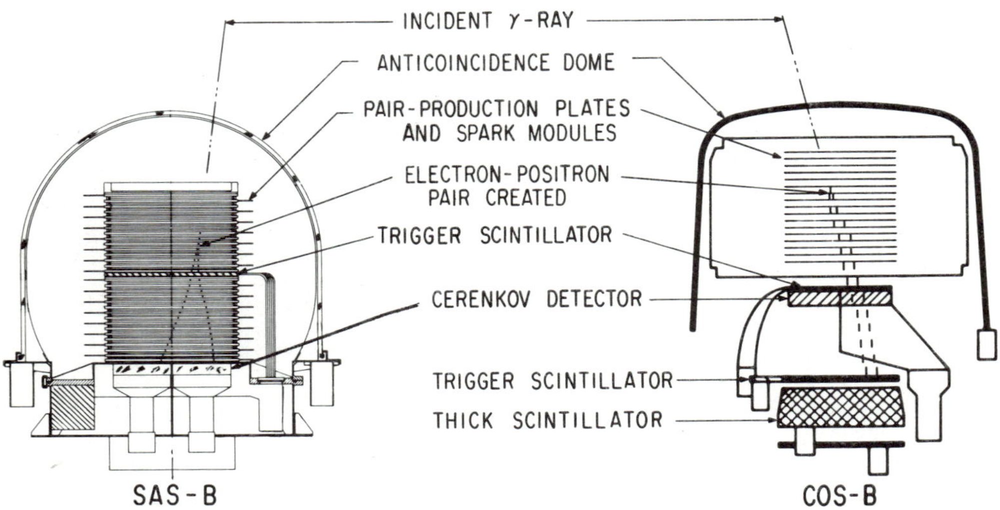


Fig. 1: Gamma-Ray Instruments (50 MeV to several GeV)

The detector triggers when a downward moving particle is noted below the spark chambers and no signal is seen in the anticoincidence dome.

The present generation of detectors achieves angular resolution of a few degrees for each photon (decreasing with energy). The accuracy in locating individual localized $\gamma$-ray emission features is limited by the number of photons detectable in reasonable exposure times to values of several tenths of a degree.

The OSO-3 experiment[4,5] was the first to detect the galactic plane, but at this point in time the SAS-2 and COS-B data constitute nearly our entire information base on gamma-ray emission from the galaxy. These two data sets have complementary aspects: COS-B, having a wider energy range and greater exposure time than SAS-2, but also a significant instrumental background, has studied localized emission features in the galactic plane with great success. SAS-2, having negligible background and broader coverage of the celestial sphere, provides the best data for study of the diffuse emission at high latitudes.[6]

GALACTIC GAMMA RAY PRODUCTION

The processes[7,8] that contribute to the diffuse galactic emission at these energies are three, illustrated in Fig. 2: 1) Collisions of GeV cosmic ray nucleons with interstellar matter producing $\pi^\circ$ mesons, which decay to $\gamma$-ray photons; 2) Electron bremsstrahlung: radiative interactions of electrons in the nuclear coulomb field; and 3) Compton interactions: Coulomb collisions of electrons >~10 GeV with ambient photons - principally 3° background radiation galactic infrared radiation, and starlight.

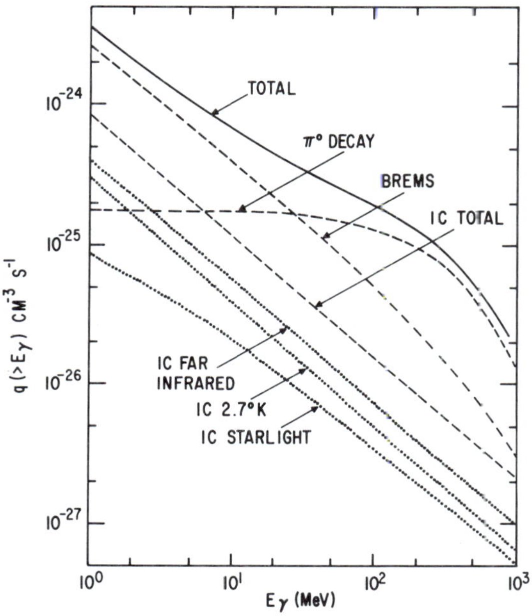


Fig. 2: Local integral $\gamma$-ray emissivity for $n_H$ = 1 atom/cm$^3$ from Worrall (1977) and Stecker (1970)

The $\pi^\circ$ and bremsstrahlung processes dominate at all energies. The key feature of these two mechanisms which has been exploited in observing the diffuse emission is their proportionality to the density of the interstellar medium. The local interstellar medium is believed to consist predominantly of neutral hydrogen, molecular hydrogen being the next most abundant component.[9] If the $\gamma$-ray emission which we see is diffuse in origin, then it ought to show a strong correlation with the neutral hydrogen observations[10,11] under the assumption that the cosmic rays are uniformly distributed.

HIGH LATITUDE OBSERVATIONS

First let us consider a region of the galaxy where we can be confident that that assumption is true: the region within 500 pc of the sun. Of course, we have no information at all on the distance at which individual $\gamma$-rays originated, but we can use the well-measured properties of the matter distribution. The scale height of neutral hydrogen is 130 pc locally. Therefore, if we confine our observations to galactic latitudes more than 12° from the plane, the diffuse matter, and thus the majority of the predicted diffuse emission, will have been generated within 500 pc of the sun. 500 pc is 3% of the galactic radius, and the order of the width of a spiral arm.

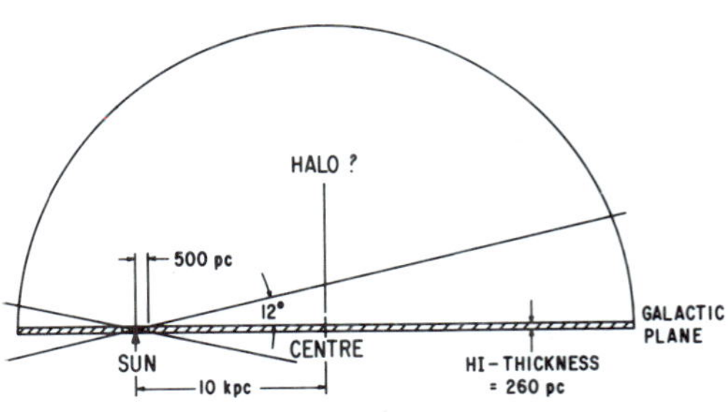


Fig. 3: Geometry of high-latitude $\gamma$-ray observations.

Our dataset is rather sparse: even with the 28-week exposure of SAS-2, only about 4000 photons were observed from latitudes outside 12°. This means that we can consider only broad-scale features of the high-latitude γ-ray distribution.

We may consider the correlation of the high-latitude γ-ray intensity with the neutral hydrogen column density[12] with the objectives of determining (1) if there exists a component of the gamma-ray intensity which is not correlated with the matter disk (i.e. an extragalactic or halo component) and (2) if the observed γ-ray intensity is consistent with observed and inferred properties of the cosmic rays.

The γ-ray intensity for high latitudes is shown plotted as a function of the 21 cm column density in Fig. 4. Here we have selected γ-rays >100 MeV, and we have divided the galaxy into broad regions of longitude (typically 60°) and latitude (10°), averaging and smoothing the 21 cm observations to match the lower resolution of the γ-ray data. We see from the figure that the γ-ray intensity in every region is consistent with a single linear correlation of the form

$$I_\gamma = A \cdot N_{HI} + B$$

This relationship also holds in the 35-100 MeV range, with different coefficients reflecting the different spectra of the two components. The slope 'A' gives directly the γ-ray photon production rate per atom of neutral hydrogen, while the intercept 'B' gives the intensity of a component which is uncorrelated with the matter, possibly originating in the galactic halo or outside our galaxy. Assuming power-law spectra, the galactic 'A' component has a spectral index $1.5 \pm 0.03$ and intensity above 35 MeV of $(6.2 \pm 1.5)\ 10^{-26}\ N_{HI}$, while for the 'B' component these figures are $2.7^{+.4}_{-.3}$ and $5.7\ 10^{-5}$ photons $cm^{-2}s^{-1}sr^{-1}$. For >100 MeV γ's the two components are equal at galactic latitudes near 30°, varying somewhat with longitude.

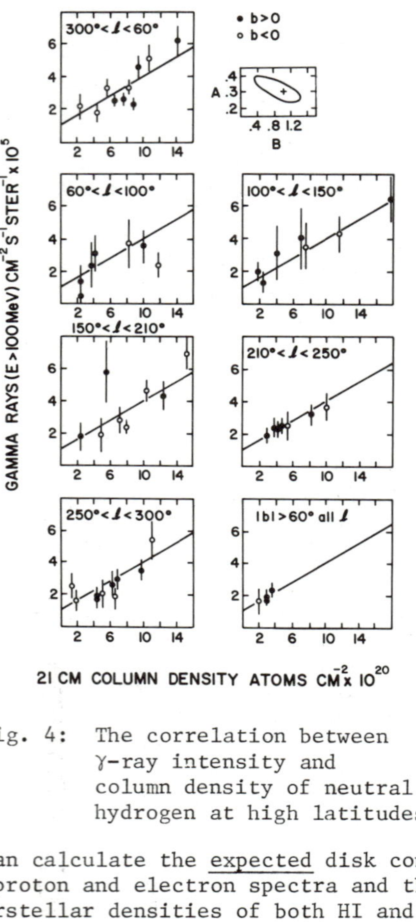


Fig. 4: The correlation between γ-ray intensity and column density of neutral hydrogen at high latitudes

We can calculate the expected disk component using measurements of the cosmic ray proton and electron spectra and the interstellar matter density. The local interstellar densities of both HI and $H_2$ have been sampled by UV absorption measurements from the Copernicus satellite[13]; the nominal value is 1.15 atoms/$cm^3$ of hydrogen in all forms. Proton spectra at the Earth have been well measured for some time and the correction for solar modulation at the energies of interest (>2 GeV) is negligible. Interstellar electron spectra, which we require to compute the bremsstrahlung and Compton components, are rather uncertain.

However, the spectrum of Cummings[14], or Daugherty et al.[15] can be used as a starting point. One important result of this calculation[12] is that a higher contribution from electrons, at least 50% more than Cummings (1973) is required to account for the spectral index. Some independent evidence for a higher electron intensity in the energy range around 1 GeV has been given by recent measurements[16] of the 5-80 MHz spectrum of synchrotron emission in the north and south galactic pole directions. When these new data are used with measurements of the high energy electron spectrum at Earth to rederive the interstellar electron spectrum,[17] a much better agreement with the requirements of the high-latitude gamma-ray observations is found. The best currently available estimates of the galactic γ-ray production rates are shown in Table 1.

TABLE 1 Source Functions In The Solar Vicinity

| SOURCE MECHANISM | VALUE OF SOURCE FUNCTION ($CM^{-3}S^{-1}$) | |
|---|---|---|
| | 10 < E(MeV) < 30 | E(MeV) > 100 |
| COSMIC RAY NUCLEON MATTER INTERACTIONS | $(0.7-1.1) \times 10^{-26}$ | $(14-22) \times 10^{-26}$ |
| ELECTRON BREMSSTRAHLUNG | $(13-39) \times 10^{-26}$ | $(4-12) \times 10^{-26}$ |
| COMPTON SCATTERING (STARLIGHT AND INFRARED) | $(1.2-3.6) \times 10^{-26}$ | $(0.4-1.2) \times 10^{-26}$ |
| COMPTON SCATTERING (3°K) | $(1-3) \times 10^{-26}$ | $(0.2-0.6) \times 10^{-26}$ |
| SYNCHROTRON RADIATION | $(0.7-2) \times 10^{-30}$ | $(0.3-0.4) \times 10^{-30}$ |
| POINT SOURCES | ? | ? |

The residual component of the high latitude gamma radiation is more difficult to interpret. However, we can immediately place some constraints on halo models for its origin by looking for anisotropies. Subtracting the disk component from the total emission in each direction, and comparing the remaining intensity from directions which pass over the galactic center to that from the anticenter directions (with $20° < |b| < 60°$) one finds a ratio of $1.10 \pm 0.18$.

If there existed an extended uniform cosmic ray halo of cylindrical geometry such as has been suggested[18], the ratio of Compton black-body γ-rays would be about 5. A uniform spherical halo of galactic dimensions gives 2.8. Models have been produced which include electron energy loss and diffusion, and give the predicted ratio as a function of energy and electron diffusion coefficient, assuming free escape into the halo. The values are around 3 to 4.[19]

If we allow the halo component to be 'diluted' by an isotropic component of extragalactic origin, then these figures imply that the halo must contribute less than 20% of the 'B' component (2σ upper limit).

## LOW-LATITUDE γ-RAY EMISSION

The γ-ray map of SAS-2[20], which includes some 14,000 photons, and that of COS-B[21] (64,000 photons) are shown in Figs. 5 and 6. The most prominent features in each of these maps are the galactic plane, and the bright localized emission regions which have been discussed in earlier papers.

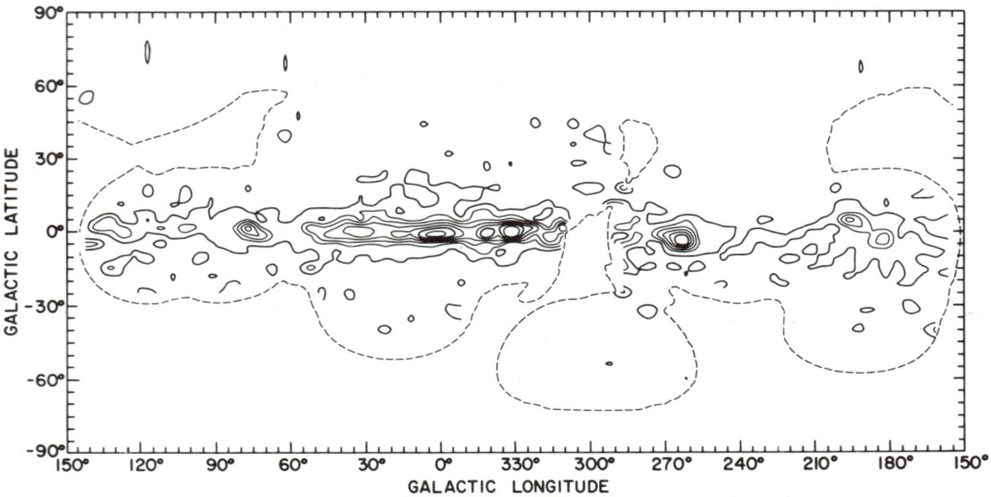


Fig. 5: The SAS-2 map of > 100 MeV γ-rays

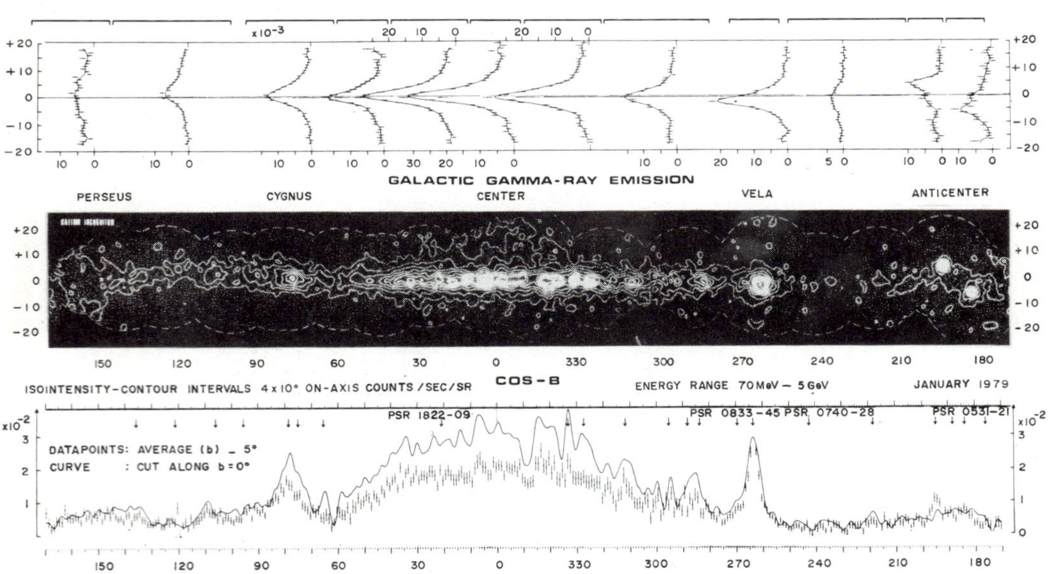


Fig. 6: The COS-B map of > 70 MeV γ-rays

The galactic plane is far more intense in directions toward the center than away from the center, which is to be expected since even γ-rays from the far side of the galaxy have only a small chance of being absorbed in their travel to our instruments. Toward the center, we are seeing emission which was produced, on the average, at the galactic center, 10 Kpc away. Thus in the galactic center latitude profile we would expect a FWHM of about 1.5°, less than our detector resolution.

The observed latitude distributions (Figure 6) towards the center show a narrow component corresponding to these distant regions of the disk, with a broader asymmetric component enhanced in the region $6^\circ < |b| < 20^\circ$. The complementary asymmetry which can be seen at negative latitudes in the anticenter direction (when the Crab and γ195 + 5 are subtracted) suggests that both these broad features are emission from the Gould's belt local cloud cluster. Another feature which can be seen is a small overall tip of the galactic plane relative to $b = 0^\circ$ in the latitude profile near $\ell = 230^\circ$: the "hat-brim effect." This effect has also been seen in HI as well as in the distribution of localized γ sources. 2/

The longitude profiles from the two experiments, as shown in Figure 7, are very similar. Here the COS-B data, shown as open circles, is superimposed on that of SAS-2, and we have chosen the scales seen on the right and left axes to match at the galactic center. There are no statistically significant disagreements at present, although we are anxious to see the COS-B data given in absolute flux units.

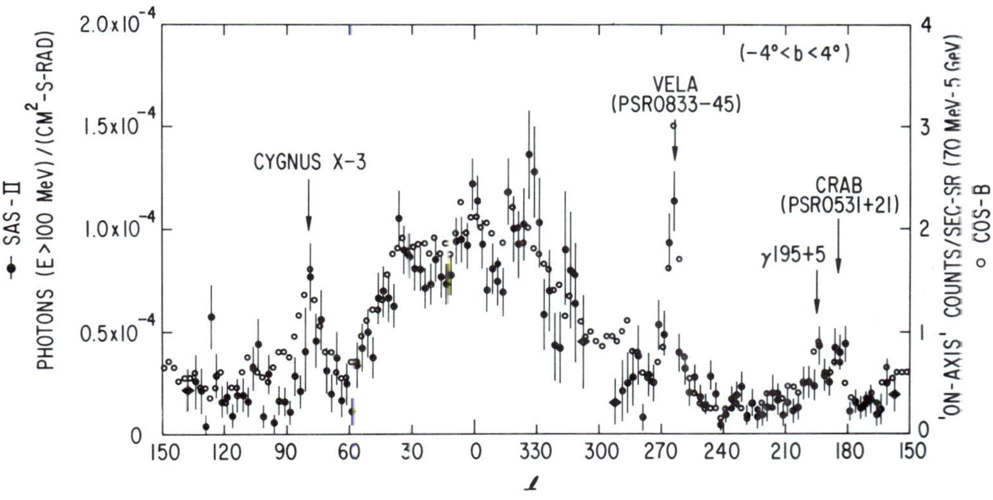


Fig. 7: SAS-2 and COS-B γ-ray intensity along the galactic plane.

2/ See Wills et al., this volume.

Various interpretations of the longitude profile have been suggested. One feature on which all agree is that the contrast between the intensity from directions toward and away from the center is stronger than the contrast which one finds in the matter distribution. This observation has been variously explained as a correlation of cosmic ray density with spiral arms,[22] the 4 kpc ring,[23] or pulsars and supernova remnants.[24] The data do not distinguish decisively among these and other similar possibilities.

The presence of a peak at the galactic center is an example of the type of observational feature which we may regard as possible, but in need of better observations.

It is quite interesting to extend the analysis of the $\gamma$-ray matter correlation to distant regions of the plane, as shown in Figure 8. As I showed for the high latitude emission, the 21 cm column density is used as the abscissa, and the line shown is that which fits the high-latitude data. Quite prominent in this figure are the excess of $\gamma$-ray intensity toward the center of the galaxy, and the deficit in regions away from the center. These deviations only appear in the points at the highest column densities, which are right in the plane, and therefore, most distant. A contribution to the enhancement toward the center is undoubtedly due to hydrogen in molecular form, which is known to be more concentrated toward the galactic center than neutral hydrogen. But quantitatively it seems most probable that cosmic ray intensity must also increase toward the galactic center and decrease at radii beyond the sun. This radial gradient of cosmic rays has been used to argue that the cosmic-ray trapping region cannot extend far above the disk of the galaxy.[25] If it did, the strong concentrations would disappear because of the lack of confining magnetic structure above the plane.

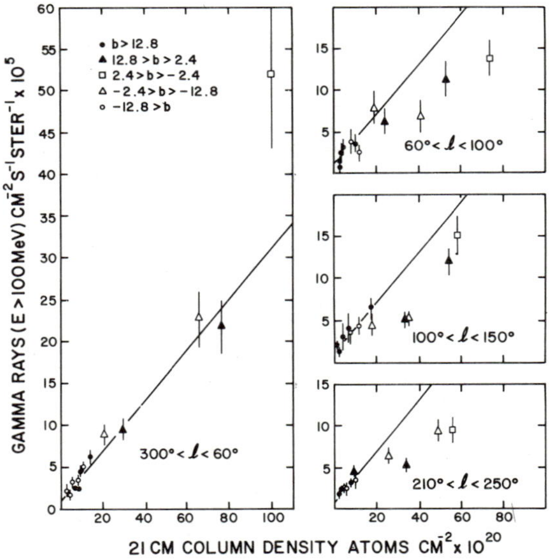


Fig. 8: The correlation between diffuse $\gamma$-ray and intensity and neutral hydrogen column density, at all galactic latitudes.

One of the most interesting questions currently being asked is "Can the longitude distribution be explained as predominantly diffuse emission, or is most of it made up of unresolved point sources?" The answer to this question must await instruments of better angular resolution, but several pieces of evidence that bear on it are worth discussing.

The argument for unresolved point sources[26,27,28] is based on the observation by COS-B of a large number (30 at present) of localized emission features which are believed to represent a partial sample of the discrete source emission within 5 kpc of the sun. The portion of the diffuse emission which can be made up depends strongly on assumptions regarding the distribution and luminosity of the sources, estimates ranging beyond 40%.

A difficulty with this debate is the lack of a precise distinction between the definitions of point source and diffuse emission. I have characterized diffuse emission as due to cosmic-ray matter and photon interactions. Between this extreme and $\gamma$-ray pulsars, which certainly are point sources, one probably has a whole spectrum of objects such as molecular clouds and cosmic-ray Fermi acceleration sites, which can appear as point-like to current $\gamma$-ray instruments, but are spatially extended and involve the diffuse processes.

The arguments in the data for the predominance of diffuse emission are as follows:

(1) At high latitudes, where the point source contribution is negligible, the diffuse $\gamma$-ray production rate is similar to that required to produce the majority of the intensity seen from the rest of the plane.[12]

(2) The spectrum of the $\gamma$ radiation from the plane has been extensively measured[29,30,31] and is shown in Fig. 9. Its shape is the same in all regions of the plane, and like the high-latitude spectrum, it requires a larger electron contribution. I have also shown the high-latitude diffuse galactic spectrum[12] and its associated uncertainty to make graphic the fact that these agree within errors. This is circumstantial evidence for predominantly diffuse origin. By contrast, of the four strongest point sources, two have spectra different than the plane, while the other two are similar.

(3) If one tries to model the longitude distribution using the matter distribution and a constant cosmic-ray density, the attempt is unsuccessful. But by invoking proportionality between cosmic rays and matter on the scale of spiral arms,[22] the distribution is fitted rather well. Similarly, the assumption that cosmic-ray sources are distributed like supernova remnants and molecular hydrogen fits the longitude data.

## OBSERVATIONS IN OTHER ENERGY RANGES

Before concluding, it is important to mention progress being made in other regions of the gamma-ray spectrum. The problems of background removal and limited angular resolution are exacerbated below 35 MeV and above several GeV; but even so, significant progress is being made. At the present the observations in the range from .5 MeV to 20 MeV have resolved the plane and its spectrum,[31,32,33] observed the brighter galactic and extragalactic point sources[34] and indicated the spectrum at high latitudes.[35]

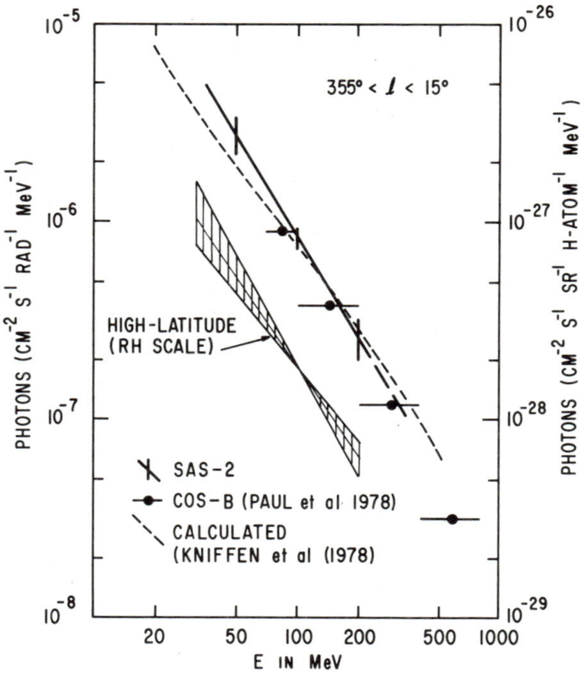


Fig. 9: Galactic γ-ray energy spectrum

The techniques include solid-state and crystal spectrometers[33] (.06-5 MeV), double Compton scattering [34,35] (1-20 MeV), and pushing the pair-production technique to lower energies.[31] At much higher energies ($10^{11}$ to $10^{14}$ eV) atmospheric Cerenkov γ-ray observations have set upper limits on the intensity from some point sources.[36,37,38] However, it appears that some new technique will be needed to make observations of the diffuse component at these very high energies.

## CONCLUSION

To summarize, observations of diffuse γ-ray emission from the galaxy have told us that (1) there are more cosmic-ray electrons in the range .1-1 GeV than we used to think; (2) the galactic halo can contribute at most 20% of the emission which is not connected with the galactic matter disk; and (3) there appears to be a radial gradient of cosmic rays in the galaxy.

Prospects for progress in observations of diffuse γ-rays are excellent, especially as the possiblity of a major γ-ray observatory in orbit approaches. Particularly exciting for diffuse galactic astronomy is the prospect of measuring the interstellar electron spectrum at low energies through its bremsstrahlung emission. This observation may impose new constraints on theories of electron acceleration and solar modulation.

REFERENCES

1. B. Rossi, High-Energy Particles, Prentice-Hall, 1952, p. 79.

2. C. E. Fichtel, R. C. Hartman, D. A. Kniffen, D. J. Thompson, G. F. Bignami, H. Ögelman, M. E. Özel, and T. Tümer, Ap. J. 198, 163 (1975).

3. K. Bennett, J. J. Burger, M. Gorisse, H. A. Mayer-Hasselwander, E. Pfefferman, P. G. Shukla, R. Stiglitz, B. N. Swanenburg, B. G. Taylor, and R. D. Wills, Ninth ESLAB Symposium, ESRO SP-106, 323 (1974).

4. W. L. Kraushaar, G. W. Clark, G. P. Garmire, R. Berken, P. Higbie, V. Leong, and J. Thorsos, Ap. J. 177, 341 (1972).

5. G. W. Clark, G. P. Garmire, and W. L. Kraushaar, Ap. J. Lett. 153, L203 (1968).

6. C. E. Fichtel, R. C. Hartman, D. A. Kniffen, D. J. Thompson, H. G. Ögelman, T. Tümer, and M. E. Özel, Tabulated data from the SAS-2 High Energy Gamma-Ray Telescope NASA TM 79650.

7. F. W. Stecker, Astrophys. Space Sci. 6, 377 (1970).

8. D. M. Worrall, Thesis, University of Durham, 1977.

9. W. B. Burton, The Structure and Content of the Galaxy and Galactic Gamma Rays, ed. Fichtel and Stecker, 1976.

10. C. Heiles, Astro. Ap. Suppl. 20, 37 (1975).

11. S. C. Simonson, III, Astron. Astrophys. 46, 261 (1976).

12. C. F. Fichtel, G. A. Simpson, and D. J. Thompson, Ap. J. 222, 833 (1978).

13. R. C. Bohlin, R. D. Savage, and J. F. Drake, Ap. J. 224, 132.

14. A. C. Cummings, Ph.D. Thesis, Caltech, Pasadena, 1973.

15. J. K. Daugherty, R. C. Hartman, and P. A. Schmidt, Ap. J. 198, 493 (1975).

16. H. V. Cane, Ph.D. Thesis, University of Tasmania (1977).

17. W. R. Webber, H. V. Cane, and G. A. Simpson, Ap. J., in press (1979).

18. V. S. Ptuskin and Ya. M. Khazan, Astrafizika, 12 #1, 81, 1976.

19. D. M. Worrall and A. W. Strong, Astron. Astrophys. 75, 229 (1977).

20. D. A. Kniffen, C. E. Fichtel, R. C. Hartman, R. C. Lamb, and D. J. Thompson, Proceedings 12th ESLAB Symposium Frascati, 45, 1977.

21. H. A. Mayer-Hasselwander, K. Bennett, G. F. Bignami, R. Buccheri, N. D'Amico, W. Hermsen, G. Kanbach, F. Lebrun, G. G. Lichti, J. L. Masnou, J. A. Paul, K. Pinkau, L. Scarsi, B. N. Swanenburg, and R. D. Wills, to be published in Proceedings of the 9th Texas Symposium in 'Annals of the New York Academy of Sciences', 1979.

22. G. F. Bignami, C. E. Fichtel, D. A. Kniffen, and D. J. Thompson, Ap. J. 199, 54 (1975).

23. F. W. Stecker, in The Structure and Content of the Galaxy and Galactic Gamma Rays, ed. Fichtel and Stecker.

24. D. Dodds, A. W. Wolfendale and E. C. M. Young, Proc. 15th Int'l Conf. Cosmic Rays, Plovdiv, 1977.

25. F. W. Stecker and F. C. Jones, Ap. J. 217, 843 (1977).

26. H. Ögelman, Nature 221, 753 (1969).

27. J. C. Higdon and R. E. Lingenfelter, Ap. J. 208, L107 (1976).

28. G. F. Bignami, P. Caraveo, and L. Maraschi, Astron. Astrophys. 67, 149 (1978).

29. R. C. Hartman, D. A. Kniffen, D. J. Thompson, C. E. Fichtel, H. B. Ögelman, T. Tümer, and M. E. Özel, Ap. J., in press.

30. J. A. Paul, K. Bennett, G. F. Bignami, R. Buccheri, P. Caraveo, W. Hermsen, G. Kanbach, H. A. Mayer-Hasselwander, L. Scarsi, B. N. Swanenburg, and R. D. Wills, Astron. Astrophys. 63, L31 (1978).

31. D. A. Kniffen, D. A. Bertsch, D. J. Morris, R. A. R. Palmeira, and K. R. Rao, Ap. J. 225, 591 (1978).

32. S. M. White, S. Moon, J. M. Ryan, R. B. Wilson, A. D. Zych, and B. Dayton, Proc. 15th Int'l Conf. Cosmic Rays, Plovdiv, 1, 100, 1977.

33. J. I. Trombka, C. S. Dyer, L. G. Evans, M. J. Beilefeld, S. M. Selzer, and A. E. Metzger, Ap. J. 212, 925 (1977).

34. V. Schonfelder, U. Graser, and J. Daugherty, Ap. J. 217, 306 (1977).

35. F. Graml, MPI - PAE/Extraterr. 151 (thesis), 1977.

36. N. A. Porter and T. C. Weekes, Smithsonian Astrophysical Observatory, Special Report 381, 1978.

37. T. C. Weeks, G. G. Fazio, H. F. Helmken, E. P. O'Mongain, G. H. Ricke, Ap. J. 174, 165 (1972).

38. J. E. Grindlay, H. F. Helmken, R. H. Brown, J. Davis, and L. R. Allen, Ap. J. 201, 82 (1975).

# LARGE-SCALE STRUCTURE OF THE MILKY WAY, AS OBSERVED BY COS-B IN HIGH-ENERGY GAMMA RAYS

R. D. Wills, on behalf of the Caravane Collaboration

*Cosmic-Ray Working Group, Huygens Laboratorium, Leiden, The Netherlands*
*Laboratorio di Fisica Cosmica e Tecnologie Relative del CNR, Milan, Italy*
*Istituto di Fisica, Università di Palermo, Italy*
*Max-Planck Institut für Extraterrestrische Physik, Garching bei München, Federal Republic of Germany*
*Service d'Electronique Physique, Centre d'Etudes Nucléaires de Saclay, France*
*Space Science Department of ESA, ESTEC, Noordwijk, The Netherlands*

ABSTRACT

ESA's satellite COS-B carries a single spark-chamber experiment for gamma-ray astronomy |1|. In the period August 1975 to February 1978 it devoted about 23 months to observations of the region of the sky along the galactic disc. These observations constitute the first complete detailed gamma-ray survey of the Milky Way.

About 64000 gamma-ray events of energies between 70 and 5000 MeV were selected and processed to derive contour maps and profiles of the gamma-ray intensity. In order not to lose fine structure in regions of high intensity the data were grouped in $0.5^{o}$ x $0.5^{o}$ bins and then carefully smoothed to make a clear presentation possible. Some details of the analytical procedure have been published elsewhere |2|. Since, as a consequence of the spectrum of the galactic gamma-rays, the spatial characteristics of a photon-flux map over the total energy range are dominated by the low-energy quanta which are measured with least accuracy in angular resolution, a sky map was generated showing the number of gamma-ray events which would have been recorded from any direction if the experiment had been pointed there (so-called 'on-axis counts').

Figure 1 shows a map of the galactic gamma-ray emission measured by COS-B together with longitude and latitude profiles. For the map a combination of grey scale and contour lines has been chosen to represent the smoothed data matrix. The smoothing applied was such that only fluctuations of angular scale smaller than the instrument angular resolution (point-spread function) were suppressed. On the other hand not all individual features in the smoothed map are statistically significant. Statistical uncertainties are indicated in the longitude and latitude profiles. In these profiles and in the map a background (determined from two observations at latitudes around $70^{o}$ and $45^{o}$) has been subtracted. The magnitude of this background, which is mainly of instrumental origin, is indicated by the shaded band in the longitude profile.

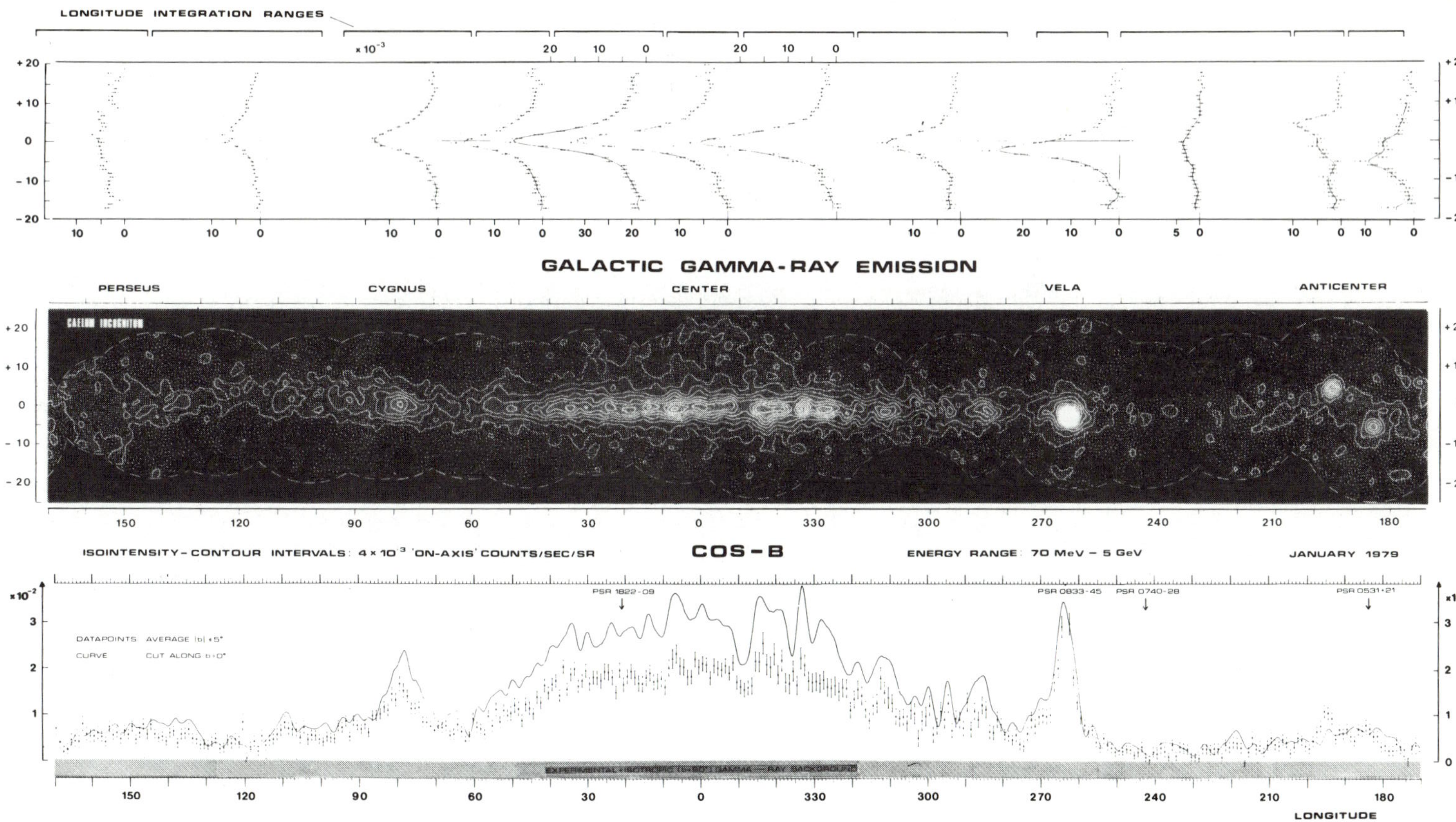


Fig. 1 Presentation in galactic coordinates of the structure of the galactic gamma-ray emission as measured by COS-B. In the map the accepted field of view is indicated by dashed circles. The latitude profiles show the data points with errors and the fitted surface (full line). The longitude ranges over which the data are averaged are indicated by brackets. In the longitude profile the data points, with errors, averaged over $\pm$ 5 degrees are shown. The full line indicates a cross-section through the fitted surface along the equator. The background is indicated by the shaded area. The map and the profiles show the parameter 'on-axis counts' $s^{-1}$ $sr^{-1}$.

The fine structure of the gamma-ray emission has been separately investigated and a number of statistically significant enhancements with the spatial characteristics of point sources have been revealed |3|.

The overall structure of the emission has been studied in more detail by selecting events corresponding to photon energies above 300 MeV for which the angular resolution is better. The highest-energy latitude profiles were used to determine the intrinsic thickness of the emissivity of the galactic disc |2|. These profiles show the two components, first noted by Fichtel et al. |4|, which correspond to galactic structures at different distances from the Sun. The broader component can be correlated well with the local ($\lesssim$ 1 kpc) matter distribution, having a broad excess at positive latitudes towards the centre of the Galaxy and a corresponding excess at negative latitudes towards the anticentre. The narrower component appears to have its origin at 4 to 6 kpc distance and shows a strong variation with galactic longitude, having features similar to those seen at radio wavelengths. The peak of the excess has a small but significant latitude displacement from the equator, corresponding to the 'hat-brim' effect seen in the HI 21 cm line.

No detailed correlation with particular features nor with any model of the galactic spiral structure has been attempted because the implied assumptions are not unique. In particular it is not clear what might be the relative contributions to the large-scale emission of truly diffuse processes (interactions of cosmic rays with gas) and of the integrated effects of unresolved sources.

References

1. L. Scarsi, K. Bennett, G.F. Bignami, G. Boella, R. Buccheri, W. Hermsen, L. Koch, H.A. Mayer-Hasselwander, J.A. Paul, E. Pfeffermann, R. Stiglitz, B.N. Swanenburg, B.G. Taylor and R.D. Wills, Proc. of the 12th ESLAB Symposium, ESA SP-124, 3 (1977).

2. H.A. Mayer-Hasselwander, K. Bennett, G.F. Bignami, R. Buccheri, N. D'Amico, W. Hermsen, G. Kanbach, F. Lebrun, G.G. Lichti, J.L. Masnou, J.A. Paul, K. Pinkau, L. Scarsi, B.N. Swanenburg and R.D. Wills, Proc. of the 9th Texas Symposium in Annals of the New York Academy of Sciences, in press (1979).

3. R.D. Wills, K. Bennett, G.F. Bignami, R. Buccheri, P. Caraveo, N. D'Amico, W. Hermsen, G. Kanbach, G.G. Lichti, J.L. Masnou, H.A. Mayer-Hasselwander, J.A. Paul, B. Sacco and B.N. Swanenburg, this volume.

4. C.E. Fichtel, R.C. Hartman, D.A. Kniffen, D.J. Thompson, G.F. Bignami, H. Ögelmann, M.E. Özel and T. Tümer, Astrophys. J., 198, 163 (1975).

# THE EFFECT OF THE DIFFUSE GAMMA RADIATION ON PULSARS

V. Radhakrishnan

*Raman Research Institute, Bangalore 560 006, India*

ABSTRACT

The possible role of the diffuse background gamma radiation in triggering the polar gap discharges in pulsars is examined. It is shown that the number of charges produced in the gap per second should suffice to continually reinitiate the breakdown in the gap. More significantly, several characteristics of pulsars that were previously not understood appear now to find a rational explanation on the basis of triggering by gamma rays.

It is well known that pulsars emit gamma rays as has already been discussed earlier at this symposium. Typical figures for the gamma ray output are $10^{33}$ to $10^{35}$ ergs/sec., about one to a hundred times the total luminosity of the sun. If we compare this with the amount of gamma radiation incident on the surface of the neutron star which we see as a pulsar, we get the ridiculously small number of $10^{3}$ ergs/sec., more than 30 orders of magnitude less than the emission from it. This is about the amount of energy that would fall on the surface of the sun due to a one-watt flash-light pointed at it from the earth. I am sure no one here would suggest that we could detect noticeable effects on the sun due to this feeble illumination. What I am going to try and persuade you to believe today is that the minute amount of diffuse gamma radiation falling on a neutron star may have more than just noticeable consequences on its behaviour as a pulsar.

Although there is no satisfactory model for the radio emission mechanism of pulsars capable of explaining all of the observed characteristics of this radiation, there is a consensus of opinion on certain processes that are believed to take place close to the magnetic poles of the rotating neutron star. Goldreich and Julian (1969) drew attention to the intense electric fields that would be generated due to the rotation of the super-strong magnetic field associated with the neutron star. Their picture of the flow of current from the surface into the magnetosphere and finally into the interstellar medium through the open field lines was modified by Ruderman and Sutherland (1975) who pointed out the inability of even such a powerful electric field to pull out ions from the very tightly bound condensed state of the stellar surface.

The binding energies that went into the Ruderman and Sutherland model have been substantially reduced by subsequent calculations (Flowers et al 1977) but their picture remains essentially unmodified. A gap is built up immediately above the magnetic pole because of the strong positive charge on the surface of the polar cap. The plasma above loses some negative charge to the polar cap and is then repelled away from the surface almost at the speed of light creating a vacuum gap with a voltage drop of the order of $10^{12}$ volts. The building up of the voltage across this gap is compared to the open circuiting of a dynamo which in this case is the homo-polar generator.

If a stray charged particle is now introduced into the gap it would be immediately accelerated to a very high energy. As it moves along the curved field lines it would emit curvature radiation and the photons could be in the MeV range without any difficulty. These gamma ray photons would produce pairs by interaction with the magnetic field as proposed by Sturrock (1971) in a earlier model for pulsars. These pairs will in turn be accelerated, and by the same process produce more pairs. Ruderman and Sutherland (1975) showed thus that the result would be an avalanche break-down of the gap and the release of a positron shower into the magnetosphere strong enough to produce the observed radiation at all frequencies through subsequent processes.

Once the polar gap, or a region of it, has been discharged as described above, a subsequent discharge can take place only by introducing one or more charged particles when the gap has been built up again to the required height. The necessity of reinitiating the discharge each time, and the

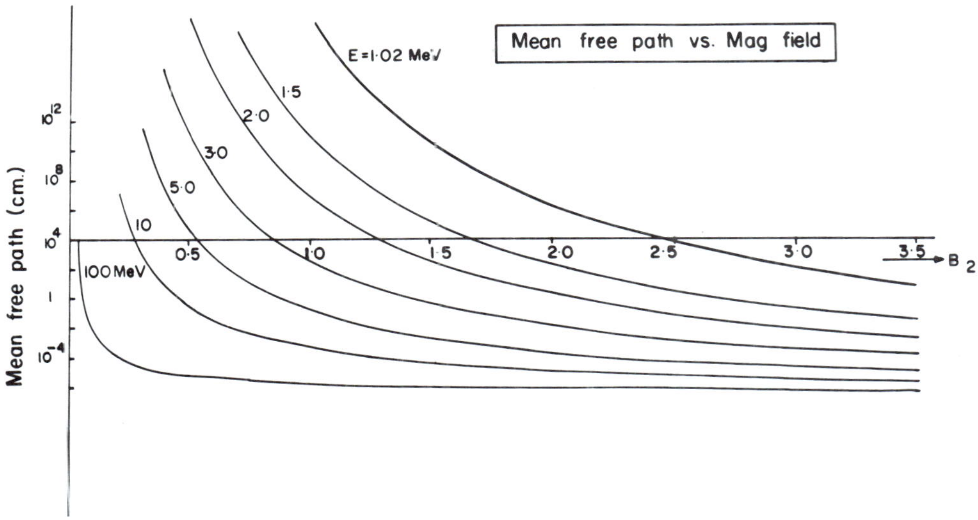


Fig.1 The mean free path for pair creation as a function of the perpendicular component of the magnetic field for gamma ray photons of various energies. The abscissa is in units of $10^{12}$G.

difficulties of so doing, have been emphasised recently by Ruderman (1978) who states that it "must continually be restarted on a time scale of order $10-10^2 \mu s$, the time it (the discharge) would take to reach either the boundary or an inner field line with no curvature". As a process for continual "restarting", he has proposed that ions are emitted from discharge heated patches on the polar cap; photons in the high energy tail of the thermal radiation from these hot patches interact with the coulomb field of these ions and make ($e^-$, $e^+$) pairs by the conventional pair production process. The pair so created then initiates the discharge as described earlier. While the above process is a possible one for maintaining the discharges, I shall assume here that it plays no role and that the operative restarting process is pair creation in the gap due to the passage of diffuse background gamma ray photons through it. The consequences predicted from this assumption will then be compared with observations to assess its validity.

The path length for pair creation of a gamma ray travelling in a magnetic field is very strongly dependent on both the energy of the gamma ray and the strength of the perpendicular component of the field. The expression given by Erber (1966) can be written to give $\ell$ the path length in cms as

$$\ell = \frac{10^{-6}}{B_\perp} \cdot e^{\left(\frac{58.7}{E B_\perp}\right)} \quad [cm] \tag{1}$$

where $B_\perp$ is the perpendicular component of the magnetic field in units of $10^{12}$ G and E is the energy of the photon in MeV. The resulting path length as a function of energy and field strength are shown in Fig. 1 from which we see that the likelihood of pair creation by any photons which enter the gap increases exponentially with both the field strength and the photon energy. However, this also creates a problem in that photons with a very short path length are unlikely to reach the gap because they must first travel through the magnetosphere in which the field is building up in strength as the photons approach the gap.

The conflicting requirements for the gamma ray photons of reaching the gap unaffected, and of pair producing while within it, can be better expressed in terms of optical depths rather than pathlengths. The probability p of a photon producing a pair within the gap is given by

$$p = e^{-\tau_{mag}} \left(1 - e^{-\tau_{gap}}\right) \tag{2}$$

where $\tau_{mag}$ and $\tau_{gap}$ are the effective "optical depths" for pair production in the magnetosphere and gap respectively. Both are functions of the photon energy, the perpendicular components of the magnetic field and the distances traversed by the photon. They can be calculated using equation (1) and the result is illustrated in Figs. 2a and 2b. For a given magnetic field strength in the polar gap, only photons lying within a narrow energy range are likely to produce pairs in the gap. Fig. 3 illustrates how the position of this photon energy window varies with the field strength in the gap.

An exact calculation of the probability of pair creation within the gap cannot be carried out without a knowledge of the field structure near the gap. The absorption coefficient in the magnetosphere will vary along the photon's trajectory with both the field strength and its angle to the trajectory. Depending on the geometry of the field structure, the net probability will also be a function of the direction of arrival of the gamma ray photon. Because of these difficulties, the values represented in Fig. 3 have been calculated on the assumption of a pure dipole geometry for the field,

Fig. 2a

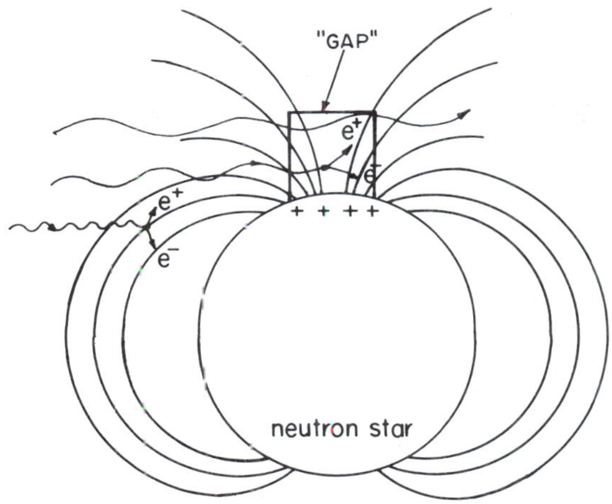


Fig. 2b

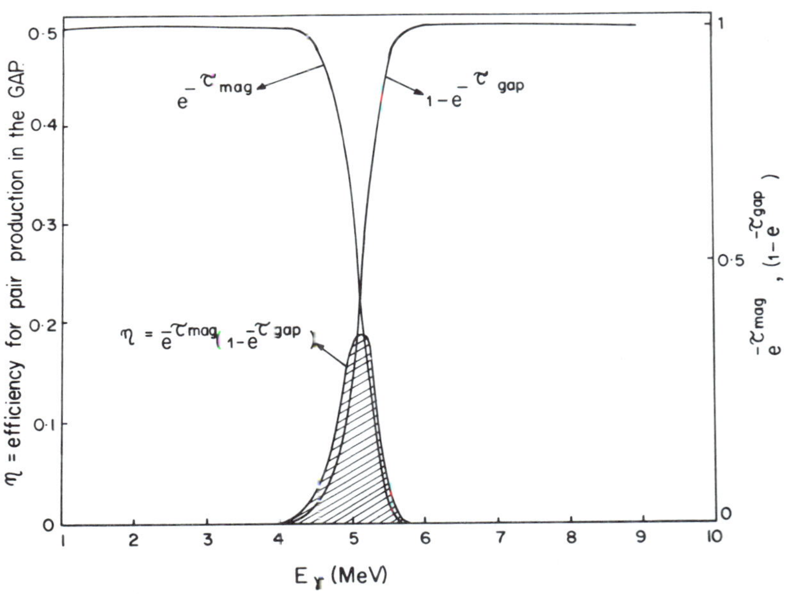


Fig. 2a For a neutron star of given magnetic field, only photons within a certain energy range are likely to produce pairs in the gap. Higher energy photons do not reach the gap and lower energy ones pass through the gap without pair-producing.

Fig. 2b The efficiency for pair production in the gap with a given magnetic field as a function of the energy of the photon. The magnetosphere acts as a low pass filter absorbing all the higher energy photons. The gap absorbs all photons reaching it with energies above a certain value. Together the combination acts as a band pass filter for pair production within the gap.

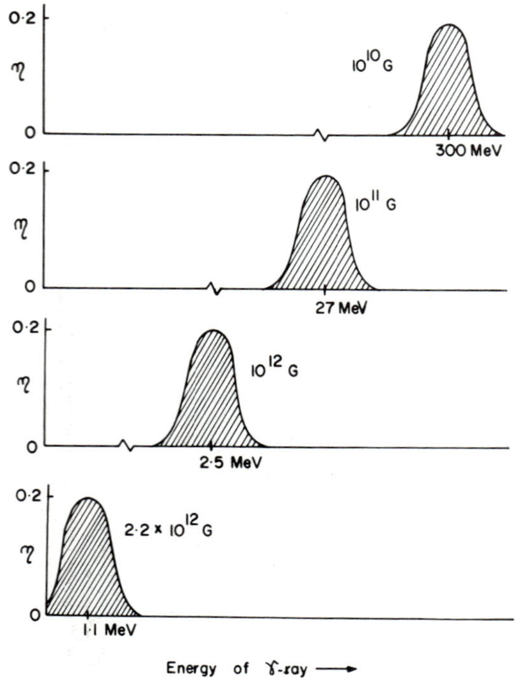


Fig. 3 The dependence of the effective window in the energy of incoming gamma rays as a function of the field strength in the gap.

a radius of 10 km for the neutron star, and a polar gap of $10^4$ cm in both height and diameter. The total number of pairs produced per second in the gap by diffuse gamma rays of all energies was then determined by combining the results illustrated in Fig. 3 with the observed differential spectrum of the background gamma radiation taken from Trombka et al (1978). The first results of such a calculation carried out by C.S. Shukre (to be published) are shown in Fig. 4 as a function of the surface field strength in the polar gap.

The peculiar shape of the plot in Fig. 4 can be understood as arising from two causes. One is the steep dependence on energy of the observed gamma ray spectrum which can be approximated as a power law of the form $n = AE^{-\alpha}$ where n is the number of photons per $cm^2$ per second per steradian per MeV, and E is the photon energy in MeV; in the region of interest $A=1.5\times10^{-2}$ and $\alpha \simeq 2.3$. The other is that gamma rays with energy less than $2m_ec^2$, although plentiful, are useless for producing pairs. Taken together, the number of pairs created within the polar gap can exceed $10^5$ per second as was already noted, albeit in parentheses, by Ruderman and Sutherland (1975).

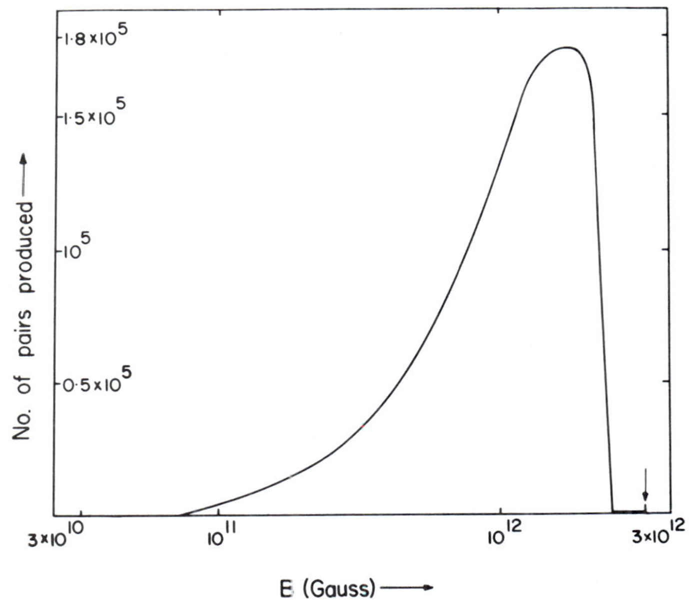


Fig. 4 The number of pairs per second produced within the gap by the gamma ray background as a function of the field strength.

This number appears to be adequate for a restarting time scale of order 10 to $10^2$ $\mu$s as required by Ruderman (1978). The inclusion of gravitational red shift effects in modifying the cut off value in the energy of gamma rays capable of producing pairs within the gap will increase the numbers shown in Fig. 4 somewhat.* As the magnitude of this increase will depend on the masses and radii of the neutron stars we have neglected this consideration in the present calculation.

The surprising new result is that such pair creation is effective only over an extremely narrow range ($\sim$ 3:1) of magnetic field strengths centred at $1.5 \times 10^{12}$ gauss. The cut off is total for fields above this range and somewhat more gradual for weaker fields, but the unambiguous implication of Fig. 4 is that discharges in the polar gap can be initiated by the gamma ray background only for neutron stars lying within this narrow "magnetic window". The obvious test is to compare the derived magnetic fields for pulsars with

*I thank A.R.P. Rau for drawing my attention to this effect.

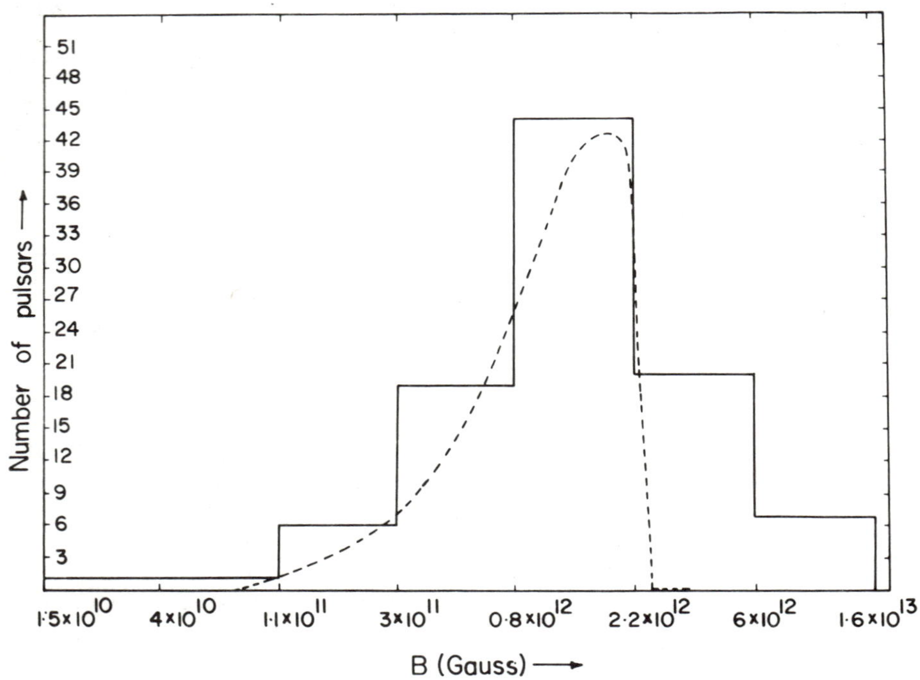


Fig. 5 Histogram of the number of pulsars with derived field strengths lying in equal logarithmic intervals of the field. The dashed line shows the magnetic window of Fig. 4 superimposed on the histogram. Field values were either taken from, or derived using the period and period derivative measurements from Taylor and Manchester (1975), Manchester et al (1976), Gullahorn and Rankin (1978) and Taylor et al (1979).

what is expected on the above hypothesis. Fig. 5 is a histogram (solid line) of such derived magnetic fields indicating a concentration at the position of the gamma ray window, shown superimposed (dashed line) on the histogram. In the histogram, we have included only those pulsars for which the period derivatives can be considered reliable.

Before drawing any conclusions from the apparent agreement of Fig. 5, I would first like to discuss the meaning of the histogram in the figure. Magnetic field strengths for pulsars are generally derived from a measurement of the period and its time derivative by applying the formula

$$B^2 = \left(\frac{3c^3}{8\pi^2}\right)\left(\frac{I}{R^6}\right)P\dot{P} \qquad (3)$$

and taking nominal values of $I = 10^{45}$g cm$^2$ and R = 10 km for the moment of inertia and radius of the neutron star. It is assumed that the braking is due to magnetic dipole radiation, or any other mechanism with the same braking rate. The energy loss rate formula derived by Goldreich and Julian (1969) for particle production and acceleration by an axisymmetric rotator is identical with that derived by Gunn and Ostriker (1969) for magnetic dipole radiation from a perpendicular rotator.

Even if no serious errors were introduced by the assumption of a standard energy loss rate, we would certainly expect a smearing of the derived values of the field due to variations in the radii and moments of inertia of the neutron stars. In any model for pulsars, the calculated values of moment of inertia and radius would be related by the equation of state used in the model. Even so, the quantity $I/R^6$ used in equation (3) should be a function of the mass, and all models admit of a considerable range in the allowed masses for neutron stars. We would certainly expect a distribution in the masses reflecting the type of parent star from which it was formed and also the conditions of formation - single star evolution, evolution in binary system etc. The spread in the derived values for B should therefore be greater than that in the true values, due to convolution with the $I/R^6$ distribution. The true values themselves would have a spread due to the variation of field strength at birth and the decay of the field with time.

The histogram in Fig. 5 shows the number of neutron stars with derived magnetic fields lying in equal logarithmic intervals; for exponential decay these intervals would represent equal intervals of time. The width and position have been chosen to make one of the intervals coincide with the magnetic window calculated from the gamma ray discharge-initiation hypothesis, and we see from the figure that about half the derived values fall in this interval. The point I wish to make is that the strong central concentration in the observed distribution would be most unlikely to expect in view of the spreading discussed above. The evidence seems compellingly in favour of the interpretation that most if not all pulsars are observed only because their field strengths have the right value for the initiation of discharges by incoming gamma rays.

The existence of a surprising clustering in the derived field values around $1.5 \times 10^{12}$ gauss has been known for years (c.f. Greenstein 1972) and has remained a puzzle all this time. It now finds a natural explanation as a selection effect for the trigger mechanism of discharges to operate. The agreement in the absolute value of the field with that calculated from pair production considerations also suggests that there are no multipole components which significantly change the field strength near the polar cap from what would be deduced from dipole braking.

The proof of the above interpretation may be difficult to establish, but the agreement shown in Fig. 5 certainly encourages me to advance the following hypotheses:

1. Most observed pulsars have a surface field strength at the magnetic poles of the right value to maximise pair production in the gap by incoming gamma rays.

2. Rotating neutron stars with <u>true</u> fields above $2.5 \times 10^{12}$ gauss will not function as pulsars unless and until their field decays to below this value. As a consequence, neutron stars with periods greater than $\sim 3$ seconds will also not function as pulsars. (This is the critical period for adequate voltage generation in the gap by a neutron star with a field of $2.5 \times 10^{12}$ gauss, according to the Ruderman and Sutherland model).

3. The magnetic field configuration is mainly dipolar and $I/R^6$ is typically $10^9$ gm cm$^{-4}$.

4. The effective energy loss rate for pulsars over their life times differs negligibly from that for magnetic dipole radiation.

5. The spread in the values of <u>derived</u> magnetic field is mainly a reflection of the spread in $I/R^6$ for the neutron stars in question.

6. If all neutron stars are created with very short periods, there is a considerable spread in the initial field strengths at birth.

7. The number period histogram of observed pulsars reflects the distribution of field strengths at birth. The observed concentration of periods around 0.6 seconds indicates that it is the most likely period a neutron star would have slowed down to when its initial field has decayed to a value within the magnetic window.

8. During the pre-pulsar phase the braking is entirely due to magnetic dipole radiation, and neither particles nor high frequency radiation are generated.

There are a number of other implications for pulsars and supernova remnants which my colleagues and I are exploring and which I do not wish to discuss in detail here. To name a few, if hypothesis 5 above is combined with a neutron star model it leads to a mass distribution for radio pulsars. The small number of pulsars observed in supernova remnants, and many of the characteristics of the remnants themselves now seem to find natural explanations if neutron stars behave as suggested above.

To sum up, many of the properties of pulsars and of supernova remnants which were difficult to understand either by themselves or in connection with other properties may find a rational explanation if the background gamma radiation does in fact play the crucial role of triggering pulsars. If it does, the most surprising aspect (to me) is that pulsars can be thought of as gamma ray boosters scattered all over the galaxy. Since most of the energy output of pulsars seems to be in gamma rays, the gain of these incoherent amplifiers is over $10^{30}$.

ACKNOWLEDGEMENTS

I am grateful to my colleagues, C.S. Shukre and G. Srinivasan, for assistance with the calculations, the preparation of the diagrams, and for numerous scientific discussions on the subject of this talk. I thank M. Vivekanand and M.M. Komesaroff for many discussions on pulsars which eventually triggered this investigation, and Rajaram Nityananda for a critical reading of the manuscript.

REFERENCES

1. T. Erber, Reviews of Modern Physics, 38, 626 (1966).

2. P. Goldreich and W.H. Julian, The Astrophysical Journal, 157, 869 (1969).

3. G. Greenstein, The Astrophysical Journal, 177, 251 (1972).

4. G.E. Gullahorn and J.M. Rankin, The Astronomical Journal, 83, 1219 (1978).

5. J.E. Gunn and J. Ostriker, Nature, 221, 454 (1969).

6. R.N. Manchester, P.A. Hamilton, W.M. Goss and L.M. Newton, Proceedings of the Astronomical Society of Australia, 3, 81 (1976).

7. M.A. Ruderman and P.G. Sutherland, The Astrophysical Journal, 196, 51 (1975).

8. M.A. Ruderman, "Pulsar Radiation Mechanisms" - Invited talk at Ninth Texas Relativistic Astrophysics Conference (Munich) (1978).

9. P.A. Sturrock, The Astrophysical Journal, 164, 529 (1971).

10. J.H. Taylor and R.N. Manchester, The Astronomical Journal, 80, 794 (1975).

11. J.H. Taylor, L.A. Fowler and P.M. McCulloch, Nature, 277, 437 (1979).

12. J. Trombka, C. Fichtel, J. Grindlay and R. Hofstadter, Science, 202, 933 (1978).

# ORIGIN OF DIFFUSE GALACTIC GAMMA RAYS

S. Hayakawa

*Department of Astrophysics, Nagoya University, Nagoya, Japan*

ABSTRACT

Theoretical interpretations of the diffuse component of galactic gamma rays with energies around 100 MeV are reviewed, on account of that these gamma rays are generated by three mechanisms, the 2$\gamma$-decay of neutral pions produced by the collision of the nuclear component of cosmic rays with interstellar matter, the bremsstrahlung of cosmic ray electrons, and the Compton scattering of cosmic-ray electrons by cosmic microwaves and starlight. Correlations of the gamma-ray intensity with various astronomical objects indicate that the gamma-ray emissivity is roughly proportional to the square of the interstellar gas density and also to the infrared emissivity. The energy spectrum of gamma rays does not show the turn-over expected from the pion decay, suggesting a large contribution of the bremsstrahlung and/or Compton scattering electrons. The contribution of the latter is estimated by reference to the starlight density which is found large through infrared observation. The contribution of localized sources to the diffuse component is discussed in connection with the star formation in dense molecular clouds.

## INTRODUCTION

Diffuse galactic gamma rays of energies around 100 MeV are regarded to be produced by the collisions of cosmic rays with interstellar matter and radiation, whose distributions are important for understanding the galactic structure. Hence the galactic gamma rays can be used as a tracer of obtaining astrophysical properties in distant regions of the Galaxy. In fact, the gamma ray intensity is correlated with a number of quantities which characterize the galactic structure, as reviewed below. Closer examination of the correlations has, however, revealed that the situation is not so simple as one has thought. A naive picture of the galactic gamma rays has been questioned in recent years particularly in the following points.

The energy spectrum of gamma rays is approximately represented by a power law in the energy range 50 MeV - 2 GeV. Turn-over of the spectrum expected for pion-decay gamma rays is not observed. This is considered to be due to the contribution of electron bremsstrahlung, but the electron intensity required for the power-law spectrum seems appreciably higher than that obtained from cosmic-ray experiments and synchrotron radio emission. A question is raised if Compton gamma rays are responsible for this gap.

Another serious question is how much diffuse gamma rays are contaminated by localized sources. Since gamma-ray detectors thus far employed have angular resolutions not good enough to resolve weak sources, a significant fraction of diffuse gamma rays may well be due to localized sources, in which the intensity of electrons relative to protons may be much higher and/or the density of radiation relative to matter may be considerably higher than in general interstellar space.

## CORRELATION OF GAMMA-RAY EMISSION WITH GALACTIC STRUCTURE

It had been naively thought that the gamma-ray intensity would be proportional to the column density of interstellar matter, under the assumption of uniformity of cosmic rays. Referring to the OSO-3 result [1], however, Hayakawa and Tanaka [2] noticed that the gamma-ray emissivity is proportional to the square of atomic hydrogen density. This implies either that the cosmic-ray density is proportional to the gas density if gamma rays are produced from diffuse matter, or that the density of gamma-ray sources is proportional to the square of the interstellar gas density if gamma rays are produced by localized sources. In favor of the latter view, they have argued that the star formation rate is approximately proportional to the square of the density of interstellar matter, and consequently that most gamma-ray sources are associated with supernova remnants (SNR). Then they have demonstrated that the longitude dependence of the gamma-ray intensity correlates with that of the radio intensity of SNR's.

The gamma-ray emissivity in proportion to the square of the gas density as suggested by Hayakawa and Tanaka [2] has been confirmed by a more refined observation by SAS-2. Kniffen et al [3, 4] have compared the longitude profile of gamma rays with that of total hydrogen (HI + HII + 2 $H_2$) and found an appreciably weaker longitude dependence of the latter. They have, however, obtained a good correlation with the square of the column density of hydrogen, as shown in Figure 1. Cesarsky et al [5] have reached similar conclusion, taking into account the spiral structure of the Galaxy.

The latitude dependence, on the other hand, indicates a linear relation. Figure 2 shows the COS-B data [6] for gamma rays with energies greater than 300 MeV in the longitude range 10° - 40°. The latitude profile is asymmetric with respect to the galactic equator, excess in the northern region, $4° < b < 10°$. Lebrun and Paul [7] demonstrated that the excess is due to dense dark clouds in the northern region and, if their contribution is taken into account, the gamma ray intensity is proportional to the column density of total hydrogen. This implies that the cosmic-ray density is uniform and equal to that in the solar neighborhood within the Gould belt region, in which the dark clouds are located.

Having remarked on the correlation between the gamma-ray emissivity and the hydrogen density, Stecker [8] has argued that the cosmic-ray density distribution is not uniform in the Galaxy but correlates with the SNR distribution as obtained by Kodaira [9]. This implies that cosmic rays are produced by SNR or pulsars and diffuse out of the galactic disk or a thin halo so as to maintain the non-uniformity of cosmic rays, as extensively discussed by Stecker and Jones [10].

The non-uniformity of cosmic rays should reveal itself in the directional distribution of the radio intensity, which is proportional to the product of the cosmic-ray electron density and some power of the magnetic field strength. On the basis of OSO-3 data[11], Schlickeiser and Thielheim [12] demonstrated that the longitude dependence of the gamma ray intensity could be accounted for if the product of cosmic-ray and hydrogen densities is correlated with the magnetic field strength.

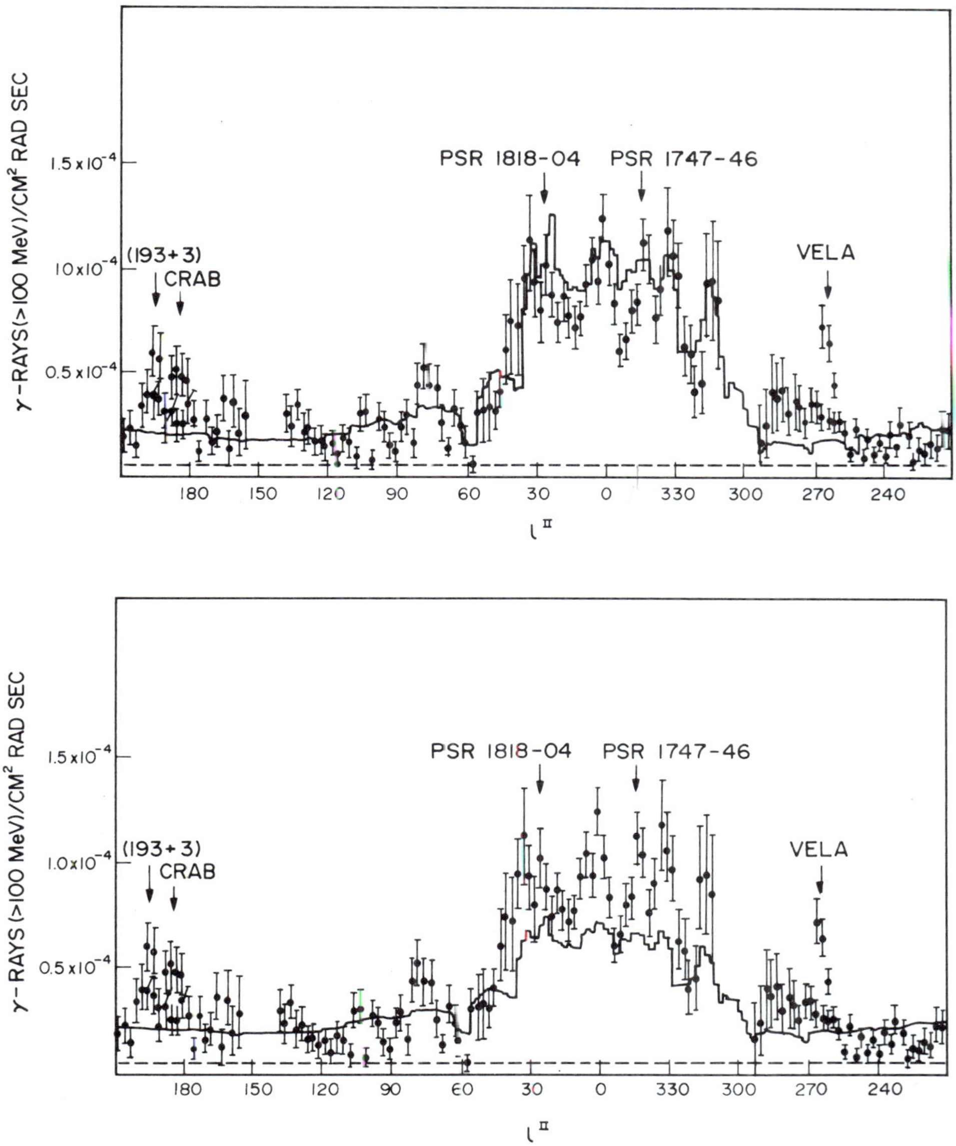


Fig.1. The longitude dependence of the intensity of gamma rays with energies greater than 100 MeV [3]. Identified sources are indicated by arrows. The dashed line represents the isotropic background. The solid line in the bottom figure represents the gamma-ray flux calculated for the uniform cosmic-ray intensity, whereas the line in the top figure gives the one for the cosmic-ray intensity proportional to the matter density.

Refferring to SAS-2 data, Paul et al [13] have shown that the distributions of the gamma-ray intensity and the synchrotron emission are similar to one another, as shown in Figure 3, and that this can be explained if the electron intensity is proportional to the square of magnetic field strength. A more quantitative study by Fichtel et al [14] gives a linear relation between the gamma ray intensity in the energy range 35 - 100 MeV, $I_\gamma(35 - 100\ MeV)$, and the radio brightness temperature at 150 MHz, $T_b(150)$, that is, $I_\gamma(35 - 100\ MeV) = A\ T_b(150) + B$, as shown in Figure 4. A correlation with the gamma-ray intensity is also found for the infrared surface brightness of the Galaxy. In Figure 5 the longitude dependence of the 2.4 μm flux observed by Hayakawa et al. [15, 16, 17] is compared with that of the gamma-ray intensity. These two also show similar latitude profiles. Moreover, the absolute flux* is so high that the starlight density estimated by reference to the near infrared intensity contributes appreciably to Compton gamma rays. Far infrared radiation observed by Serra et al. [18] and by Maihara et al. [19] also shows a similar longitude dependence, though the longitude ranges covered are limited.

There are so many quantities which are correlated with the gamma-ray intensity. Some of them may be related only indirectly, or may arise from a common origion which is yet to be disclosed.

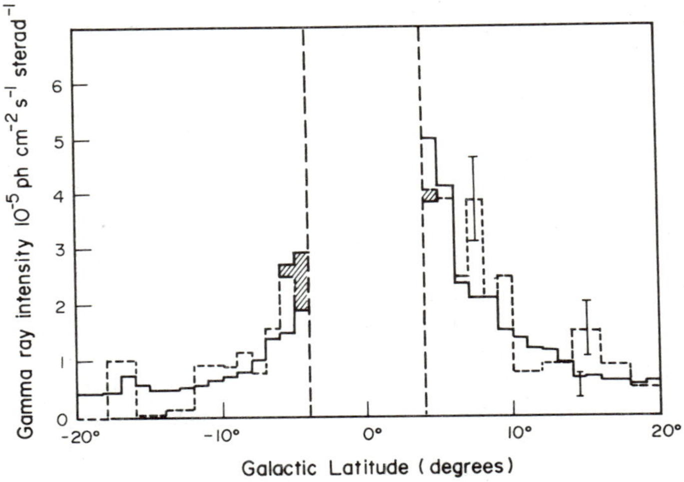


Fig.2. The latitude profile of the gamma-ray intensity in the longitude range 10° - 40°. The solid line indicates the calculated profile for the constant cosmic-ray intensity in the region including dark clouds. The hatched area represents the contribution of gamma rays from the region $|b| < 4°$. After Bennett et al [6].

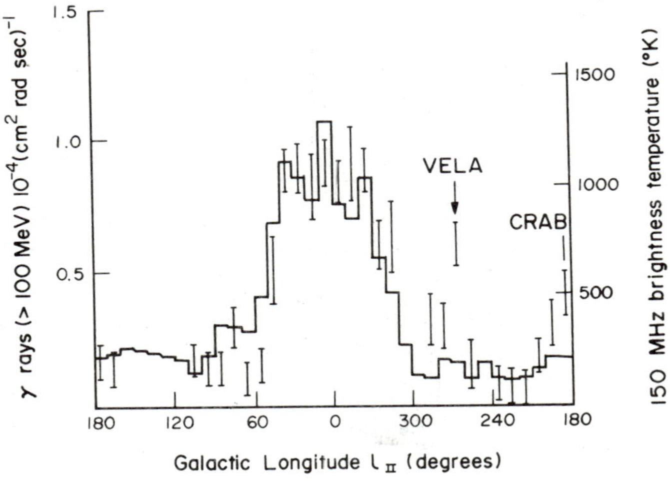


Fig. 3. The longitude dependences of the gamma-ray intensity (> 100 MeV) and the radio brightness temperature at 150 MHz. After Paul et al. [13].

* This was not properly quoted by Bignami and Piccinotti [20].

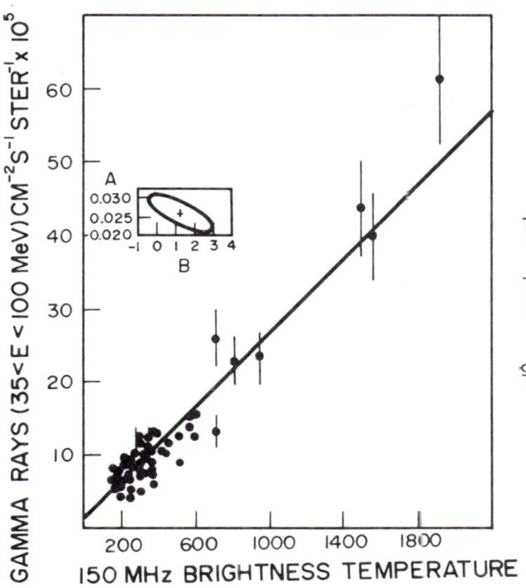


Fig.4. Correlation between the gamma ray intensity (35-100 MeV) and the radio brightness temperature at 150 MHz. The insert represents the contour line of 1$\sigma$ above minimum $\chi^2$ for $I_\gamma = A\ T_{150} + B$. After Fichtel et al. [14].

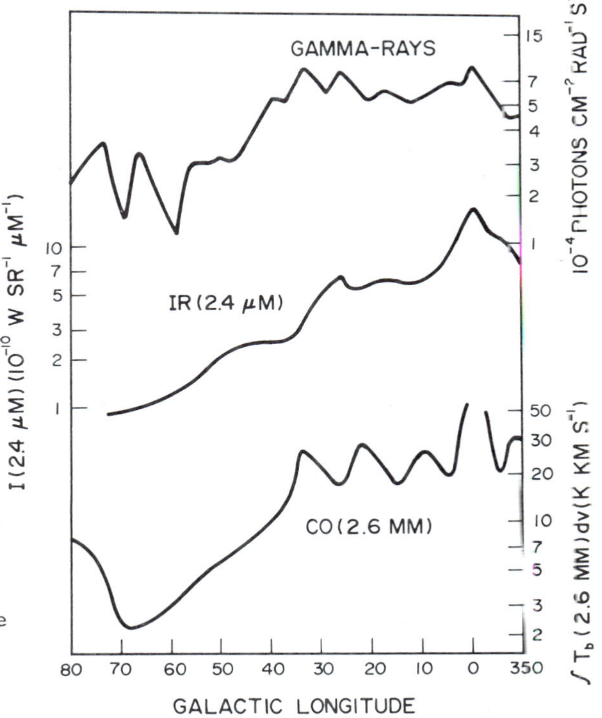


Fig.5. The longitude dependences of the gamma-ray intensity (>100 MeV), the infrared surface brightness (2.4 μm) at the equator [17], and the CO column density [21].

## ENERGY SPECTRUM

The energy spectrum of diffuse gamma rays is essentially the same in all directions, if the isotropic component which is supposed to be of extragalactic origin is subtracted. At low latitudes the differential spectrum is approximately represented by $E^{-2}$ at energies above 150 MeV and becomes somewhat less steep at lower energies [22]. At high latitudes Fichtel et al. [23] have derived a spectrum of $E^{-1.6}$ in the energy range 35 - 200 MeV, after subtraction of the isotropic component which shows a much steeper spectrum. The increase in the spectrum towards low energy requires a contribution of cosmic-ray electrons which emit gamma rays by bremsstrahlung and Compton scattering, since the spectrum of gamma rays from neutral pions turns over at low energies.

The production spectra of respective component in the solar neighborhood calculated by Cesarsky et al. [24] are given in Figure 6. The dashed curves ($\pi$) represent the spectra of pion decay gamma rays calculated by Badhwar and Stephens [25] with estimated error of 30% and are least ambiguous. The dot-dashed curves (C) refer to the spectra due to Compton scattering and are also not ambiguous, since electrons responsible for this process are of high energies and accordingly are little modulated. However, the gamma-ray intensity strongly depends on the distribution of electrons away from the galactic plane. The bremsstrahlung spectrum is model dependent, as shown by several solid lines, since electrons of energies around 100 MeV are subject to strong solar modulation. The curve B2 is regarded as standard, whereas the curve B1 gives an upper bound. The observed spectrum can be accounted for by the sum of π and B1, if the contribution of Compton scattering is neglected.

A considerable contribution of Compton gamma rays is assumed by Fichtel et al. [14], adopting the scale height of electron distribution to be 750 pc, as indicated by synchrotron radio emission, whereas the scale heights for atomic and molecular hydrogen are 120 pc and 50 pc, respectively. Then the contribution of Compton gamma rays is found comparable to that of bremsstrahlung, if the standard electron spectrum is assumed. The contributions of respective components in two energy bands for the pole region of N(HI) = $2.4 \times 10^{20}$ atoms $cm^{-2}$ are given in Table 1. The spectra of respective components and their sum are given in Figure 7, in which the difference between the scale heights of cosmic-ray electrons and interstellar matter is taken into account.

Table 1 shows that the contribution of electrons is dominant in the low-energy band, whereas pion decays give a major contribution in the high-energy band. The observed intensity is consistent with the predicted one, if the high side of the electron intensity is adopted. The contribution of Compton gamma rays is somewhat different at low galactic latitudes, since the contribution of the 2.7 K radiation is relatively small

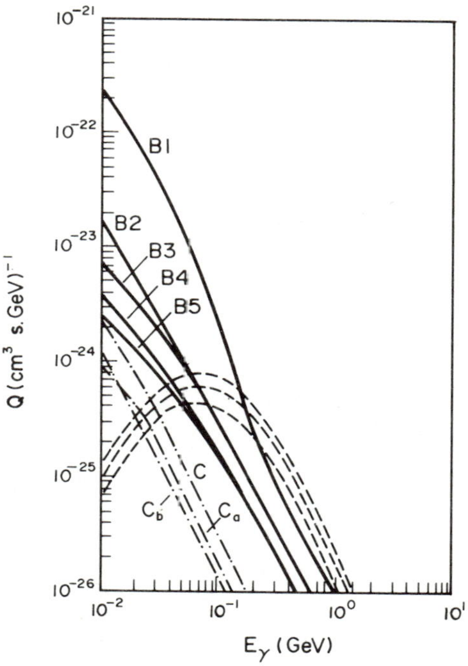


Fig.6. The production spectra of gamma rays by neutral pion decay (π), bremsstrahlung (B) and Compton scattering (C). Uncertainty in π-decay gamma rays is indicated by two dashed lines. The spectra of bremsstrahlung gamma rays calculated with different model spectra cf electrons are indicated by B1 - B5, in which B1 and B4 give the upper and lower bounds, respectively, whereas B2 gives the standard spectrum which we refer to hereafter. $C_a$ and $C_b$ represents the spectra of Compton gamma rays due to the 2.7 K radiation and starlight in the solar neighborhood, respectively. After Cesarsky et al. [24].

TABLE 1 Gamma Ray Intensities Predicted and Observed ($10^{-5}$ photons $cm^{-2}$ $sr^{-1}$ $s^{-1}$) for the Pole Region, $N_H$ = $2.4 \times 10^{20}$ H atoms $cm^{-2}$. After Fichtel et al [14]

| | 35 - 100 MeV | > 100 MeV |
|---|---|---|
| $\pi^\circ$ —— $2\gamma$ | 0.08 - 0.12 | 0.32 - 0.48 |
| Bremsstrahlung | 0.14 - 0.33 | 0.08 - 0.20 |
| Compton | 0.08 - 0.32 | 0.05 - 0.20 |
| Total predicted | 0.30 - 0.77 | 0.45 - 0.88 |
| Observed | 0.64 ± 0.16 | 0.88 ± 0.22 |

compared with that of starlight. In fact, Strong et al [26] have explained the gamma-ray spectrum for the galactic plane of 110° < ℓ < 140° by a superposition of pion decay and bremsstrahlung gamma rays, assuming the proton and electron intensities 0.4 times and 4 times the solar neighborhood value, respectively, in order to achieve agreement between the calculated and observed spectra.

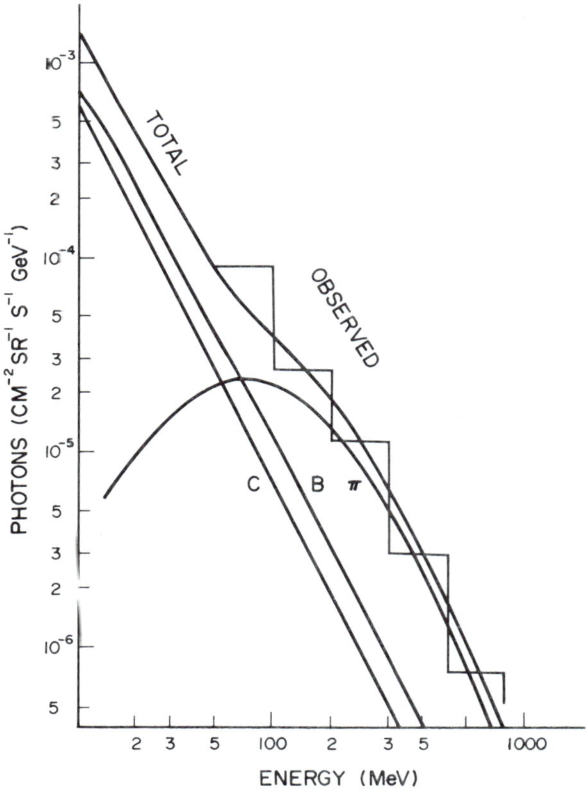


Fig.7. The differential energy spectra of gamma rays produced by three mechanisms. Curves π and B are adopted from the median line and B2 of Figure 6, whereas curve C is obtained by increasing $C_b$ in Figure 6 by a factor 6. The thick line represents the sum of three components, and the staircase is the spectrum observed by Paul et al. [22].

Too large an electron intensity can be get rid of , if Compton gamma rays are properly taken into account. Bignami and Piccinotti [20] have assumed that the starlight density is proportional to the mass density, whereas the far infrared density is computed by reference to the distribution of dust whose density is proportional to the total hydrogen density. Then they have obtained the radiation density of about 3 eV cm$^{-3}$ at 5 kpc from the galactic center, which is compared with about 1 eV cm$^{-3}$ in the solar neighborhood. An increase of factor three is comparable to that in the total hydrogen density as obtained by Gordon and Burton [27]. Hence the relative contribution of Compton gamma rays in the inner part of the Galaxy would be smaller than that given in Table 1, since the contribution of 2.7 K radiation is relatively small in the inner part. The intensity of Compton gamma-rays at ℓ = 30° would be twice larger than at ℓ = 90° and only a small fraction of that of bremsstrahlung.

A different way of estimating the radiation density results in that Compton gamma rays contribute as much as bremsstrahlung also in the inner part of the galactic plane. The near infrared spectrum observed for the longitude range 30° - 50° indicates that stars of the surface temperature of about 3000 K are responsible for near infrared emission. This allows us to derive the total emissivity from the 2.4 μm emissivity. Since these stars are regarded as red supergiants born in molecular clouds, there are more blue supergiants in the clouds. Using the luminosity ratio of blue to red supergiants, which is about four for galactic clusters of ages as great as $10^7$ years, we obtained the total emissivity

$$Q(\text{all stars}) \simeq 30\ q(2.4\ \mu m), \tag{1}$$

in proportion to $q(2.4\ \mu m)$, the 2.4 $\mu m$ emissivity per $\mu m$.

Since these stars distribute with the scale height $z_s \simeq 50$ pc perpendicular to the galactic plane and their density decreases with the scale length $R_s \simeq 3$ kpc from the 5 kpc ring in the galactic plane, the radiation density at z from the galactic plane near the 5 kpc ring is approximately given by

$$\varepsilon_s \simeq \frac{4z_s}{c} Q \left(\ell n\frac{R_s}{z_e} + \gamma\right), \tag{2}$$

where $z_e \simeq 750$ pc is the scale height of cosmic-ray electrons and $\gamma = 0.577$ is Euler's constant.

Assuming $q(2.4\ \mu m) = 5 \times 10^{25}\ W\ pc^{-3}\ \mu m^{-1}$ [16], we obtain

$$\varepsilon_s \simeq 12\ eV\ cm^{-3}. \tag{3}$$

This is 4 times greater than the radiation density near the 5 kpc ring as obtained by Bignami and Piccinotti [20].

The emissivity of $Q \simeq 5 \times 10^{-29}\ W\ cm^{-3}$ is three times greater than that estimated from $L_H(FIR) = 2 \times 10^{-30}$ W per H atom and $n_H = 8\ cm^{-3}$, if all radiation from stars is converted to far infrared radiation through the thermal emission of dust. This would give the far infrared flux at $\ell = 30°$, $b = 0°$ over the all wavelength range to be $6 \times 10^{-8}\ W\ cm^{-2}$ [28]. This is comparable to the flux obtained by Rouan et al. [29], but appreciably higher than that of Serra et al. [18].

If the radiation density in the 5 kpc ring is about ten times that in the solar neighborhood, as given in Equation (3), the relative contributions of respective components of gamma-rays in the direction of $\ell \simeq 30°$ and $b \simeq 0°$ are nearly the same as those in the polar region, as given in Table 1. In the pole direction, 1/4 of Compton gamma rays arise from stellar radiation, whereas 3/4 of them from the 2.7 K radiation. Since the former increases 10 times, the total Compton gamma rays increase by a factor of 3.3, provided that the cosmic-ray density were uniform. Both pion and bremsstrahlung gamma rays increase in proportion to the surface density of interstellar hydrogen, which increases by a factor of about 3 from the solar neighborhood to the 5 kpc ring. Thus all components increase nearly by the same factor, thus the spectrum being independent of direction, nearly independent of the cosmic-ray density. The gradient of the cosmic-ray density is needed to explain the absolute emissivity of gamma rays, which is high in the inner region of the Galaxy.

## CONTRIBUTION OF LOCALIZED SOURCES

We have thus far tried to explain the energy spectrum of gamma-rays under the assumption that they are of interstellar origin. If gamma-rays are produced mainly in localized sources, the spectrum may be explained in different ways.

Ögelman [30] suggested that gamma rays could be produced by X-ray sources, extending the X-ray spectrum by $E^{-2}$. However, such a power law holds up to gamma-ray energies only for special sources. Hayakawa and Tanaka [2] proposed a model such that gamma rays are produced by the nuclear component of cosmic rays in the general medium of SNR, whereas radio emission takes place in a thin shell of SNR. The gamma ray emissivity relative to the radio yield requires a thin shell with thickness small compared with the radius, a high proton to electron ratio and/or a high matter density in SNR.

As the number of discrete sources resolved has increased, the theory of localized source origin has revived. Taking into account the resolving power of existing gamma-ray detectors and the surface density of discrete sources thus far resolved, Bignami et al. [31] have demonstrated that a substantial part of diffuse galactic gamma rays can be accounted for by a superposition of unresolved sources, if a constant source luminosity is assumed. This may well be the case towards the inner region of the Galaxy, but the contribution of sources is considered to be insignificant in the pole and anticenter regions.

A little different approach has been adopted by Rothenflug and Caraveo [32]. They have found the log N - log S relation represented by slope -1, which is expected for the distribution in a thin disk. The log N - log S relation for gamma-ray sources at $60° < \ell < 200°$ can be reproduced by a luminosity function $dN/dL \propto L^{-1}$ for $5 \times 10^{34} < L < 10^{36}$ erg $s^{-1}$, as shown in Figure 8. The longitude distribution of gamma-ray flux can be fitted with the SNR-like radial distribution of sources as obtained by Kodaira [9]. Figure 9 demonstrates that the model may account for a substantial part of gamma rays in the inner part of the Galaxy, whereas the genuine diffuse component is dominant in the outer part. In this model nearly a half of gamma-rays emitted from the Galaxy are attributed to localized sources. Using extreme models, they have found the contribution of localized sources to lie between 30% and 60% of the gamma-ray luminosity of the Galaxy, which is estimated to be $6 \times 10^{38}$ erg $s^{-1}$ for energies about 100 MeV.

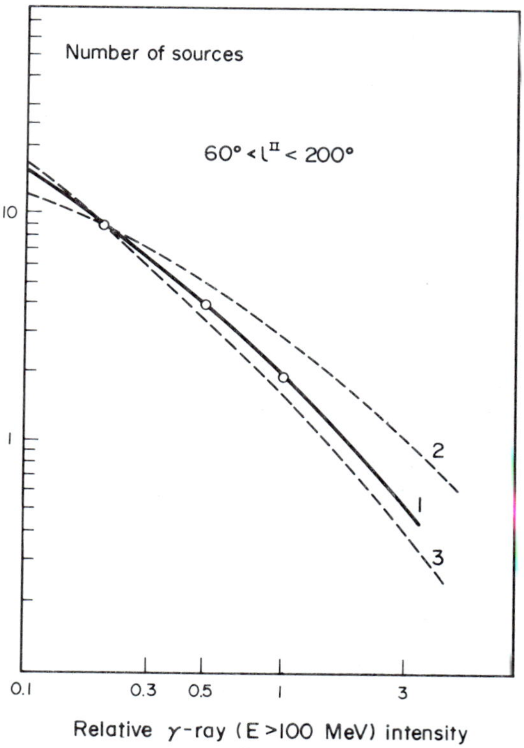


Fig.8. Log N - Log S relation for gamma-ray sources. Curve 1 represents the relation for the luminosity function given in the text, whereas curves 2 and 3 give the relations for higher and lower luminosities, respectively. After Rothenflug and Caraveo [32].

A question concerning the energy spectrum arises if a substantial part of gamma rays are generated by localized sources. Since the spectra of individual sources are different from each other, the fact that the spectra in different directions are essentially the same [22, 14] requires either that sources of different spectra are mixed with a ratio independent of position, or that a single kind of sources dominate over others. Since the first alternative seems artificial, the second one is considered more plausible.

Although SNR's were thought to be candidates [2], no gamma rays have been observed from SNR's, except from pulsars such as the Crab and the Vela. Montmerle [33] has thus suggested that gamma-ray sources are not isolated SNR's but SNR's belonging to OB associations, named SNOB. This hypothesis may be supported by the observation that among 11 unidentified gamma-ray sources 5 can be attributed to SNOB's, and that among 19 SNOB's 9 coincide with gamma-ray peaks in addition to 5 resolved sources.

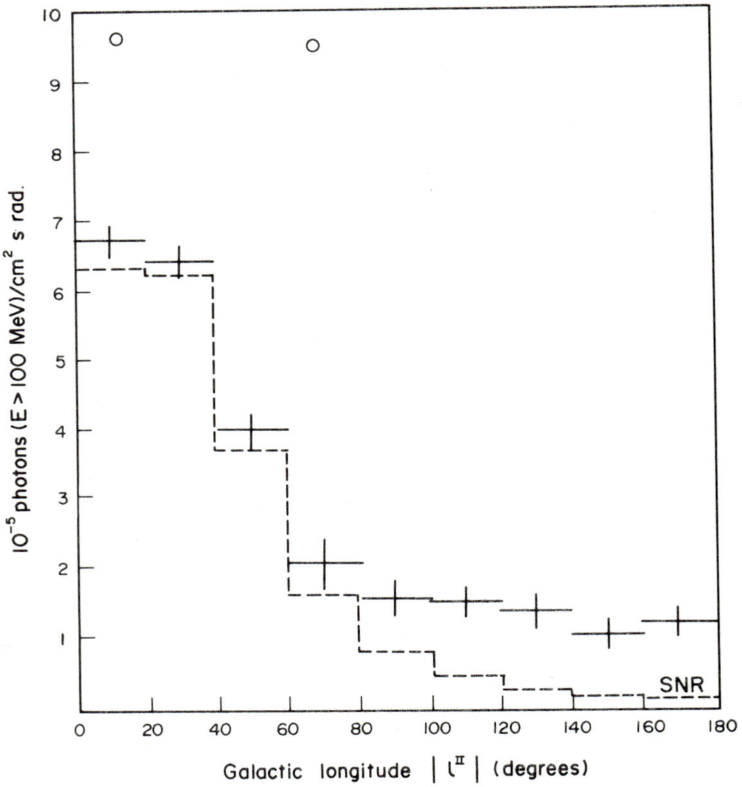


Fig.9. The longitude dependences of the intensity of "diffuse" gamma rays and of the gamma-ray intensity from localized sources whose distribution is the same as that of SNR. After Rothenflug and Caraveo [32].

The SNOB model has the following theoretical background. OB stars inject charged particles to be accelerated by the shock wave of SNR's as worked out by Bell [34], Injected particles with velocities appreciably higher than the shock velocity are necessary for acceleration to relativistic energies. The electron-proton ratio increases as energy decreases in the non-relativistic region of protons, which favors to increase the contribution of electron bremsstrahlung relative to pion decay. Moreover, protons lose their energy in the proximity of a SNOB by pion production, so that the relative intensity of secondary electrons is high. Compton gamma-rays are also of considerable contribution, since the radiation density is high in dark clouds in which supergiants emit a high flux of radiation and radiationin is trapped because of a large optical depth.

The SNOB model indicates that the gamma-ray source spreads over a dark cloud in which both OB stars and supernovae are born. Electrons accelerated at SNOBs emit both gamma rays and radio waves. Their relative yields observed can be accounted for if the magnetic field strength and the hydrogen density are of the orders of 10 μG and $10^3$ H atoms cm$^{-3}$, respectively. Red supergiants are responsible for near infrared emission, while light emitted by blue stars heats up dust which then emits far infrared radiation. Correlations of gamma-ray emission with radio and infrared emission are due therefore to their common origin. However, only a small fraction of dark clouds can possess SNOB's. If the average gamma-ray luminosity of a SNOB is 5 x $10^{35}$ erg s$^{-1}$ and a half of galactic gamma rays are produced by SNOB's, there are about 600 active SNOB's, among which only 25 have been observed as SNOB's and gamma-ray sources, and the rest are hidden in dark clouds. If the supernova rate is one in 30 years, the active period of SNOB would be about 2 x $10^4$ years. Since the average age of a dark cloud is considered to be $10^7$ years, about one third of dark coulds are capable to generate SNOB's. This is not inconsistent with the mass function and the total stellar mass in a cloud.

REFERENCES

1. G.W. Clark, G.P. Garmire and W.L. Kraushaar, Astrophys. J. Lett. 153, L203 (1968).

2. S. Hayakawa and Y. Tanaka, in: Non-Solar X- and Gamma-Ray Astronomy, ed. by L. Gratton, 1970, p. 374.

3. D.A. Kniffen, C.E. Fichtel and D.J. Thompson, The Structure and Content of Galactic Gamma Rays, NASA CP-002, 301 (1976).

4. D.A. Kniffen, C.E. Fichtel and D.J. Thompson, Astrophys. J. 215, 765 (1977).

5. C.J. Cesarsky, M. Cassé, and J.A. Paul, Astron Astrophys. 60, 139 (1977).

6. K. Bennett, G.F. Bignami, R. Buccheri, W. Hermsen, G. Kanbach, F. Lebrun, H.A. Mayer-Hasselwander, J.A. Paul, G. Piccinotti, L. Scarsi, F. Soroka, B.N. Swanenburg and R.D. Wills, Proc. 12th ESLAB Symp. ESA SP-124, 83 (1977).

7. F. Lebrun and J.A. Paul, Astron. Astrophys. 65, 187 (1978).

8. F.W. Stecker, Phys. Rev. Lett. 35, 188 (1975).

9. K. Kodaira, Publ. Astr. Soc. Japan 26, 255 (1974).

10. F.W. Stecker and F.C. Jones, Astrophys. J. 217, 843 (1977).

11. W.L. Kraushaar, G.W. Clark, G.P. Garmire, R. Borken, P. Higbie, C. Leong, and T. Thorses, Astrophys. J. 177, 341 (1972).

12. R. Schlickeiser and K.O. Thielheim, Astron. Astrophys. 34, 167 (1974).

13. J.A. Paul, M. Cassé and C.J. Cesarsky, Astrophys. J. 207, 62 (1976).

14. C.E. Fichtel, G.A. Simpson and D.J. Thompson, Astrophys. J. 222, 833 (1978).

15. S. Hayakawa, K. Ito, T. Matsumoto, T. Ono and K. Uyama, Nature 261, 29 (1976).

16. S. Hayakawa, K. Ito, T. Matsumoto and K. Uyama, Astron. Astrophys. 58 325 (1977).

17. S. Hayakawa, T. Matsumoto, H. Murakami, K. Uyama, T. Yamagami and J.A. Thomas, Nature 279, 510 (1979).

18. G. Serra, J.L. Puget, C.E. Ryter and J.L. Wijnbergen, Astrophys. J. Lett. 222, L21 (1978).

19. T. Maihara, N. Oda and H. Okuda, Astrophys. J. Lett. 227, L129 (1979).

20. G.F. Bignami and G. Piccinotti, Astron. Astrophys. 59, 233 (1977).

21. W.B. Burton and M.A. Gordon, Astron. Astrophys. 63, 7 (1978).

22. J.A. Paul, K. Bennett, G.F. Bignami, R. Buccheri, P. Caraveo, W. Hermsen, G. Kanbach, H.A. Mayer-Hasselwander, L. Scarsi, B.N. Swanenburg and R.D. Wills, Astron. Astrophys. 63, L31 (1978).

23. C.E. Fichtel, R.C. Hartman, D.A. Kniffen, D.J. Thompson, H.B. Ögelman, M.E. Özel, and T. Tümer, Astrophys. J. Lett. 217, L9 (1977).

24. C.J. Cesarsky, J.A. Paul and P.G. Shukla, Astrophys. Space Sci. 59, 73 (1978).

25. G.D. Badhwar and S.A. Stephens, Proc. Int. Cosmic Ray Conference (Plovdiv) 1, 198 (1977).

26. A.W. Strong, A.W. Wolfendale, K. Bennett and R.D. Wills, Month. Not, Roy. Astr. Soc. 182, 751 (1978).

27. M.A. Gordon and W.B. Burton, Astrophys. J. 208, 346 (1976).

28. S. Hayakawa and T. Matsumoto, Presented at COSPAR at Tel Aviv. (1976).

29. D. Rouan, P. Léna, J.L. Puget, K. de Boer and J.J. Wijnbergen, Astrophys. J. Lett. 213, L35 (1977).

30. H. Ögelman, Nature 221, 753 (1969).

31. G.F. Bignami, P. Caraveo and L. Maraschi, Astron. Astrophys. 67, 149 (1978).

32. R. Rothenflug and P. Caraveo, Astron. Astrophys. in press.

33. T. Montmerle, preprint. (1979).

34. A.R. Bell, Month. Not. Roy. Astr. Soc. 182, 147, 443 (1978).

# GAMMA RAYS AND THE LOCAL ORIGIN OF COSMIC RAYS

M. Cassé and J. A. Paul

*Section d'Astrophysique, Centre d'Etudes Nucléaires de Saclay, France*

## ABSTRACT

Supersonic stellar winds are quite common among the brighest and/or youngest stars such as OB stars, T-Tauri stars, Herbig-Haro objects and related objects. A supersonic wind must inevitably trigger a shock wave in the surrounding interstellar medium (ISM). The shock wave generated at the boundary between the stellar wind and the ISM would be able (in certain conditions) to accelerate pre-injected particles as suggested by Jokipii in the case of the solar wind.

This process may have a bearing on the production of high-energy gamma rays in cosmogonically active regions of the Galaxy. The close proximity of a young (mass-losing) star to the Rho Oph molecular cloud is invoked to explain the enhanced emission of this region inferred from the COS-B satellite observations. The local character of the observed cosmic rays (CR) is quite naturally explained in this context by the high concentration of OB stars (potential cosmic-ray sources) in the Gould Belt.

## INTRODUCTION

In a recent paper [1], we have proposed that the shocked region at the boundary between the stellar wind and the surrounding medium is, in certain conditions, the site of acceleration of charged particles with a power-law energy spectrum, extending a suggestion proposed by Jokipii [2] for the solar wind. This mechanism would operate up to $\simeq 10^{15}$ eV. We emphasize here the consequences of this mechanism on the local high-energy gamma-ray emission and particularly on the emission of the Gould Belt and the Rho Oph cloud.

Astronomical observations have revealed that we do not sit in an ordinary place. A recent, expanding structure of young stars and dense clouds surrounds us : the Gould Belt (Stothers and Frogel [3]; Frogel and Stothers [4]; and references therein). It is a belt inclined about 20° to the galactic plane, with a diameter less than 1 kpc, in which most of the nearest OB stars, young clusters, associations as well as dark clouds are concentrated [5, 6, 7]. This local structure has been recently recognized as a source of high-energy gamma rays produced in all likehood by the interaction of CR with the ISM [8, 9]. Furthermore, a bright spot in this structure coincides with the Rho Oph complex [9], which is the nearest aggregate of gas, dust and young stars pertaining to the Gould Belt. A substantial enhancement of the CR density in the Rho Oph cloud, with respect to the CR density in the solar vicinity, is required to account for the gamma-ray flux observed in this direction [1, 9, 10, 11]. Is stellar-wind-induced CR acceleration responsible for these local features ?

Since the major unknown of the mechanism is the efficiency of the energy transfer between the wind and the particles, the best that we can do is to compare the mechanical power released by the local mass-losing stars to that required (i) to maintain the observed CR energy density and (ii) to account for the high-energy gamma-ray luminosity of the Rho Oph cloud.

## THE GOULD BELT

A compilation of the stellar catalogs [12, 13, 14] reveals the existence of at least 44 OB stars with $M_{bol} \leq -6$ in the Gould Belt ,i.e. within 400 pc of the Gould Belt center ( $\ell = 180°$, $b = -16°$, distance = 212 pc ; from Stothers and Frogel [3]). Their luminosity distribution is given in table I, together with their estimated mechanical power.

TABLE 1 Luminosity Distribution And Estimated Mechanical Power Of 44 OB Stars With $M_{Bol} \leq -6$ Located In The Gould Belt

| $M_{Bol}$ range | Number of stars | Estimated mass-loss rate per star * in $10^{-6} M_{\odot}\ yr^{-1}$ | Total mechanical power in $10^{37}$ erg $s^{-1}$ |
|---|---|---|---|
| - 10 to - 11 | 2 | 25 | 6.3 |
| - 9 to - 10 | 5 | 10 | 6.3 |
| - 8 to - 9 | 5 | 4 | 2.5 |
| - 7 to - 8 | 11 | 1.6 | 2.2 |
| - 6 to - 7 | 21 | 0.6 | 1.6 |

Note to table I *: A wind velocity $\simeq 2\ 10^3$ km $s^{-1}$ has been adopted.

Among these 44 stars, 18 have been observed by Copernicus [15] of which 17 show evidence for mass loss. The estimated total mechanical power liberated by these 44 stars is $\simeq 2.10^{38}$ ergs $s^{-1}$, i.e.$\simeq$ 10 times the power required to maintain the local CR density in a sphere of 400 pc radius, taking a confinement time of $2.10^7$ yr [16]. PMS objects and T-Tauri stars have not been taken into account in the power balance. Then, as far as power is concerned, the situation appears confortable and a moderate efficiency for the conversion of stellar wind energy into particle energy ( < 0.1) is sufficient to account for the observed CR energy density.

## INTERPRETATION OF THE HIGH-ENERGY GAMMA-RAY LUMINOSITY OF THE RHO OPH CLOUD

According to Black and Fazio [17], a nearby dense cloud would yield a gamma-ray flux resulting from the interaction between CR and the cloud material, given by $F_\gamma (\geq 100\ MeV) = 10^{-6} M\ \epsilon_\gamma / d$ , where $F_\gamma$ is the gamma-ray flux in photon $cm^{-2}\ s^{-1}$, M the mass of the cloud in $M_\odot$, $\epsilon_\gamma$ the gamma-ray emissivity in $10^{-25}$ photon ($\geq$ 100 MeV)$s^{-1}$ (H-atom)$^{-1}$ and d the distance of the cloud in pc.

In the Rho Oph case, d is quite well known, ranging between 150 and 170 pc [18]. The mass estimate is less precise : from visual extinction measurements, Encrenaz, Falgarone, and Lucas [19] derived a total mass M = 2000 $M_\odot$ using the standard extinction law $N_H = 2.10^{21} A_v$, where $N_H$ is the total mass column-density (in H-atom $cm^{-2}$) and $A_v$ the visual extinction in magnitude. However, Bohlin, Savage, and Drake [20] have shown, on the basis of the Copernicus UV observation of the Rho Oph star, that in the Rho Oph region, the standard extinction law may not apply. On the line of sight toward Rho Oph, they derived $N_H = 15.4\ 10^{21}$ E(B-V), where E(B-V) is the color excess in magnitude. As $A_V$/E(B-V) = 4.3 in the Rho Oph region [21] the standard extinction law must be replaced by $N_H = 3.5\ 10^{21}\ A_V$, leading to a cloud mass M = 3600 $M_\odot$. Taking into account the uncertainty on $A_V$ itself which is of the order of 20% [19] and on the extinction law, the Rho Oph mass might range between 2000 and 4000 $M_\odot$.

If the CR intensity and energy spectrum in the Rho Oph cloud are taken as observed in the solar vicinity, the emissivity $\epsilon_\gamma$ in the cloud must be equal to that derived from local observation : $1.5 \leq \epsilon_\gamma \leq 2.5$ [22]. Under this assumption, the expected gamma-ray flux of the Rho Oph cloud ranges between $10^{-7}$ and $4.10^{-7}$ photon ($\geq$100 MeV) $cm^{-2}s^{-1}$. A source with such a flux is certainly not detectable by the COS-B gamma-ray telescope, especially in a region of high unresolved background (the Gould Belt).

Nevertheless the Rho Oph cloud is within the error circle of the gamma-ray source discovered by COS-B at $\ell$ = 353°, b = 16° [9]. The flux of this source is certainly above $10^{-6}$ photons ($\geq$100 MeV) $cm^{-2}\ s^{-1}$. Moreover, from the soft X-ray observations [23], Bignami and Morfill [11] have shown that there is no clump of ISM between the sun and the cloud able to contribute significantly to the gamma-ray flux observed toward Rho Oph. As no other gamma-ray source candidate (as e.g. radio pulsar, SNR) lies in the COS-B error circle, the natural interpretation of the observational material is that the gamma-ray source results from CR interacting with the Rho Oph cloud, the CR density within the cloud being substantially enhanced (by a factor ranging between 2 to 10) with respect to the CR density in the solar vicinity. Then, we conclude in agreement with Simpson [10] and Bignami and Morfill [11] that relativistic particles must be accelerated within the Rho Oph cloud. In the absence of SNR in the Rho Oph cloud and its vicinity [23] we propose that this acceleration is the result of the stellar wind mechanism presented in [1].

The total CR power required to sustain the gamma-ray emission is of the order of $6\ 10^{33}$ erg $s^{-1}$ [1], i.e. 8 times the gamma-ray luminosity $L_\gamma$ ($\geq$100 MeV) amounting to $8.10^{32}$ erg $s^{-1}$. None of the early type stars embedded in the cloud is brighter than $M_{bol}$ = - 6 [24, 25], but the B1 III star HD147165, about 5 pc away, is expected to release $\simeq 10^{36}$ erg $s^{-1}$ in the form of stellar wind. From geometrical arguments, a fraction of $\simeq$ 4% of this power would be transferred to the Rho Oph cloud. This amount ($\simeq 4.10^{34}$ erg $s^{-1}$) appears more than sufficient to sustain the gamma-ray emission since other presumed mass-losing stars, like the 11 T-Tauri stars embedded in the cloud [25, 26, 27] could also contribute to the particle acceleration. In the absence of a SNR, stellar wind acceleration appears as a promising mechanism to inject in the cloud interior the high-energy protons responsible for the gamma-ray emission.

## CONCLUSION

We have presented arguments in favour of the existence of local CR acceleration sites : the shock regions around the most luminous stars in the Gould Belt. Since these stars are generally not located within dense clouds but at their fringes, most of high-energy particles they generate in the intervening medium could contribute to the observable CR, whereas those accelerated by stars located within dense clouds may not completely leak out as in the Rho Oph case [1]. This strengthens the idea of a local origin of the observed CR based on the similarities between the elemental and isotopic abundances in the pre-accelerated material and in the local galactic medium [28, 29]. A direct consequence of our suggestion is that the observed CR protons and nuclei are of local origin. Consequently, the CR distribution is predicted to be somewhat patchy in the galactic disc. In this context, high-energy gamma rays are likely to be sensitive tracers of cosmogonically active regions ; we expect that closeby active regions, and more precisely the Orion region [30] are detectable gamma-ray sources.

## REFERENCES

1. M. Cassé and J.A. Paul, to be published in Ap.J. (1979).
2. J.R. Jokipii, Ap.J., 152 (1979).
3. R. Stothers and J.A. Frogel, A.J., 79, 456 (1974).
4. J.A. Frogel and R. Stothers, A.J., 82, 890 (1977).
5. P.B. Lucke, Astr.Ap., 64, 367 (1978).
6. G.S. Rossano, A.J., 83, 241 (1978).
7. D.C.B. Whittet, Astr. Ap., 72, 370 (1979).
8. C.E. Fichtel, R.C. Hartman, D.A. Kniffen, D.J. Thompson, G.F. Bignami, H.B. Ogelman, M.E. Ozel and T. Tümer, Ap.J., 198, 163 (1975).
9. H.A. Mayer-Hasselwander, K. Bennett, G.F. Bignami, R. Buccheri, N. D'Amico, W. Hermsen, G. Kanbach, F. Lebrun, G.G. Lichti, J.L. Masnou, J.A. Paul, K. Pinkau, L. Scarsi, B.N. Swanenburg and R.D. Wills, to be published in Proc. of 9th Texas Symp. (1979).
10. G. Simpson, Workshop on Particle Acceleration Mechanism in Astrophysics, La Jolla (1979).
11. G.F. Bignami and G.E. Morfill, submitted to Astr. Ap. (1979).
12. J.R. Lesh, Ap.J. Suppl., 17, 371 (1968).
13. J.R. Lesh, Astr. Ap. Suppl., 5, 129 (1972).
14. R.M. Humphreys, Ap.J. Suppl., 38, 309 (1978).
15. T.P. Snow and D.C. Morton, Ap.J. Suppl., 32, 429 (1976).
16. H. Garcia-Munoz, G.H. Mason and J.A. Simpson, Ap.J., 217, 859 (1977).
17. J.M. Black and G.G. Fazio, Ap.J.(Letters), 185, L 7 (1973).
18. D.C.B. Whittet, M.N.R.A.S., 168, 371 (1974).
19. P.J. Encrenaz, E. Falgarone and R. Lucas, Astr. Ap., 44, 73 (1975).
20. R.C. Bohlin, B.D. Savage and J.F. Drake, Ap.J., 224, 132 (1978).
21. I.G. van Breda, I.S. Glass and D.C.B. Whittet, M.N.R.A.S., 168, 551 (1974).
22. C.J. Cesarsky, J.A. Paul and P.G. Shukla, Ap. and Space Sci., 59, 73 (1978).
23. K.M.V. Apparao, S. Hayakawa and D.R. Hearn, submitted to Ap. and Space Sci.(1979).
24. G.G. Fazio, E.L. Wright, M. Zelik II and F.J. Low, Ap.J.(Letters), 206, L 169 (1976).
25. J.H. Elias, Ap.J., 224, 453 (1978).
26. G.H. Herbig and N.K. Rao, Ap.J., 174, 401 (1972).
27. A.E. Rydgren, S.E. Strom and K.M. Strom, Ap.J. Suppl., 30, 307 (1976).
28. M. Cassé and P. Goret, Ap.J., 221, 703 (1978).
29. M. Cassé, Workshop on Particle Acceleration Mechanisms in Astrophysics, La Jolla (1979), invited paper.
30. H. Reeves, in Protostars and Planets , Univ. Ariz. Press, Tucson, 1978, p. 368.

# DYNAMICS OF THE GALAXY

S. A. Stephens

*Tata Institute of Fundcmental Research, Homi Bhabha Road, Bombay 400 005, India*

## ABSTRACT

The quasistationary spiral structure seen in many external galaxies and the spiral pattern of the Galaxy deduced from the 21 cm observations have raised the question of the stability of such patterns against galactic rotation. In this regard, the spiral density wave theory and the stabilisation of density waves by the gas are discussed. The development of shock resulting from the propagation of density wave, the resultant distribution of gas in spiral arms and the formation of stars by the compressed gas are summarised. The stability of gas-field system in the Galaxy, consisting of gas, magnetic field and cosmic rays is examined. It is inferred that thermal instabilities can induce condensation of gas in the size range .01 to .1 pc, while cloud formation with large sizes can be achieved through Parker instabilities. A brief description of the hydrostatic equilibrium of the gas-field system perpendicular to the galactic plane is given. It is shown that cosmic rays extend well beyond the gas disk and the distribution of cosmic rays is wider than that of radio emission in the Galaxy.

## INTRODUCTION

It has been customary to include this subject in Gamma Ray Astronomy Meetings. The main reason is that a large part of the observed gamma rays seems to be resulting from the interaction of cosmic rays with interstellar gas. In this paper I will restrict myself to the theoretical results concerning the structure and dynamics of the Galaxy, relevant to Gamma Ray Astronomy. Though progress in this field has been rather slow over the last few years, may review papers have appeared [1-5] in the literature. Therefore my main aim is to look at the subject with the interest of the Gamma Ray Astronomer in mind and briefly bring out the essence of the theoretical basis and the results arising therefrom in this field. I have broadly divided the subject into two. The first deals with the spiral pattern of the Galaxy and the implications of density wave theory. The second concerns with the dynamics of the gas-field system with specific reference to cosmic rays.

## DENSITY-WAVE THEORY

The presence of the spiral structure in our galaxy suggests that the Galaxy has no axial symmetry and it has not reached a steady stage. However, the most prominent constituents of the spiral arms, namely, the bright young stars and the gas represent only a small fraction of the total mass [6]. The bulk of the matter in the Galaxy is in the form of old stars, which form the basic symmetric disk. Clearly, spiral structure can be conceived as a perturbation over the stable symmetric disk. The concept of density wave was advocated by Lindblad [7] as the dynamical basis for the spiral structure. By studying the orbits of individual stars, he showed that a spiral gravitational field would tend to produce a density redistribution, which in turn supports the imposed field. In particular, if the spiral structure has two arms, the resonance between the epicycle motion of stars and the density wave can extend over a substantial portion of the galactic disk. Both the symmetric and spiral fields are due to the collective gravitational effects of stars and gas. Lin [8] realised that a study of the collective behaviour might allow one to describe the spiral density wave. Density wave could be initiated by certain forms of dynamical instability (peculiar to such gravitational system) which evolves sufficiently slowly so that a "Quasi Stationary Spiral Structure" can be maintained. This theory successfully explained why spirals do not wind up despite differential rotation and provided a physical explanation to the observed deviation of systematic motion of clouds from circular motion.

The persistence of the waves remained to be the greatest problem of the density wave theory. Toomre [9] showed that the density waves, when combined into wave packets forming spiral arms, cannot be sustained because the wave packets propagate radially inwards until the inner Lindblad resonance, where they disappear due to resonance effects. 1/ This only implies that fresh waves must somehow be created either continuously or intermittently. Further, numerical experiments showed that the self gravitating disks do not form long-lived spiral structure. Instead they are prone to large scale two-armed instabilities, heat up the disk and finally end up as a finite amplitude wave in the form of a bar [10]. The response of gas to the spiral wave is non-linear in character. This leads to the formation of shocks and the gas dissipates energy at the expense of the wave [11]. This damping argues in favour of trailing density waves, which is consistent with the observed trailing of spiral structure. The interaction of gas with the density wave stabilises the amplitude of perturbation at some finite value so that quasistationary spiral structure can be realised. Further, the observed coolness of the stellar disk population could be understood.

Lin developed an analytical theory of density wave by assuming that the perturbed potential is in the form of tightly wrapped spiral [8]. This forms the basis of the extensive work carried out in this field of research. Without going into any detail, I briefly state here some of the features of this theory as summarised by Dekker [2]. Spiral density waves are intrinsically generated by a gravitational instability that is inherent to cold stellar disks. This instability is propagated by the stars that resonate with the wave. These resonant stars

---

1/ Lindblad resonances occur at radii at which the pattern speed equals $\Omega(r) \pm k/2$, where $\Omega(r)$ is the rotational speed and $k$ the epicycle frequency at a distance $r$ from the centre.

absorb angular momentum at the expense of the non-resonant stars. As a result the non-resonant stars are slowed down with respect to the rotational velocities they possessed in the unperturbed disk. This is expressed by adding to the total angular momentum of the stars a negative term. Thus the density wave has negative energy and the wave grows as more angular momentum is lost to the resonant stars. During this process, angular momentum has to be transported by the wave from the inner regions to the outer parts of the disk. Such a transport can only be effected by trailing density waves [12]. The growth of the wave amplitude is limited by the non-linear response of the gaseous component of the disk. Equilibrium is reached when the angular momentum gained by the wave from the gas is lost to the resonant stars. The final amplitude of the wave is not expected to exceed about 10% of the local mean gravitational field. This is compatible with the value of 5% as derived from the observations. Shocks are formed and cause the gas to pile up at the bottom of the gravitational potential well of the wave, resulting in the subsequent formation of young stars and H II regions. These bright objects form the visible outline of the stellar density wave structure.

In order to illustrate some of the aspects mentioned above, we show in Figure 1 a photographic simulation of a stellar density wave of 5% amplitude superposed on a stellar disk similar to our Galaxy [3]. One can notice from this figure that the resulting background wave pattern of old to moderately old stars is hardly visible. Such a background pattern would be certainly difficult to detect in real galaxies. Thus, the observed spiral pattern is not due to old and moderately old stars, which constitute the bulk of the matter in the galaxy, but due to young bright stars, dust and gas.

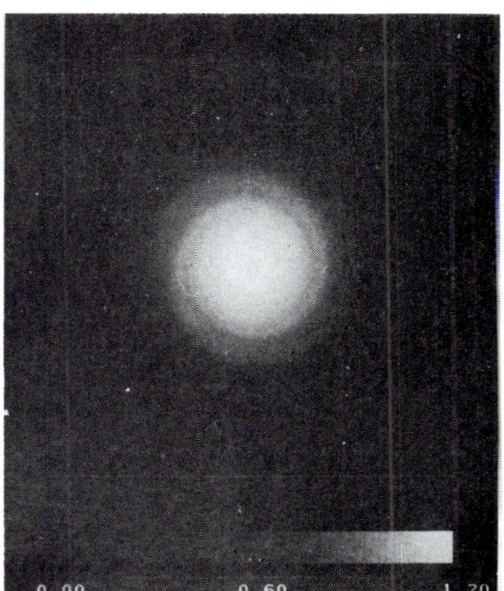

Fig. 1: Photographic simulation of stellar density wave of 5% amplitude in a model galaxy (from Roberts, ref. 3).

The spiral galaxy M81 has been studied in great detail to test the predictions of density wave theory. The detailed analysis made by Visser [13] is summarised here. Figure 2 is a photograph of M81. The dashed lines shown in this figure are the theoretical spiral pattern calculated using linear stellar density wave theory for pattern speed of $\Omega_p = 18$ km $s^{-1}$ $kpc^{-1}$. After determining the amplitude of the wave from an analysis of the azimuthal brightness variation, the theoretical gas flow has been computed with non-linear wave theory. The shock front thus determined is shown by the solid lines. The observed dust lanes seem to be in general agreement with the calculated positions of the shock front. In order to compare the velocity fields from model calculations with hydrogen observations, the model field should be smoothed to the beam of the radio telescope. Such a smoothing has been carried out by Visser [13] and the results are shown in Figure 3. This figure is a radiograph of M81 in 21 cm hydrogen

Fig. 2: Photograph of M81 in which the theoretical spiral pattern is shown by dashed lines and the shock front is indicated by solid lines (from Visser, ref. 13).

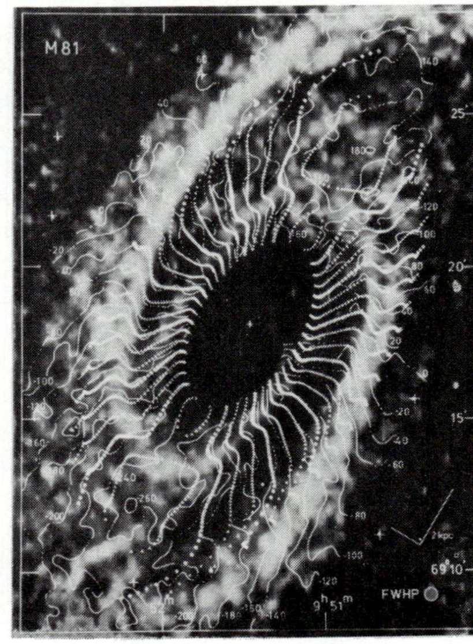


Fig. 3: Radiograph of the distribution of gas in M81, in which the calculated velocity fields are indicated by symbols and the measured fields are shown by full and dashed lines (from Visser, ref. 13).

line showing the distribution of hydrogen density. In this figure the smoothed velocity fields at 25" resolution from model calculations are shown by various symbols. These can be compared with the observed velocity fields which are indicated by full and dashed lines. The kinks in the consecutive contours at the inner edge of the spiral arms indicate the shock front. Clearly, one can notice from this figure that the model velocities are in good agreement with the observations, indicating the success of density wave theory.

## GALACTIC SHOCK AND GAS DISTRIBUTION

As mentioned earlier, the non-linear response of interstellar gas to the stellar density wave lead to the formation of shock front along the arms of the background pattern [3]. The surface density response of a particular component in a galaxy is roughly proportional to the inverse square of the characteristic speed. The effective acoustic speed of the interstellar gas is typically about one-fourth of the R. M. S. random velocity of the disk stars. Therefore, a mild spiral gravitational field which induces only a small fractional variation of the disk stars (Fig. 1) can drive a large density response in the interstellar gas towards the formation of shock front. The shock that forms is a hydromagnetic shock wave because the magnetic field is embeded in the gas. Figure 4 shows the distribution of gas along a typical stream line. One notices a rather narrow peak in the density. This is induced by shock at the potential well, which corresponds to the centre of the background spiral arm. The adjacent figure

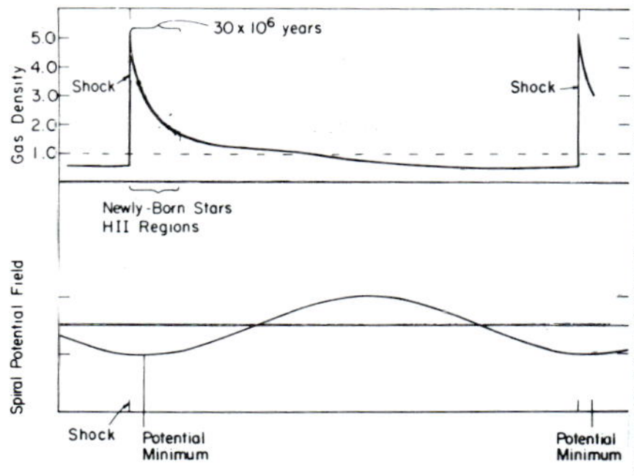


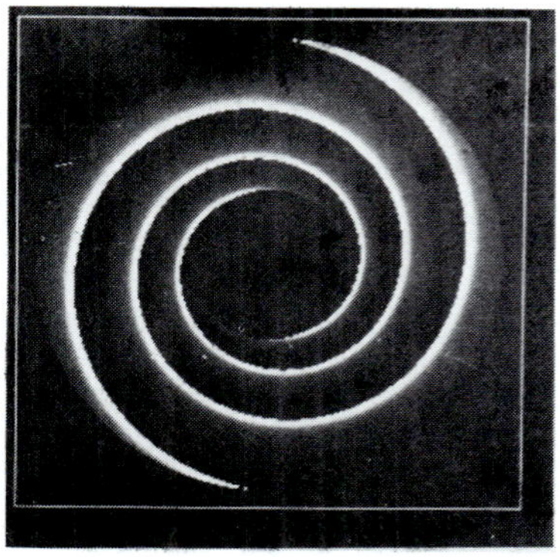

Fig. 4: Distribution of gas along a typical stream line and a photographic simulation of the same over the face of a model galactic disk (from Roberts, ref. 3).

illustrates a photographic simulation of the distribution of gas over the face of a model disk; here the ridge of the light distribution traces out the shock and the gas density ridge. Newly formed stars and H II regions result from the compression of the gas are shown in Figure 4. The enhancement of the density could be by a factor 3 or 4 depending upon the strength of the shock. The shock propagates over a period of about 30 million years across a spiral arm.

The shock wave may trigger the gravitational collapse of gas clouds leading to the formation of stars and other tracers along the spiral arm. The two phase model of the interstellar gas provides a favourable physical state to effect this through compression and phase transitions during the flow of the shock [14]. Inside the shock layer, the transient pressure tends to achieve a value higher than the maximum pressure consistent with the thermal stability of the hot-intercloud medium, and this forces transition of some of the hot gas into cold cloud phase. In the interarm region, the pressure tends to drop below the minimum pressure consistent with the thermal stability of cold medium and this forces the reverse transition resulting in the evaporation of a few clouds; this helps to maintain a constant pressure environment for the remaining clouds. The above picture tells us that clouds adjust to the changes of the ambient pressure in the environment. The increase in the pressure across the galactic shock occurring in the hot-intercloud phase is in turn transmitted to the cold clouds. This triggers the gravitational collapse of some of the dense clouds, leading to the formation of stars.

Figure 5 gives an overview of the temporal sequence of physical phenomena across a spiral arm expected from the density wave theory. One can notice that, with the formation of galactic shock, the gas density peaks around a narrow front. The magnetic field which is frozen into the gas and the dust particles are also compressed. Therefore, a shock, a sharp H I peak, a narrow dust lane, and the strongest magnetic fields are all expected to lie on a narrow lane on the inside edge of the bright optical arms of young stars and H II regions. This simple picture could be complicated by local processes relating to the interactions of clouds with intercloud medium and by instabilities resulting from the

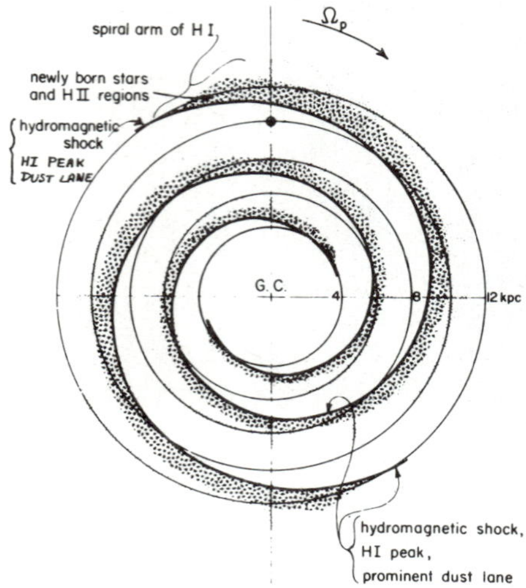


Fig. 5: Temporal sequence of shock related phenomena in a model galaxy (from Roberts, ref. 3).

strong perturbation in the gas-field system due to the passage of shock.

The strength of the shock and the degree of compression of the gas vary as the square of ratio $w_\perp/a$ where $w_\perp$ is the total velocity component of the gas normal to the spiral arm and a is the effective acoustic speed of the interstellar gas [3]. $w_\perp$ oscillates along a stream line about its unperturbed value $w_{\perp o}$ because of the passage of the spiral density wave. Shocks form when $w_\perp$ achieves transonic values. When $w_{\perp o} > a$, most of the gas on a stream line moves at supersonic speeds and as a result shocks tend to be strong giving rise to narrow regions of high gas compression. On the other hand, if $w_{\perp o} < a$, most of the gas on the stream line moves subsonically and the shocks tend to be weak resulting in broad regions of relatively low gas compression. This helps in differentiating between spiral structures with narrow filamentary arms and with broad massive arms. The galaxy M81 comes under the former category while M33 comes under the latter [15]. Similar study carried out on 24 external galaxies shows [3] that the two fundamental parameters that govern $w_{\perp o}$, the potential strength of shock and the geometry of spiral pattern are (i) the total mass of the galaxy divided by the co-rotation radius ($GM/r_c$) and (ii) the concentration of mass towards the centre ($r_{.5m}/r_c$).

We can readily apply some of these principles to our galaxy. In the inner parts of the Galaxy, the value of $w_{\perp o}$ increases as r decreases, and therefore potentially strong shocks and high gas compressions are possible as far as the inner Lindblad resonance at 4 kpc. In this region the frequency of compression $2(\Omega - \Omega_p)$ is also large due to the large differential rotation inherent in the inner parts of the disk. With increasing galactocentric distance the shock strength and the frequency of compression decrease. Close to the solar circle, the intrinsic frequency $\nu$ of the density wave satisfies the relation $\nu^2 - \alpha^2 = n^2$, where $\alpha$ is the intrinsic acoustic speed and $n = \pm 2, \pm 3, \ldots\ldots$ Such ultra harmonic resonances lead to secondary arm structure, as noticed in many external galaxies. Near the co-rotation radius (around 15 kpc) $w_{\perp o} < a$ and only broad regions of weak compression are possible. Therefore conditions for the formation of new stars and molecules may be rather unfavourable in the outer parts of the Galaxy in constrast to the inner parts. These qualitative assessments [3] fit well with the observed molecular density distribution.

The large scale spiral structure of the Galaxy has been constructed by Simonson [16] from the velocity profiles of 21 cm line radiation, through a model fitting approach based on density-wave kinematics. Figure 6 shows the grand design of

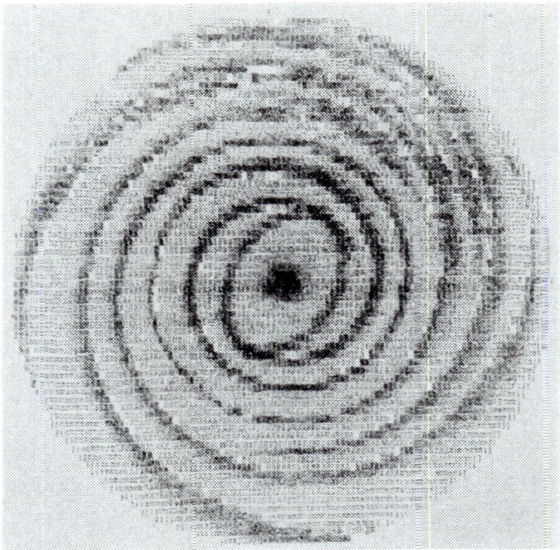

Fig. 6: Spiral structure of the Galaxy derived by Simonson (ref. 16).

the Galaxy derived from such an analysis. This synthetic optical photograph shows a basically two-armed spiral pattern with a pitch angle of $6^o$ - $8^o$ between 4 kpc dispersion ring and solar circle. Near the solar circle, two additional arms originate and the pattern outside is multiple-armed. This pattern of the spiral structure with the expected density contrast between the arm and the inter-arm regions, is widely used in the study of longitudinal distribution of gamma rays.

## INSTABILITIES IN THE GAS-FIELD SYSTEM

So far in our discussion, we have neglected the effect of magnetic field and cosmic rays because they do not play an important role in the grand design. However, they do produce instabilities in the gas-field system leading to local changes in the distribution of gas. One should not also neglect the thermal instability which is an important dynamical effect responsible for the inhomogeneous state of interstellar gas. Heating of the interstellar gas by cosmic rays or by other radiations leads to thermal instability [17]. The temperature of the gas tends to either rise to stable values around $10^4$K or sink to low values in the neighbourhood of $10^2$K. Since these two phases co-exist under pressure equilibrium, the gas density in the hot phase is low and that in the cold phase is high. Detailed study of the ionisation and thermal balance of the interstellar medium have been carried out using time-independent and time-dependent models [18]. The major sources of heating and ionisation considered in such studies are low energy cosmic rays around 2 MeV from supernova outbursts and soft X-rays below 1 keV. These agents can ionise atoms to high ionisation states. Thermalisation of hydromagnetic turbulences generated by high energy cosmic rays has also been suggested as a mechanism [5] for heating without associated ionisation.

Whatever may be the exact cause, the presence of thermal instabilities is evidenced by observation, that the interstellar medium is in two phases described above. The ultra-violet study of the interstellar gas [19] indicates that significant ionisation of H I clouds is by radiation from hot stars. High ionisation states are not observed, as expected on the basis of the gas being heated mainly by cosmic rays or X-rays. Thus the role of these radiations as heating agents is questioned. The observations also suggest [19], the existence of high temperature coronal gas of very low density at about a million degrees. Such high temperature gas could possibly result from successive supernova explosions and this may occupy much of interstellar space. Thermal instabilities can produce condensation of clouds in the size range .01 - .1 pc, since the wavelength associated with these instabilities are small [20]. Of course, cloud can grow when hot intercloud gas condenses on its surface. However, the observed large cloud complexes can be attributed to Parker instability which is associated with large wavelengths. In order to examine this further a brief description of Parker instability is essential.

The interstellar medium is a composite fluid composed of thermal gas, cosmic rays and magnetic field [17]. Cosmic rays constitute a very high temperature gas with a pressure at least as large as that of thermal gas. They make a large contribution to the speed of sound, to the hydrostatic pressure supporting the interstellar medium and to the propagation of interstellar disturbances. The magnetic field contributes the usual magnetic stresses to the composite fluid and in addition ties the two gases together. One can derive the dispersion relation for a plane wave with the horizontal wave number $k_y$ parallel to the field lines and vertical wave number $k_z$. From this it can be shown [17] that instability occurs when the ratio of the specific heats $\gamma$ of the composite fluid satisfies the relation

$$\gamma < 1 + \beta + \alpha \left[ \tfrac{1}{2} - 2(k_y^2 + k_z^2)\Lambda^2 \right] + 0(\alpha^2, \alpha\beta, \beta^2),$$

where $\Lambda$ is the scale height of the gas and, $\alpha$ and $\beta$ are respectively the magnetic and cosmic ray pressures relative to the gas pressure. The effect of $k_y^2$ and $k_z^2$ is to stabilise the system and the system is stable against short wave lengths because of the tension in the magnetic field. For long wave-lengths, one cannot avoid the instability.

The physical nature of the instability is rather simple and is shown in Figure 7. Any perturbation which depresses the magnetic lines of force at one point and raises them at another, upsets the equilibrium. The thermal gas slides downward along the lines of force into the depression. The increase of the gas density in the magnetic field valleys is primarily due to efficient drainage along the field lines, rather than due to compression perpendicular to the galactic plane. The instability grows as the field expands upward in the raised portion. Cosmic rays, which maintain an uniform pressure along the lines of force, have pressure above the ambient value in the raised portion of the lines of force, where they tend to inflate and weaken the field further. The instability is opposed by the gas perturbation and by tension in the magnetic field. From a detailed numerical analysis [20] it is shown that the scale height of the gas in the final state is larger by a

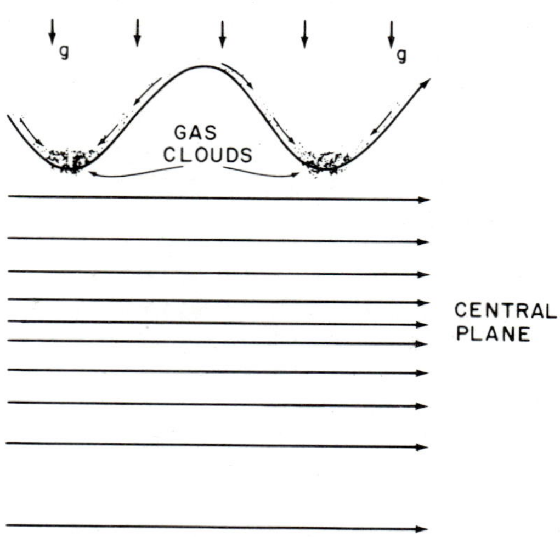


Fig. 7: A schematic drawing of the gas-field system perpendicular to the galactic plane undergoing inflation (from Parker, ref. 17).

factor about 2 in the region of condensation than in the corresponding stratified initial state; at the wings of the condensation the opposite is true. Since Parker instability is associated with long wave-lengths, final condensation involving $10^6 M_\odot$ can be produced and this instability can be triggered by stellar density wave.

The inflation of field lines due to cosmic rays will not continue for ever because as the magnetic field lines expand, the volume of various flux tubes increases causing the cosmic ray pressure to decrease [21], unless there is copious production of cosmic rays in the flux tube. Therefore, except for the initial stages of the instability, the cosmic ray pressure gradients will decrease faster with z than the magnetic forces. If the magnetic field lines are held down by the gas clouds, the quasi-steady number of cosmic rays in each flux tube will not cause an unlimited inflation of the field lines. This is because as the inflation grows the pressure exerted by cosmic rays in the space between the gas clouds would not exceed the tension of the field lines resulting from the curvature. Final state is reached when the confining tension of the field lines becomes equal to the expansive pressure resulting from the magnetic field gradients and cosmic rays.

If the inflation does not continue, how then do the cosmic rays escape out of the galactic plane ? One possible reason is that as the tension of the field lines exceeds certain values, the ambipolar diffusion of the magnetic field lines from the gas condensation becomes effective. As a result the tension in the field lines reduces and this causes further inflation. Secondly, cosmic rays traverse to the surface of the gas disk through magnetic flux tubes which emerge out. Since the gas density is low in these regions cosmic rays are scattered efficiently by the hydromagnetic waves created by them [22]. When scattering is efficient, then a transfer of a major fraction of cosmic ray energy to the gas takes place. This accelerates the gas to about 100 km $s^{-1}$ and the cosmic rays rise with a mean velocity of about 200 km $s^{-1}$. Such a large internal pressure at the weak spots can certainly let the cosmic rays escape without much difficulty.

In the direction perpendicular to both y and z, the instability can grow as $k_x$ increases. The physical origin of the increase of instability with $k_x$ is, when $k_x^2 > k_y^2$, the raised portions expand in the x direction into the room available by the depression of the lines on either side. The energy is thus lower than when $k_x^2 = 0$. It has been shown [17] that for large values of y, the gas which is displaced in z above or below a cloud is powerfully attracted towards the cloud, and the gas which is displaced in x are repelled. Thus the properties of dynamical forces indicate that the interstellar gas must be in a continually chaotic state, with a strong tendency to clump, but with no evident equilibrium state toward which to relax until the density becomes sufficiently high for the self-gravitation to take over.

It is interesting to see whether cosmic rays can penetrate into the dense clouds, which form the major component of gas in the Galaxy. This has important consequences in the study of gamma ray production in the Galaxy since most of the observed gamma rays have been attributed to the interaction of cosmic rays with interstellar gas. This subject has recently attracted the attention of a few people [23-25]. Here I summarise the results from the recent work of Cesarsky and Volk [25]. It is assumed that dense clouds are threaded by the general interstellar magnetic field. Resonant diffusion in the cloud can enhance the energy losses of fast particles and diminish their flux in clouds. However, it is difficult to create effective wave fields which can diffuse resonantly, corresponding to particle energies above 10 MeV/nucleon. Instabilities caused by streaming of cosmic rays at the border of clouds due to self-generated waves could also inhibit the cosmic-ray penetration; this effect could stop cosmic rays below about 50 MeV/nucleon. If particles can be trapped in quasistatic magnetic mirrors which may arise at local condensations, the trapped particles will be depleted by energy losses before they are released by adiabatic field changes. Estimates show that such depletion could be as large as a factor of two at relativistic energies. One should be careful before using these numbers because the assumption regarding the configuration of magnetic field may not be a common feature in the Galaxy. In these studies it is further assumed that cosmic ray sources are outside the dense clouds. However, if cosmic ray sources are associated with dense clouds, the clouds would shine in gamma rays due to the large density of cosmic rays inside the clouds.

## HYDROSTATIC EQUILIBRIUM OF GAS-FIELD SYSTEM

From the above description of the dynamic state of the gas-field system, it appears that the observed configuration of the interstellar gas represents one particular phase of the dynamic evolution of a given initial state under a specific set of perturbing forces. It is also clear from the study of the dynamical state of such complex system, that it is possible to obtain only qualitative description of the system. In particular, if one wants to know the distribution of the components, such as cosmic rays and magnetic field perpendicular to the galactic plane, it is difficult to obtain them from the final states resulting from instabilities. On the other hand, the study of the hydrostatic equilibrium of the gas-field system (inspite of the fact that this cannot be considered to be stable) perpendicular to the galactic plane is found to be useful [17]. Therefore, a brief description of this aspect is presented in this section.

Hydrostatic equilibrium of the gas-field system perpendicular to the galactic plane is determined by the balance between the gradient of the total internal pressure of the system and the gravitational force per unit area in that direction. The total internal pressure includes that due to gas, magnetic field and the cosmic rays. It is generally assumed, following the work of Parker [17], that the distributions of all three forms of pressure are similar. Using this and knowing the mean velocity of the gas and the gravitational acceleration one can deduce the distribution of gas. It is found that the gas distributions derived from such studies are inconsistent with the observation at large heights from the plane. It is quite obvious that by using the distribution of gas as the only observational test, no further information about the distribution of the individual components could be obtained. However, it was realised that the observed non-thermal radio spectrum towards the galactic poles could be effectively used to set constraints on the distribution of magnetic field and cosmic rays. A detailed study has been made by Badhwar and Stephens [26] on this basis.

The total internal pressure of the gas-field system perpendicular to the galactic plane can be written as $Q(z) = \int_z^\infty \rho(z)\, g(z)\, dz$, where g is the gravitational acceleration and $\rho$ the gas density. One can determine g from the galactic model and make use of the observed gas distribution to determine Q as a function of z. It is found that the distribution of pressure due to magnetic field and cosmic rays is different from that of thermal gas. The integrated radio emissions observed perpendicular to the galactic plane cannot be accounted for, if only the distribution of the observed gas is made use of, but requires the presence of an additional component of gas in the Halo. The required distribution of halo gas should have a scale height probably of the order of a few kpc with a density of about one-hundredth of an atom/$cm^3$ in the galactic plane. The radio emissivity deduced in the direction of the Halo suggests that the Galaxy does not have a new spherical halo, but the emissivity extends beyond the gas disk with a scale height of about half that of the halo gas. From the distribution of the total internal pressure one can derive the distribution of cosmic rays if one knows the distribution of magnetic field. For a simple assumption that the distribution of pressure due to cosmic rays is similar to that of magnetic field, it is found that the equivalent thickness of cosmic ray disk is about a few kpc and the magnetic field extends well beyond the galactic disk. One other deduction made from this study [26] is that clouds are perhaps not dynamically coupled to the gas-field system. If this is true, then most of the observed gamma rays in the Galaxy should be due to unresolved gamma ray sources.

The calculated parameters for one of their models [26] in which it is assumed that the halo gas has a scale height of about 2 kpc are shown in Figure 8. Curve A in this figure shows the gas pressure, curve B the total internal pressure and curve C the radio emissivity at 408 MHz. The distribution of radio emissivity obtained from this calculation is similar to that deduced from the observed latitude profiles [27]. One can make use of such models to evaluate the gamma ray intensity, at galactic latitudes $|b| > 0$. In the anticentre region, where the gas extends well above the central plane of the Galaxy a knowledge of the distribution of cosmic rays perpendicular to the galactic plane is important. Further, if one wants to calculate the gamma ray contribution through inverse compton process, such a knowledge is essential, because the distribution of stellar photons varies slowly with z and the black body photons at 2.7°K are universal. The stability of such horizontal equilibria has been studied [28] and it is found that the system becomes unstable in times $\sim 10^8$ years.

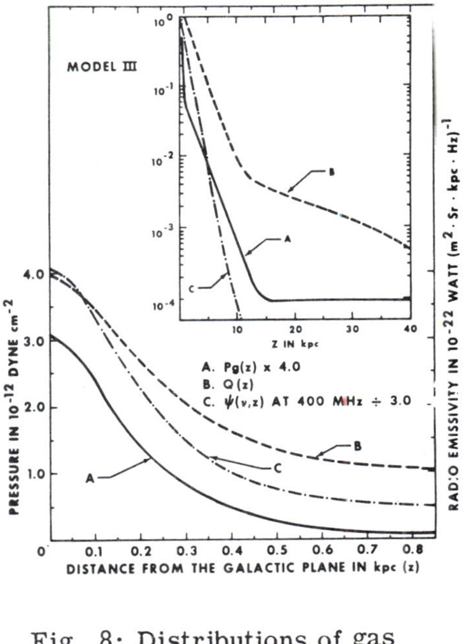


Fig. 8: Distributions of gas pressure, total internal pressure and radio emissivity at 408 MHz on the basis of hydrostatic equilibrium of the gas-field system (from Badhwar and Stephens, ref. 26).

## REFERENCES


1. A. Toomre, Ann. Rev. Astron. Ap. 15, 437 (1977).
2. E. Dekker, Phys. Report, C24, 315 (1976).
3. W.W. Roberts, Proc. NASA International Gamma Ray Symposium, GSFC, (NASA X-662-76-154), p. 128 (1976).
4. E.N. Parker, Proc. NASA International Gamma Ray Symposium, GSFC, (NASA X-662-76-154), p. 320 (1976).
5. D.G. Wentzel, Ann. Rev. Astron. Ap. 12, 71 (1974).
6. J.H. Oort, in "Galactic Structure", Ed. Blaauw and M. Schmidt, U. Chicago Press, p. 455 (1965).
7. B. Lindblad, Stockholm Obs. Ann. 22, 3 (1963) and the reference therein.
8. C.C. Lin, Ann. Rev. Astron. Ap. 5, 453 (1967).
9. A. Toomre, Ap. J. 158, 899 (1969).
10. R.A. Miller and K.M. Prendergast, Ap. J. 151, 699 (1968).
11. A.J. Kalnajs, Ap. Letters 11, 41 (1972); W.W. Roberts and F.H. Shu, Ap. Letters 12, 49 (1972).
12. D. Lynden-Bell and A.J. Kalnajs, M.N.R.A.S. 157, 1 (1972).
13. H.C.D. Visser, I.A.U. Symposium No. 77, Ed. Berkhuijsen and Wielebinski (Reidel Pub. Co.), p. 105 (1978).
14. F.H. Shu, V. Millione, W. Uebel, C. Yuan, D.W. Goldsmith and W.W. Roberts, Ap. J. 173, 557 (1972).
15. W.W. Roberts, M.S. Roberts and F.H. Shu, Ap.J. 196, 381 (1975).
16. S.C. Simonson, Ap. J. 208, 893 (1976).
17. E.N. Parker, Sp. Sci. Rev. 9, 651 (1969).
18. A. Dalgarno and R.A. McCray, Ann. Rev. Astron. Ap. 10, 375 (1972).
19. L. Spitzer and E.B. Jenkins, Ann. Rev. Astron. Ap. 13, 133 (1975).
20. T. Ch. Mouschovias, Ap. J. 192, 37 (1974) and 206, 753 (1976).
21. T. Ch. Mouschovias, Astron. Ap. 40, 191 (1975).
22. J. Skilling, Ap. J. 170, 265 (1971).
23. J. Skilling and A.W. Strong, Astron. Ap. 53, 253 (1976).
24. R.L. Brown and A.P. Marcher, Ap.J. 185, L 7 (1977).
25. C.J. Cesarsky and H.J. Volk, Astron. Ap. 70, 367 (1978).
26. G.D. Badhwar and S.A. Stephens, Ap. J. 212, 494 (1977).
27. J.E. Baldwin, Proc. NASA International Gamma Ray Symposium, GSFC, (NASA X-662-76-154), p. 206 (1976).
28. M. Lachieze et al., 15th Int. Cosmic Ray Conf., Plovdiv, Vol. 2, p. 219 (1977).

SESSION 5

# Cosmic-rays and Extragalactic Gamma-rays

*Chairman:* **S. L. MANDELSHTAM**

# GALACTIC GAMMA RAYS AND THE ORIGIN OF COSMIC RAYS

R. Cowsik

*Tata Institute of Fundamental Research, Bombay, India*

## ABSTRACT

High energy gamma rays in the galaxy are generated primarily through the interactions of cosmic ray nuclei and electrons with matter and radiation fields present in a variety of astrophysical objects and in the interstellar medium. Because of this close generic relationship, the observations of both discrete galactic gamma-ray sources and the diffuse galactic gamma-ray emission provide significant clues as to the nature and the distribution of cosmic ray sources. In addition, since the non-thermal radio emission of the galaxy arises from the synchrotron radiation of relativistic cosmic ray electrons spiralling in the galactic magnetic field, radio astronomy and gamma ray astronomy complement each other in the information they provide on cosmic rays in the galaxy. In this paper we argue that cosmic rays leaking from gamma-ray sources must contribute significantly to the density of cosmic rays in the galaxy. Conversely, the sources of cosmic rays, whatever they may be, are also likely to be the gamma ray sources.

Since high energy nuclear interactions of cosmic rays also generate the rare elements like Li, Be and B, observations of the (Li + Be + B)/(C + O) ratio, both in cosmic rays and also overall in the galaxy, place important constraints on the relative luminosities of the cosmic ray and gamma ray sources. We discuss several possible models for the sources with reference to the aforementioned constraints. We also critically review problems such as cosmic ray gradients, varying electron to proton ratios and so forth that are apparently implied by the gamma ray observations.

## INTRODUCTION

The importance of astronomy in the gamma-ray domain in providing clues regarding the origins of cosmic rays was realised very early. Thus the subject of gamma-ray astronomy can be said to have begun in the early 1950's, from theoretical speculations that the interactions of high energy particles with matter and radiation in galactic and extragalactic objects should give rise to detectable fluxes of gamma-rays [1-3]. These early estimates provided the motivation for a number of balloon-borne experiments that were carried out over the next decade without, however, any positive results (see review in [4]). This was partly because of the severe experimental difficulties involved and in particular the complex subtraction procedures involved in separating out a small possible celestial flux from the huge background of gamma rays generated in the Earth's atmosphere itself.

Moreover, some of the theoretical estimates had been too optimistic and had predicted fluxes from celestial objects, far in excess of what was determined experimentally later [3]. It was only in 1968 that Clark et al. [5] achieved the first unambiguous detection of celestial gamma rays with a satellite-borne experiment. Their directional detector on board the OSO-3 satellite detected gamma rays with energies above 50 MeV that were highly anisotropic in their arrival directions, such that there was a $\sim 10\sigma$ excess contribution along the galactic equator and an additional pronounced excess in the direction of the galactic center [6]. The very limited angular resolution of the instrument, however, precluded any conclusions from being drawn about whether the radiation arose in unresolved discrete sources or was truly diffuse in nature. Further, the crude spectral data did not enable any definite identification of the mechanism/s of gamma ray generation. In succeeding years, a number of balloon experiments were conducted by several experimental groups with inconclusive and sometimes contradictory results (see review in [7]). The next definite progress in the field was achieved with the launch of the SAS-2 gamma ray satellite in 1972 by the GSFC group. With improved angular and spectral resolution, this experiment was able to demonstrate convincingly that the gamma radiation is, in fact, confined to a thin galactic disc and in addition identified several point sources of gamma rays that stand out against the 'diffuse' disc emission [8,9]. More recently, the ESA gamma ray satellite COS-B, launched in 1975, has furnished excellent data on galactic gamma ray emission and has significantly increased the number of identified point sources of gamma rays [10,11].

The extent to which unresolved discrete sources contribute to the galactic gamma ray emission is still the subject of considerable debate [11-16] and one must await observations with still higher angular resolution for a proper clarification of this important question. The genuinely diffuse part of the radiation is believed to arise from cosmic ray interactions with interstellar matter as well as interactions of cosmic ray electron with interstellar photon fields [12,17-19]. The longitudinal distribution and energy spectrum of the galactic gamma-ray emission have been interpreted by various authors as suggesting a non-uniform cosmic ray density in the galaxy [12,21], a varying electron to proton ratio in cosmic rays [20] and several such other alternatives.

In this paper we discuss in some detail the knowledge that has been gained on the high energy processes that are going on in the galaxy through the studies in the areas of cosmic rays and gamma ray astronomy. We start by first discussing the close generic relationship that exists between high energy gamma rays and cosmic rays and how the luminosity of the galaxy in these two forms are closely related to each other. We then discuss the diffuse emission of gamma-rays and show the general consistency of these observations with a variety of cosmic ray observations and observations in the radio band. Then the emission from discrete sources is discussed. Here again, we shall show how the early predictions of the typical gamma-ray luminosity of the cosmic ray sources have been confirmed by the recent observations. Finally, we discuss some constraints on the models of the sources and suggest a possible scenario in terms of which the generation of cosmic rays and high energy radiation from the galaxy could be understood.

## THE GENERIC RELATIONSHIP BETWEEN GAMMA RAYS AND COSMIC RAYS

### Gamma-ray Observations

Let us first briefly review the gamma-ray observations as available to date. In high energy gamma rays ($E > 100$ MeV) the galaxy appears as a narrow line source with a width at half maximum of $5.5^\circ$ ($-2.5^\circ \leqslant b^{II} \leqslant 3^\circ$) towards $l^{II} = 0^\circ$ and $10^\circ$ ($-3^\circ \leqslant b^{II} \leqslant 7^\circ$) towards $l^{II} = 120^\circ$ [10]. The line source strength of the galaxy as obtained by COS-B is shown in Fig.1. In addition to the broad

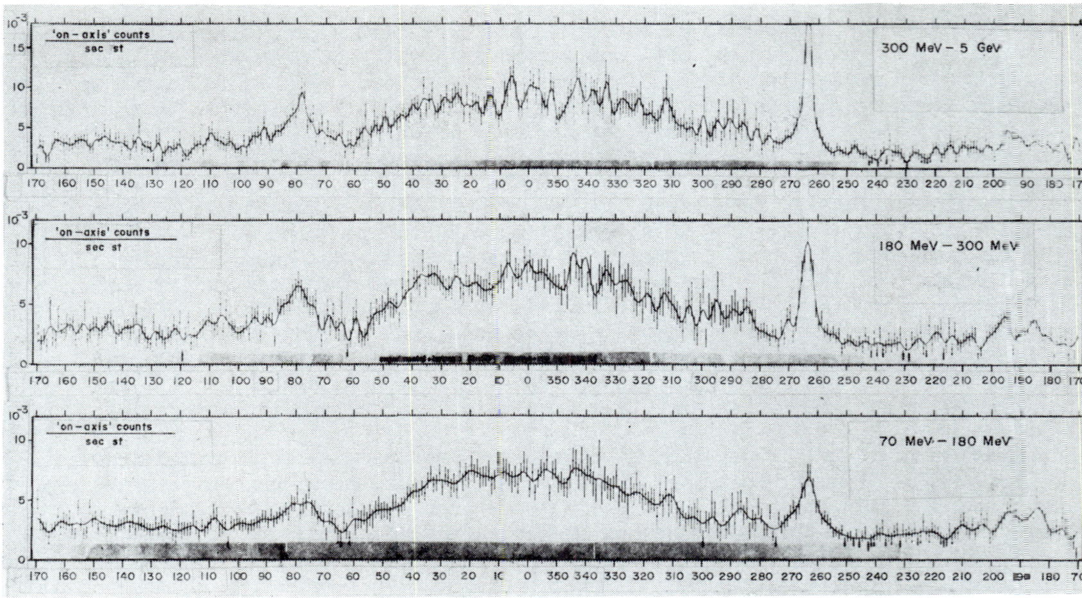


**Fig.1** Profile of gamma-ray intensity along the galactic equator

enhancement in the neighbourhood of the galactic center, regions of high gamma-ray emissivity also appear to occur where the line of sight tangetially intersects the galactic spiral arms that are delineated by HII-regions and appear as intensity maxima in the radio continuum and neutral hydrogen maps [22] . In addition, there are many apparently point sources of emission and 22 sources have been seen along the galactic plane by COS-B [23] . Because of the limited angular resolution, unambiguous identification of these sources with known astronomical objects have been possible only for 4 sources that have been identified with the pulsars PSR 1822 - 09, PSR 0532 + 21, PSR 0740 + 28 and PSR 0833 - 45. The energy spectrum of the gamma rays is about the same in all directions and is close to a power law of differential index 1.7 ± 0.1 [24,14] .

The longitudinal distribution of the gamma-ray intensity can be unfolded, assuming azimuthal symmetry, to yield the total luminosity of the galaxy in gamma rays as $\sim 10^{42}$ photons . $s^{-1}$ , above 100 MeV, or about $4 \times 10^{38}$ erg . $s^{-1}$ [25] . As mentioned before, there is some dispute over what fraction of the luminosity is due to unresolved discrete sources. Some estimates constrain the source contribution to lie between 15% and 30% [19] , whereas other calculations indicate that as much as 40% of the total luminosity could be due to unresolved sources [15] . The reverse problem, namely that of estimating the contribution to the galactic emission of known types of sources such as pulsars or X-ray objects, is also not settled. For example, estimates of the pulsar contribution have been obtained by Strong et al. to be only 10% [26] whereas Higdon and Lingenfelter believe that a much larger contribution is possible [16] . Extrapolation of the brightness of X-ray sources into the gamma ray domain have also yielded rather conflicting values for the extent of their contribution [27,28] .

## Cosmic Ray Observations

Let us now briefly consider some basic aspects of the high energy particle population of the galaxy and the various interactions they undergo with matter and radiation fields. The cosmic radiation is composed mostly of nucleons (~86% protons, ~13% helium nuclei, ~1% heavier nuclei) with about 1% electrons. In energy, cosmic rays extend from a few tens of MeV to over $10^{20}$ eV, following an essentially power law spectrum. The energy densities of these particles in interstellar space is then ~0.5 eVcm$^{-3}$ for nucleons of energy greater than 1 GeV/n and ~0.004 eVcm$^{-3}$ for electrons of energy greater than 1 GeV [29]. Whereas the energy density of nucleons of energy below 1 GeV/n is rather small, the energy density of electrons in this energy range is equal to or somewhat larger than that for E > 1 GeV. Thus the total energy density in electrons with energies greater than 100 MeV is ~0.01 eVcm$^{-3}$. For comparison, the energy density of starlight is ~0.4 eVcm$^{-3}$, that of the 2.7 °K universal blackbody radiation is 0.24 eVcm$^{-3}$ and that of the galactic magnetic field is ~0.5 eVcm$^{-3}$. From arguments involving the interactions of electrons with the universal 2.7 K photons, it can be definitely shown that the cosmic ray electrons of galactic origin. However, for the nucleons the situation is not as simple and at least the highest energy particles are quite likely to be of extragalactic origin.

Information about the propagation of cosmic ray nucleons may be obtained from the presence of such elements as Li, Be and B in cosmic rays, that have extremely low universal abundances and therefore are presumably absent in the sources of cosmic rays. Their finite abundances in the observed cosmic rays is then attributed to interactions of heavier nuclei such as C and O with interstellar matter subsequent to their acceleration in the sources. The determination of the ratio (Li + Be + B)/(C + O) then leads to an evaluation of the amount of interstellar matter traversed by the cosmic ray nucleons during their propagation in the galaxy (Fig.2). Further, measurements of the abundances of radioactive isotopes such as $Be^{10}$, indicate that cosmic ray nucleons have an effective lifetime of ~ 6 x $10^6$ years before being leaked into the intergalactic space.

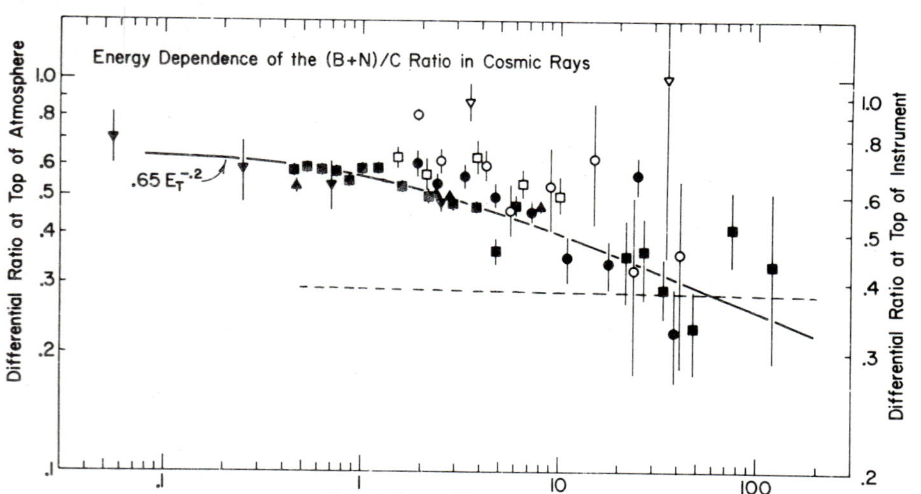


Fig.2 Ratio of the flux of secondary cosmic-ray nuclei to that of the primaries. The broken line indicates the spallation in the source region and the rest is due to spallation in the interstellar medium.

The galaxy is threaded by a large scale magnetic field which is determined, from measurements of the Faraday rotation of pulsar signals, to have a regular component of ~ 3 $\mu$g together with a random component of the same order. Cosmic ray

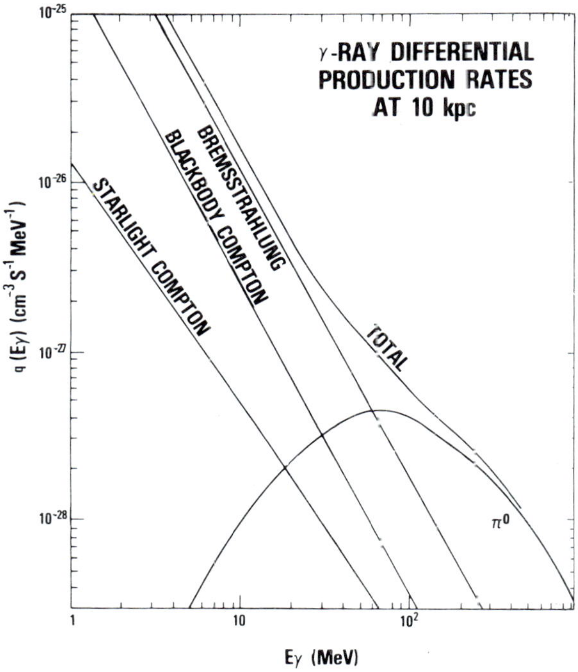


Fig.3 Local galactic gamma-ray differential production rate for a total gas density of 1 atom per cm$^3$ and starlight radiation density of 0.44 eVcm$^{-3}$ .

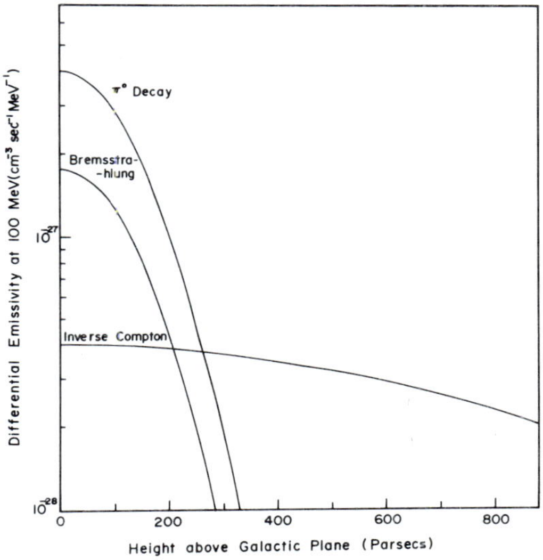


Fig.4 Dependence of the specific emissivity in gamma-rays due to various processes as a function of altitude above the galactic plane.

electrons spiralling along the magnetic field lines emit radio waves by the synchrotron process. The maps of the radio emission of the galaxy can then give us information on the region of confinement of relativistic electrons and the current interpretation is that they occupy a disc with a scale height of $\sim$750 pc [30] . (In contrast, the distribution of atomic and molecular hydrogen in the galaxy can be represented by much narrower discs, for example, by gaussian functions with dispersions of 50 pc and 120 pc for molecular and atomic hydrogen respectively [31] ). Then adopting the volume of the cosmic ray confinement region to be $\sim 2 \times 10^{67}$ cm$^3$ and the containment lifetime to be $\sim 2 \times 10^{4}$ seconds, we obtain the 'luminosity' in cosmic ray nucleons to be $\sim 8 \times 10^{40}$ erg.s$^{-1}$ and that in cosmic ray electrons to be $\sim 1.6 \times 10^{39}$ erg s$^{-1}$.

It is clear from the earlier discussions that high energy gamma-rays are produced by cosmic rays through the three processes: (a) decay of neutral pions which are produced through nuclear interactions with ambient matter, (b) bremsstrahlung of the electron component and (c) through the inverse Compton scattering of various photon fields in the galaxy. The spectrum of gamma-rays generated by cosmic rays through these processes is shown in Fig.3 [17] . The relative amplitudes of the three spectra correspond roughly to their production rates in the interstellar medium close to the solar system. Because of the unknown amount of the large solar modulation of the low energy cosmic-ray electrons, their bremsstrahlung spectrum is very uncertain [34] . Since the gamma-ray telescopes integrate the source function along their look direction, we show in Fig.4 the possible dependence of source strength in these processes as a function of height [7] above the galactic plane. The bremsstrahlung and the $\pi^{o}$ components essentially follow the gas distribution falling off with a scale height of ~120 pc. On the other hand the scale height of the star-light photons exceeds ~1.5 kpc and that of the relativistic electrons is ≳ 750 pc as suggested by nonthermal radio observations [30] ; thus the Compton-generated source strength drops off much more slowly. Telescope sampling regions of high 'z' would be sensitive to these Compton gamma rays. The variations of these source functions over the galactocentric distance is less well known. But even if one assumes that the density of cosmic rays is roughly uniform through out the disc, substantial increases in all these components are expected at the locations of spiral arms, particularly at the one lying at ~4 kpc. These increases are caused by increases in the density of gas due to molecular clouds and increases in the photon densities due to the presence of young OB associations and HII regions. The increased matter density at the 4 kpc arm and its relevance to gamma-ray astronomy and possible cosmic ray gradients has been extensively discussed in the literature. But the importance of the increase in the density of the photons is being realised only recently [35] . The overall increase in the density of stars as one approaches the galactic centre is discussed in [18].

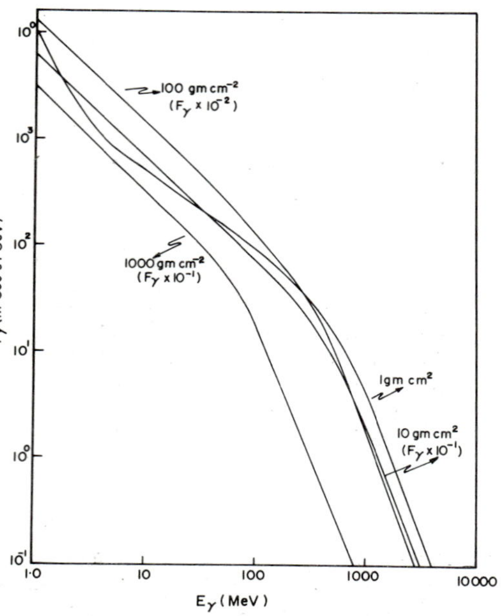


Fig.5 The gamma ray flux generated by cosmic rays in targets of various thicknesses.

In Fig.5 we show the spectrum of gamma rays generated when cosmic rays go through slabs of matter of various thicknesses. It is seen that production rate per gramme is roughly constant upto slab thicknesses of ~100 gm.cm$^{-2}$ but the rate drops exponentially beyond that with absorption length of ~100 g.cm$^{-2}$. Thus, if one wants to make an efficient gamma-ray source, then the shroud of matter around a source of high energy particles should have a depth of ~ 100 g cm$^{-2}$ or less.

General Energetic Considerations

The total luminosity of the galaxy in various forms of highly energetic particles and quanta are summarised in the Table 1 below:

TABLE 1 Galactic Luminosities

| | |
|---|---|
| Cosmic-ray nuclei | $= L_{nuc} \sim 8.10^{40}$ erg s$^{-1}$ |
| Cosmic-ray electrons | $= L_e \sim 1.6 \times 10^{39}$ erg s$^{-1}$ |
| Diffuse gamma-rays | $= L_{\gamma,d}$ $2 \times 10^{38}$ erg s$^{-1}$ |
| Discrete sources of gamma-rays | $= L_{\gamma.s}$ $2 \times 10^{38}$ erg s$^{-1}$ |

The electrons are capable of transferring almost all their energy to the photon component either through the inverse Compton process or through bremsstrahlung. On the other hand only that part of the nuclear component which is above the pion production threshold ($\sim 1$ GeV/n) is effective and through repeated collisions the nucleon component can transfer roughly 15% of its energy to the high energy gamma-rays. The uncertainties in the observations are about a factor of two, perhaps towards the lower luminosities. All the same it is clear energetically that either the cosmic-ray electrons or the nucleons could generate the gamma-ray emission from our galaxy.

Now we will make use of the fact that all cosmic rays go through about 4 g/cm$^2$ of matter. The interaction mean free path for the pion producing collisions by nucleons is $\sim 60$ g cm$^{-2}$ and in these collisions about 15% of the energy of the nucleonic component is transferred to gamma rays. Thus the nucleonic component generates a gamma ray luminosity.

$$L_{\gamma,\mathrm{nucl}} = \frac{4}{60}\, 0.15 \times L_{\mathrm{nuc}} = 8 \times 10^{38} \text{ erg s}^{-1}$$

Similarly the electrons have a radiation length of $\sim 60$ g cm$^{-2}$ where they transfer $\sim 70\%$ of their energy to the gamma-ray component; thus

$$L_{\gamma,e} = \frac{4}{60}\, 0.7\, L_e = 8 \times 10^{37} \text{ erg s}^{-1}$$

It is also interesting to note that in $\sim 10^7$ years the electron component would also generate, through Compton scattering of the galactic starlight fields, a luminosity also comparable to the galactic diffuse emission. Thus keeping in mind the uncertainties in the various numbers this rough balance between the total luminosity in gamma-rays and that expected purely from the cosmic ray observations establishes the close generic relationship that exists between these two components.

## DISCUSSION OF THE EXTENDED EMISSION

Having set up a general framework for the consideration of the interactions of high energy particles in the galaxy that lead to the generation of gamma rays, we now proceed to a discussion of the observational data on the galactic emission. To obtain clues as to the relative importance of the various processes, within the context of our current knowledge about the structure and content of the galaxy, we must look at two fundamental observational parameters, the energy spectrum and the angular distribution.

As mentioned before, the energy spectrum of galactic gamma rays is the same in all directions in the galactic disc ($|b| < 10^\circ$) and follows a power law with a differential index of $1.7 \pm 0.1$ between a few tens and a few hundreds of MeV [14,24]. In addition, the spectrum derived for the galactic component of the high latitude ($|b| > 10^\circ$) radiation has an index of $1.5 \pm 0.3$ and is therefore within errors, the same as that for the radiation from the disc [33]. We have seen that in this energy range, the inverse-compton gamma rays would have a slope of 1.8, the bremsstrahlung gamma rays would probably have a slope closer to 1.6, and pion decay gamma rays would have a characteristic hump-shaped spectrum with a turnover at $\sim 70$ MeV. Also, in the context of the similarity of the energy spectrum at both high and low galactic latitudes, it is pertinent to note that bremsstrahlung and pion decay gamma rays arise in a thin matter disc of scale height $\sim 120$ pc [31], whereas inverse compton gamma rays are generated in the much thicker electron disc of scale height $\sim 750$ pc [30]. From these two considerations, it appears plausible that gamma rays from pion decay and bremsstrahlung [17,34] may not necessarily be the dominant contributors to the galactic emission everywhere, and that inverse-compton gamma rays probably play not an insignificant role.

Let us now see what the observed angular distribution tells us. On being unfolded from the longitude distribution to give the variation in intensity as a function of galactrocentric distance, the radial distribution obtained is seen to be similar to that of molecular hydrogen and of supernova remnants as shown in Fig.6 [31] . This has led to the suggestion that the major contribution to the

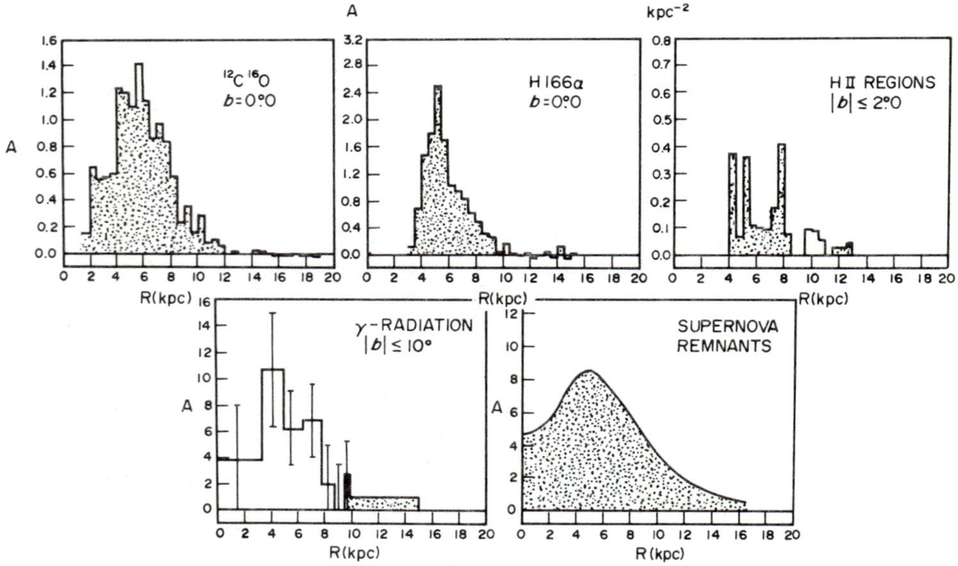


Fig.6 The similarity of gamma-ray production rates and other galactic constituents like molecular clouds supernova remnants, etc.

gamma-ray luminosity is from the decay of pions formed in interactions of cosmic ray nucleons, leaking out of supernova remnants, with the excess matter present in their neighbourhood [17] . Alternatively, however, it may be noted that the radial distribution of HII regions is also the same as that of the gamma ray emissivity [31] with the obvious implication that the inverse-compton process may be important. Of course, the most important consideration in this respect is the extent of the contribution due to discrete sources. For example, the enhancements in gamma-ray emissivity along directions that tangentially intersect spiral arms may be interpreted to mean that the cosmic ray density is correlated with the matter density leading to quadratic enhancements in emissivity in the spiral arms [12] . Alternatively, this may be more simply interpreted as indicating that unresolved discrete sources, that might be making a significant contribution to the galactic luminosity, are population-II objects that naturally have a higher density along spiral arms. Such arguments have been recently summarised by Pinkau [35] who has pointed out that there are two possible approaches towards an interpretation of the data on the angular distribution of the gamma ray emissivity. One may either argue that the point source contribution stays at a comparatively low level through the galaxy and hypothesize variations of the gamma-ray emissivity through cosmic-ray-matter interactions in terms of correlations with molecular hydrogen distribution or spiral arms, for example. Alternatively, one may conjecture that gamma-ray point sources are correlated with the stellar birth rate (as deduced from the radial variation of emissivity of the galactic disc in lyman alpha photons [32] ) and this requires that the cosmic ray produced gamma-ray emissivity through interactions with matter) stays at a low and perhaps even constant level throughout the galaxy. In the context of the second hypothesis, the large number of sources, that are expected to lie within the rather wide viewing angle of the gamma-ray telescopes flown so far, may account for the relative constancy of the energy spectrum at all galactic longitudes. An interesting way to distinguish between the two hypotheses would be to compare their predictions with the observations, not merely in terms of the absolute intensities but also with respect to the level of Poissonian fluctuations, which would tend to be larger if the discrete sources were important.

## The Galactic Halo

Another aspect of the interpretation of the angular distribution data has been the attempt to resolve a long outstanding question, namely, the extent of the region of confinement of cosmic rays. Stecker and Jones [36] have considered detailed models of diffusion halos of various sizes perpendicular to the galactic plane and compared the implified gamma-ray longitude distributions with the SAS-2 data, for goodness of fit. Assuming the sources of cosmic rays to be supernova remnants or pulsars, cosmic-ray nucleon halo models with scale heights greater than 3 kpc are found to provide a poor fit to the observations. The main thrust of their argument is that unless the containment disc is rather thin one would not be able to generate the sharp cosmic rays gradients needed to generate the strong luminosity variations implied by the gamma-ray data. Though the scale heights of ~ 3 kpc seem to be in agreement with the radio data one should be rather cautious in accepting these arguments till the role of the discrete sources, which have been neglected in the present analysis, is made clear.

## Possible Gradient Towards The Anticentre

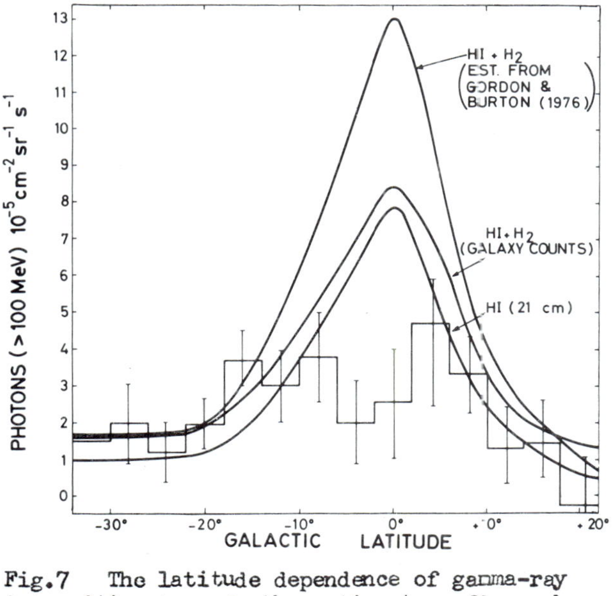


Fig.7 The latitude dependence of gamma-ray intensities towards the anticentre after substraction of the contribution of discrete sources.

There have been several analysis of the gamma-ray sky maps in the literature which claim to have established a strong gradient in cosmic rays towards the anticentre [20] . In fact the claim is that the nucleonic component decreases substantially by a factor more than four but the electron component increases by a similar factor. In view of the tremendous importance of this result to cosmic-ray astrophysics let us examine this result carefully. We show in Fig.7 the latitude distribution of gamma-rays integrated over the longitudes $l^{II}$ = 160° - 210° obtained from COS-B data after major substractions of the contribution of point sources in this region. Also on the same figure the expected latitude distribution as estimated by

Strong et al. [37] is plotted and it is the large discrepancy between the two that has led to the claims regarding the cosmic ray gradients stated earlier. The discrepancies as seen in Fig.7 are clear enough but it is not at the moment very clear how much of this discrepancy is due to an over estimation of the contribution of the sources, over estimation of the column densities of interstellar matter and a true gradient in the cosmic ray densities. In Fig. 8 we show the the 21-cm map at these longitudes [38]. Notice that contour lines which are 10% apart are either nearly parallel to the $b^{II}$ axis or very sparsely distributed around $b^{II} = 0$ indicating that the hydrogen itself is not concentrated very sharply at the galactic plane. Further, the spectra of gamma-rays arriving from different longitude bins is also analysed in terms of the relative contribution of the $\pi^o$-decay gamma-rays and the bremsstrahlung gamma-rays which are used as probes of the nuclear and electronic components of the cosmic rays respectively. Basically the conclusions of Strong et al. [37], that the contribution of $\pi^o$-decay gamma rays to be observed spectrum is small, is not new [18]. But here they have a more extensive analysis of the longitudinal dependence of the $\pi^o$-contribution. Their analysis is summarised in Table 2. Here 'f' represents the factor by which the (e/p) ratio in cosmic rays changes from its value of ~1-2% measured locally. In view of the fact that the observed spectra imply very small $\pi^o$- contribution it is more convenient to show also the value of $f^{-1} = g$, representing the variation of the (p/e) ratio.

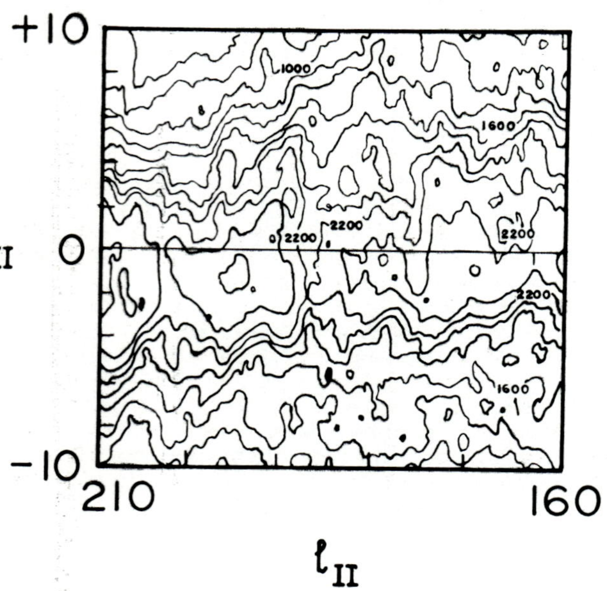


Fig.8 The 21-cm map towards the anticentre. Notice the weak dependence of temperature on galactic latitude.

TABLE 2 Distribution of e/p Over The Galaxy

f = e/p enhancement factor ; g = $f^{-1}$ = p/e variation

| $l^{II}$ | $10^o - 40^o$ | $60^o - 90^o$ | $110^o - 140^o$ | $305^o - 335^o$ | All longitudes |
|---|---|---|---|---|---|
| f | $15^{+\infty}_{-8}$ | $4.4^{+4}_{-1.5}$ | $11^{+\infty}_{-6}$ | $17^{+\infty}_{-9}$ | $11^{+11}_{-3}$ |
| g | 0.07 ± 0.07 | 0.22± 0.12 | 0.09 ± 0.99 | 0.06 ± 0.06 | $0.09^{+0.03}_{-0.05}$ |

Looking at Table 2, particularly at the g values, it is easy to decipher what is happening. The so called nominal bremsstrahlung spectrum of Shukla and Cesarsky [34] is so close to the observed spectral shape that if we make the assumptions (a) that the nominal spectrum is right and (b) that the contributors to the observed spectrum are exclusively the bremsstrahlung with standard spectrum and the $\pi^0$- decay gamma-rays, then we are led to the conclusion that the contribution of the $\pi^0$'s is negligible. Notice that if we assume that the density of the nucleonic component is roughly uniform over the galaxy, these results would mean that one needs a production rate about 10 times as much as the production rate of Shukla and Cesarsky [30]. Since there is no firm experimental basis any way for the so called standard spectrum no definitive conclusions can be drawn.

Summarising the discussions of the diffuse galactic emission, we can make the following observations: Observations of (Li + Be + B)/(C + O) ratios in cosmic rays do indicate that the cosmic-ray interactions can generate the overall gamma-ray luminosity. But in order to interpret the observed gamma-ray spectra as super positions of $\pi^0$-decay and bremsstrahlung contributions we need rather steep electron spectra at energies below $\sim 1$ GeV, i.e. spectra with logarithmic slopes greater than 2. It looks as though we have made very little progress in establishing cosmic ray gradients; other than the absence of a prominent peak in the latitude distribution at $l^{II} \sim 180$ there is very little else to go on.

## THE SOURCES

The nature of the cosmic ray and gamma ray sources will now be discussed under three separate headings: (1) the observational aspects and the attempts at identification, (2) the implications of cosmic ray studies as to the nature of gamma ray sources and (3) the implications of observations of gamma ray astronomy and radio astronomy as to the nature of the cosmic-ray sources. Having made these inter-relationships clear, we finally present a schematic model for the sources.

### The Observational Aspects of Gamma-ray Sources

There are now some 24 discrete sources of high energy gamma-rays that have been seen by the gamma-ray telescopes with a flux level $\gtrsim 10^{-6}$ photons $cm^{-2}s^{-1}$ at $E \gtrsim 100$ MeV. Firstly one should realise that this list may not exhaust all the gamma-ray sources in our neighbourhood in the galaxy. It is possible that even a relatively bright source could have been missed if it lies low on the galactic plane within $\sim 40^\circ$ in longitude from the direction of the galactic centre. The rather intense diffuse emission could easily obscure such a source. It is also quite likely that there are indeed a large number of sources with luminosity levels below the detection threshold and their integrated emission could be contributing significantly to the so called diffuse-component. We shall return to these weak sources later. Due to limited angular resolution, unambiguous identification of the bright sources has been possible only in a very limited number of cases, four of which are pulsars and two possibly extragalactic objects. A more ambitious attempt at identification is due to Montmerle [39] . He suggests that gamma-rays sources are closely linked with young extreme population I objects and perhaps association of supernova remnants and O B stars, the so-called 'SNOB', he also has looked at the possible gamma-ray emission from Vela and other supernova remnants and says that the COS-B data indicate that SNR's as a class are not strong gamma ray sources. This conclusion naturally leads him to suggest that SNOB's could be the major source of galactic cosmic rays if $\pi^0$ decay contributes dominantly to the gamma-ray flux. We shall return later to this point when we discuss source models. But at this stage it is worthwhile noting the following rough estimates regarding the absolute luminosity of these sources. With observed flux levels at $\sim 2 \times 10^{-6} cm^{-2} sec^{-1}$ , using typical distances of $\sim 1$ kpc these sources emit $2.10^{38}$ gamma-rays/sec above 100 MeV. Their spectra are not too different from that of the 'diffuse emission' $E^{-1.7}$ corresponding to a mean energy of ~ 400 MeV so that a total emissivity of $2.10^{38}$ $s^{-1}$ corresponds to ~ $10^{35}$ ergs/sec per source. There may be about a thousand sources of this luminosity yielding a nett luminosity of ~ $10^{38}$ erg . $s^{-1}$. From our previous discussions on the generic relationship between cosmic rays and gamma-rays, we see that this luminosity is about 0.1% of the luminosity in the nucleonic component of cosmic rays of the galaxy. We will now see how cosmic ray observations do indicate that the high energy particles do transfer roughly this fraction of their energy to the gamma rays before escaping from the sources.

## CONSTRAINTS DERIVED FROM THE COSMIC RAY OBSERVATIONS

Careful experimental work by several groups [40-44] has established that the relative abundances of cosmic ray nuclei are dependent on rigidity in the range ~ 10-100 GV/C. The main observational feature of interest here is that the ratios of the fluxes of secondary nuclei like Li, Be, B to that of their progenitors like C and O are decreasing functions of rigidity as shown in Fig.2. In the so called nested leaky box model for the cosmic rays, it has been argued that the most likely explanation of this and several other features is that cosmic rays are stored briefly around the sources before they are released into the interstellar medium [45,46]. The mean residence time of cosmic rays in the source region (the inner box) is a decreasing function of their rigidity so that the low rigidity cosmic rays are spalled more effectively at lower energies leading to the observed steeper spectra of the secondary cosmic ray nuclei. One would also be able to reproduce the same result if a few of the cosmic ray sources are heavily shrouded by gas of high density and if the cosmic ray spectra in these sources have somewhat steeper slope (This is a kind of an inversion of the closed galaxy model of Peters and Westergaard [47] ). The point which is relevant here is that cosmic rays do traverse about 2-3 g.$cm^{-2}$ of matter in the sources before they escape into the interstellar medium and that during this traversal they will generate gamma rays. After escaping from the sources the cosmic rays spend about ~ $10^7$ years in an extended galactic volume during which time they pass through an additional 2-3 g. $cm^{-2}$ of interstellar matter.

Since about half of the total matter traversed by the cosmic rays in the sources and half in the diffuse interstellar matter, the sources and the diffuse emission should contribute about equally to the total galactic emission as estimated by Bignami et al. [15] but for the Compton-component. And this is not a mere fortuitous coincidence. These arguments based on the cosmic ray residence times were given early in 1975 [46] , before the COS-B telescope had observed any of the sources!

In fact the early predictions went even further: It is interesting that one can estimate typical flux from a nearby gamma ray source. As we remarked before the rate at which gamma rays are generated is given by

$$G_s = L_\gamma \frac{\tau_s}{\tau_G + \tau_s}$$

where $\tau_G$ and $\tau_S$ are the cosmic ray path lengths (g $cm^{-2}$) in the interstellar matter and the source respectively. If there are $\nu$ sources generating gamma-rays distributed roughly uniformly throughout the galaxy then the luminosity per source is roughly equal to

$$g_s = G_s/\nu$$

The mean distance to the nearest of these sources is

$$\langle d \rangle = \sqrt{\frac{\pi r_G^2}{\nu^2}}$$

where $r_G \simeq 4.5 \times 10^{22}$ cm, the radius of the galaxy. Thus the flux from the nearest sources becomes

$$f \simeq \frac{g_s}{4\pi \langle d \rangle^2} = \frac{L_\gamma}{4\pi^2 r_G^2} \frac{\tau_s}{\tau_G + \tau_s} \simeq \frac{10^{42}}{4\pi^2 (4.5 \cdot 10^{22})^2} \frac{3}{3+3}$$

$$\simeq 3 \cdot 10^{-6} \ cm^{-2} s^{-1}$$

which is consistent with the subsequent observations of COS-B.

But how many gamma-ray sources are indeed there in the galaxy? We cannot directly answer this question but observations of the spectrum of the cosmic ray electrons yield useful clues to estimate this number. The high energy electrons, because of their high rate of energy-loss through the synchrotron and the Compton processes in the galactic disc, are sensitive indicators of the effective time they take to reach us from the cosmic ray sources. Whereas the high energy electrons can reach us only from the nearest of the sources, the low energy electrons, because of their smaller rate of energy loss, could reach us even from the more distant sources. Of course, the contribution of the very distant sources to the electrons observed near the solar system even at low energies would become progressively smaller because of the leakage of the electrons from the sides of the galactic disc. Cowsik and Lee [48] have discussed this problem in detail and we reproduce here their results in Fig.9. Notice that one needs sources situated

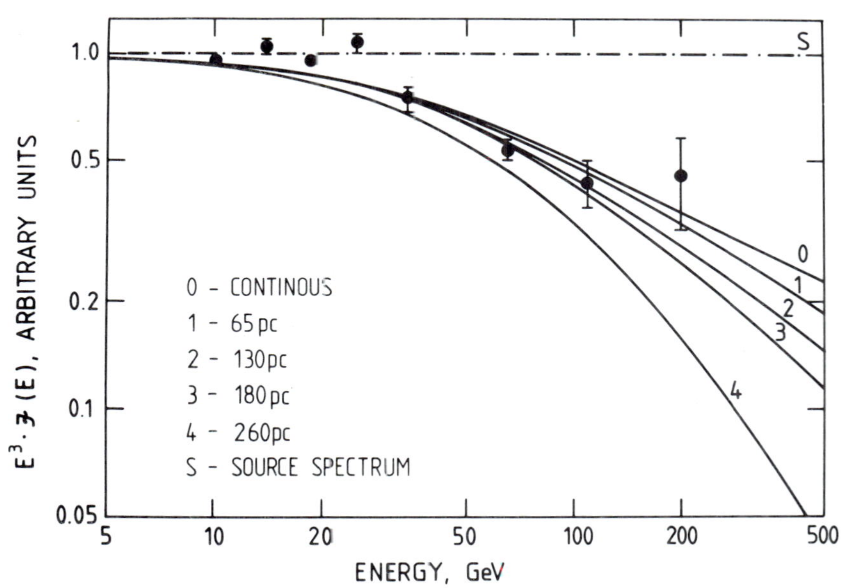


Fig.9 The spectrum of cosmic-ray electrons expected from a distribution of sources with various mean spacings compared with the observations.

within a few hundred parsecs from the solar system to reproduce the high energy tail of the electron spectrum. If the mean density of the sources is roughly uniform over the galactic plane it would mean that there should be at least $\pi (r_G / r_S)^2 \gtrsim 3 \times 10^4$ sources distributed over the whole galaxy. Each of these need accelerate only $\lesssim 3.10^{34}$ erg . $s^{-1}$ in high energy electrons or $\lesssim 3.10^{36}$ erg $s^{-1}$ in all cosmic rays, on the average. Now from our earlier discussion it is clear that these sources together contribute roughly one half of the total gamma-ray luminosity of the galaxy. Thus the mean luminosity of the sources in gamma-rays is

$$\langle L_S \rangle \lesssim \frac{10^{42}}{3 \cdot 10^4} \frac{1}{2} = 1.6 \cdot 10^{37} \ \gamma \ s^{-1}$$

This is a factor of ~30 below the typical source luminosity observed by COS-B in gamma rays. This would mean that about only one in 100 of the cosmic ray sources could be an intense gamma-ray source and the rest will have to be very weak so as to maintain the average luminosity specified above. So the distribution in luminosities of the cosmic ray sources would have qualitatively the behaviour shown in Fig.10. Since the gamma-ray luminosity increases roughly linearly with the grammage traversed in the source it means that the distribution shown in Fig.10 also refers to the distribution of the mean residence times in the different cosmic ray sources. (Of course this distribution is not to be confused with the distribution in path lengths that the individual cosmic rays go through). Since the typical distance to a cosmic ray source is ~100-200 pc the most probable distance of the brightest gamma-ray source, which is the hundredth cosmic ray source, is ~1-2 kpc as seen by the COS-B telescope. Thus the predictions based on the cosmic ray observations are consistent with the observations in gamma-ray astronomy.

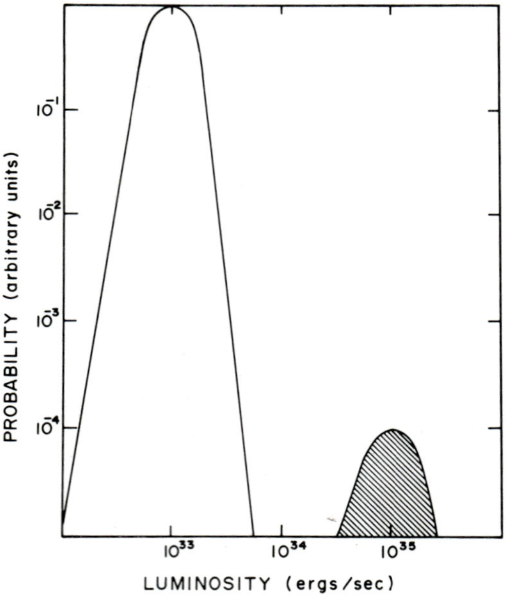


Fig.10 Distribution of gamma-ray luminosities of cosmic ray sources expected on the basis of the nested leaky-box model.

What are the implications of these considerations to the field of cosmic rays? Firstly the observations of the gamma-ray sources indicate that at least in some of the sources the cosmic rays traverse through substantial amount of matter. We can now make use of the observation that the elements Li, Be and B are extremely rare, in nature and the spallogenic calculations of Audouze, Meneguzzi and Reeves [49] to show that cosmic rays have to escape even from the gamma-ray sources. The reason for this is that cosmic rays can at best transfer ~15-20% of their energy to gamma-rays of E ~100 MeV but they are very effective in spalling the heavy nuclei even after their energy has been degraded below the pion production threshold. For this reason and for the economy of high energy particle generation in our galaxy it is most likely that the high energy particles leaking out of the gamma-ray sources also form part of the galactic cosmic rays. We now therefore have a situation in which cosmic rays are being generated by a very large number of sources. In 99% of these sources cosmic rays leave without substantial breakup passing through only $\sim 10^{-2}$ g $cm^{-2}$ . In rare cases $\sim 1\%$ the source of cosmic rays are enveloped in a heavy cloud of gas and cosmic rays come out in a highly spalled state. The composition of cosmic rays predicted by such a model is not very different from the so called closed model of the galaxy of Peters and Westergaard [47] though the model suggested by the gamma-ray observations is closer to the cannonical leaky-box models for the cosmic ray production and propagation.

COMPTON AND SYNCHROTRON EFFECTS IN THE SOURCES

The luminosity of the galaxy in electrons is about one hundredth of all cosmic rays and about two or three times the gamma-ray luminosity of the galaxy as a whole. Therefore the cosmic ray electrons in the sources should not transfer more than $\sim 30\%$ of their energy on the average to the gamma-rays of $E > 100$ MeV through Compton interactions. Most of the energy in the cosmic ray electrons resides in the energy range $\sim 1$-3 GeV or Lorentz factors $\gamma \approx 2 \times 10^3 - 6 \times 10^3$ . The average energy of the scattered photon $E_\gamma \approx \gamma^2 \epsilon_{ph} \cdot \frac{4}{3}$ . Thus in order that we do not generate too much Compton gamma rays we must have

$$\sigma_{Th}\, n_{ph}\, c\, T \left(\frac{4}{3}\gamma^2 \epsilon_{ph}\right) < 0.3\, \gamma\, mc^2$$

which gives

$$T\, n_{ph}\, \epsilon_{ph} < 3 . 10^{14} \quad s.eV.cm^{-3}$$

In the close proximity of population I objects and in HII regions $\epsilon_{ph} \approx 10$ eV (corresponding to Ly-$\alpha$ photons whose density seems to correlate well with the gamma-ray emissivities) and $n_{ph} \approx 10$-100. Thus the cosmic ray electrons should not spend a time $T > 10^{12}$ s in these regions on the average. On the other hand an accidental or genetically related coincidence of a cosmic ray source with a region of high photon density would produce a rather powerful Compton gamma-ray source. Thus one may expect OB-star association to be Compton gamma-ray sources. Some of the weak sources of $\pi^0$ gamma rays which populate the large peak at about $10^{32}$ erg $s^{-1}$ in Fig.10 could indeed be more powerful as Compton sources. They will have at least $10^{48}$ erg in all cosmic rays or $\sim 10^{46}$ erg in electrons. Their Compton luminosity would be $10^{33} - 10^{34}$ erg in high energy gamma rays. In fact all the cosmic ray sources could also be Compton gamma-ray sources probably very weak depending on the value of the product $T\, n_{ph} \epsilon_{ph}$ where T is the effective time spent near the source, $n_{ph}$ is the mean photon density per $cm^3$ in this region and $\epsilon_{ph}$ is the mean photon energy. It would not be surprising if these weak Compton sources should contribute importantly to the total galactic emission.

There would be relatively strong synchrotron emission at radio wavelengths from these numerous sources and this emission may contribute significantly to the total galactic non-thermal radio emission. Observations of our own galaxy and that of many external galaxies [32] does indicate that the radio spectrum is very weakly dependent on the height above the galactic plane. In fact Allen [50] after studying the high resolution specture synthesis observations, remarks "energetic electrons are produced not only by explosions of type II SN near spiral arms but also by some other sources which are distributed through out a galaxy more or less as an older stellar population". This conclusion is identical to the one drawn by Lee and myself [48] from the study of the galactic cosmic ray electron spectrum.

Thus the sources of cosmic rays could be objects like different pulsars, flare stars, magnetic white dwarfs, X and gamma-ray bursters and so-forth. Our personal preference would be for condensed objects in close binary systems. These objects would become intense gamma-ray emitters when they are encompassed either by a dense matter cloud or in a strong UV photon field. If the typical luminosity of $\sim 10^{36}$ erg/source estimated by us is correct, this would also be the upper limit to the absolute gamma-ray luminosity of such sources.

## EPILOGUE

In this paper we have presented a preliminary study of the relationship that exists between the sources of cosmic rays and gamma rays in the galaxy. The study is decidedly sketchy but it brings into focus areas where much theoretical and experimental work needs to be done. The tentative conclusions that can be drawn are listed below:

1. The galactic gamma-ray luminosity is clearly due to the interactions of cosmic rays with matter and radiation fields.
2. The specific contributions of the various physical processes like $\pi^0$ production, bremsstrahlung and inverse Compton scattering are not yet clear.
3. There seems to be variations in the cosmic ray density over the galactic disc thus making extragalactic origin of cosmic rays very improbable.
4. There seems to be a gradient towards the anticentre direction but one should exercise much caution; further studies are required to confirm this exciting result.
5. Cosmic ray observations predict a large number of sources distributed over the galaxy and the gamma ray fluxes from a typical source is also correctly predicted.
6. There seems to be the possibility of a very large number of sources distributed over the galaxy which could simulate the diffuse emission. If this is indeed true one can understand the apparent insensivity of the measured gamma-ray spectra on the galactic co-ordinates.
7. High resolution radio aperture synthesis also indicates the presence of a large number of sources distributed in a thick disc similar to the older stellar population.

## ACKNOWLEDGEMENTS

I wish to thank Dr. V. Radhakrishnan and S. Sarkar for many very fruitful discussions. This talk was originally scheduled to have been given by Prof. I. Shklovsky.

## References

1. S. Hayakawa, Prog. Theoret. Phys. 8, 571 (1952).
2. G.W. Hutchinson, Phil. Mag. 43, 847 (1952).
3. P. Morrison, Nuovo Cimento 7, 858 (1968).
4. G.G. Fazio, Ann. Rev. Astron. & Astrophys. 5, 481 (1967).
5. G.W. Clark, G.P. Garmire and W.L. Kraushaar, Astrophys. J.(Lett), L-203 (1968).
6. W.L. Kraushaar, G.W. Clark, G.P. Garmire, R. Borken, P. Highbie, C. Leong and T. Thoros, Astrophys. J. 177, 341 (1972).
7. E.L. Chupp, Gamma Ray Astronomy, Reidel, Dondr echt (1976)
8. D.A. Kniffen, R.C. Hartman, D.J. Thompson and C.E. Fichtel, Astrophys. J.(Lett) 186, L-105 (1973).
9. C.E. Fichtel, R.C. Hartman, D.A. Kniffen, D.J. Thompson, G.F. Bignami, H.B. Ogelman, M.E. Ozel and T. Tumer, Astrophys. J. 198, 163 (1975).

10. K. Bennett et al., Caravanne Collaboration, in Recent Advances in Gamma-ray Astronomy, ESA SP-124, 83 (1977).
11. W. Hermsen et al., Caravanne Collaboration, Nature 269, 494 (1977).
12. D.A. Kniffen, C.E. Fichtel and D.J. Thompson, Astrophys. J. 215, 765 (1977).
13. H.B. Ogelman, C.E. Fichtel, D.A. Kniffen and D.J. Thompson, Astrophys. J. 209 584 (1976).
14. J.A. Paul et al., Caravanne Collaboration, Astron. Astrophys. 63, L-31 (1978).
15. G.F. Bignami, P. Caraveo and L. Maraschi, Astron. Astrophys. 67, 149 (1978).
16. J.C. Higdon and R.E. Lingenfelter, Astrophys. J. (Lett) 208, L-107 (1976)
17. F.W. Stecker, Astrophys. J. 212, 60 (1977)
18. R. Cowsik and W. Voges, 14th Intern. Cosmic Ray Conf., 1, 74 (1975)
19. C.J. Cesarsky, M. Casse and J.A. Paul, Astron. Astrophys. 60, 139 (1977)
20. A.W. Strong, A.W. Wolfendale, K. Bennett and R.D. Wills, Mon. Not. R. Astr. Soc. 182, 751 (1978).
21. F.W. Stecker and F.C. Jones, Astrophys. J. 217, 843 (1977)
22. Y.M. Georgelin and Y.P. Georgelin, Astron. Astrophys. 49, 57 (1976)
23. W. Hermsen et al., Caravanne Collaboration, 6th Eur. Cosmic Ray Symp. Kiel (1978)
24. R.C. Hartman, D.A. Kniffen. D.J. Thompson, C.E. Fichtel, H.B. Ogelman, T. Turner and M.E. Ozel, NASA Tech. Mem. 79651 (1978)
25. A.W. Strong and D.M. Worrall, J. Phys. A. 9, 823 (1976)
26. A.W. Strong, A.W. Wolfendale and C. Dahanayake, Mon. Not. R. Astr. Soc. 179 69 p (1977).
27. M.J. Coe, J.J. Quenby and A.R. Angel, Nature 274, 343 (1978)
28. R.J. Prothroe, A.W. Strong, A.W. Wolfendale, K. Bennett and R.D. Wills, Nature 277, 542 (1979).
29. A.W. Wolfendale, in Origin of Cosmic Rays, Reidel, Dordrecht, (1975) p.1.
30. J.E. Baldwin, in The Structure and Content of the Galaxy and Galactic Gamma Rays, NASA CP-002 (1976) p.189.
31. W.B. Burton, A. Rev. Astron. Astrophys. 14, 275 (1976).
32. P.G. Mezger, Astron. and Astrophys., in Press (1979)
33. C.E. Fichtel, G.A. Simpson and D.J. Thompson, Astrophys. J. 222, 833 (1978).
34. P.G. Shukla and C.J. Cesarsky, in Recent Advances in Gamma-ray Astronomy, ESA SP-124, 1977, p.131.
35. K. Pinkau, Nature, 277, 17 (1979).
36. F.W. Stecker and F.C. Jones, Astrophys. J. 217, 843 (1979).
37. A.W. Strong, A.W. Wolfendale, K. Bennett and R.D. Wills, Proc. 12th ESLAB Symposium, p. 167 (ESA-SP-124, Frascati, 1977).
38. W.B. Burton, The Structure and Content of the Galaxy and Gamma Rays, p.177 (GSFC Preprint X-662-76-154, Greenbelt, 1976).
39. T. Montmerle, Astrophys. J. 231, 95 (1979)
40. L.H. Smith, A. Buffington, G.F. Smoot, L.W. Alvarez, M.A. Wahlig, Ap. J. 180 987 (1973).
41. V.K. Balasubramanian and J.F. Ormes, Ap.J. 186, 109 (1973)
42. E. Juliusson, Ap. J., 191, 331 (1974)
43. J.H. Caldwell, Ap. J., 218, 269 (1977).
44. M. Garcia-Munoz, G.M. Mason and J.A. Simpson, Ap. J. 202, 265 (1975)
45. R. Cowsik and L.W. Wilson, Proc. 13th Intern. Conf. on Cosmic Rays, 1, 500 (1973).
46. R. Cowsik and L.W. Wilson, Proc. 14th Intern. Conf. on Cosmic Rays, 2, 659 (1975).
47. B. Peters and N.J. Westergaard, Danish Space Research Institute Report (1976).
48. R. Cowsik and M.A. Lee, Ap. J. 228, 297 (1979).
49. H. Reeves, Ann. Rev. Astron. Astrophys. 12, 437 (1974)
50. R.J. Allen and R. Sancisi, Astron. and Astrophys. 62, 397 (1978); Also Prv. Comm. (1979).

# LONGITUDINAL DISTRIBUTION OF GAMMA RAYS AND THE VARIATION OF MAGNETIC FIELD IN THE GALAXY

S. A. Stephens

*Tata Institute of Fundamental Research, Homi Bhabha Road, Bombay 400 005, India*

## ABSTRACT

An analysis has been carried out to understand the longitudinal distribution of gamma rays and radio emission. It is inferred from the analysis that the large scale variation of the density of cosmic rays in the Galaxy is of the form $\rho_{CR}(w) = \rho_{CR}(\odot)\exp(1-w/w_\odot)$ upto $w_\odot$ and the density remains nearly constant in the outer parts. It is also found that the variation of magnetic field is exponential in nature with a scale length of about $2w_\odot$ over the entire extent of the Galaxy. There is no evidence for the proton to electron ratio in cosmic rays to vary over the Galaxy.

## INTRODUCTION

Many attempts have been made in the past to understand the longitudinal distribution of gamma rays as obtained by SAS-II data; these have been summarised recently by Fichtel [1] and Pinkau [2]. Most of these attempts have taken into account the structure and/or the dynamics of the Galaxy. Each one of these models reproduces the gamma ray observation in a satisfactory manner. It is not possible to uniquely identify one model to be preferred over the others, because the line of sight integration washes out the details in the theory. In this paper a simple approach is made to understand the consistency between Cos-B data and the radio data at 408 MHz, and to deduce the large scale variation of cosmic rays and magnetic field in the Galaxy.

## THEORETICAL APPROACH

The gamma ray line intensity $I_\gamma(>E_\gamma, l)\Delta\theta$ of energy greater than $E_\gamma$ in the direction of galactic longitude $l$ using a telescope having acceptance angle $\Delta\theta$, is given by

$$I_\gamma(>E_\gamma, l)\Delta\theta = \int_0^L \frac{\varepsilon_\gamma(w)}{4\pi} ds \int_{-b}^{b} \eta_{CR}(z) \sum_i n_i(w)\,\eta_i(z)\,db \qquad \text{.. (1)}$$

Here, $z = s \tan b$, where $b = \Delta\theta/2$ is the galactic latitude, $n_i$ the gas density in terms of hydrogen atoms in the galactic plane for the ith component, $\eta_{CR}$

and $\eta_i$ the distribution functions respectively for cosmic rays and gas components perpendicular to the galactic plane, w the galactocentric distance and L the distance along the line of sight. In the above equation $\varepsilon_\gamma$ is the gamma ray emissivity per hydrogen atom and is a function of cosmic ray density $\rho_{CR}(w)$. The cosmic ray density may vary on a galactic scale independent of the gas density and there could also be local variation depending upon the gas density. In order to separate these, we define $\varepsilon_\gamma(w) = \varepsilon_\gamma'(w)F_{CR}(w)$, where $F_{CR}(w) = \rho_{CR}(w)/\rho_{CR}(\theta)$ so that $\varepsilon_\gamma'(w)$ takes care of the local variation. One can now calculate the mean emissivity using equation (1) along any longitude and we define

$$\langle\varepsilon_\gamma'(w)\rangle_l = 4\pi I_\gamma(>E_\gamma, l)\,\Delta\theta \Big/ \Big\{\int_0^L F_{CR}(w)\,n(w)\,ds \int_{-b}^{b}\eta_{CR}(z)\sum_i f_i\,\eta_i(z)\,db\Big\} \quad ..\ (2)$$

where $n(w) = \sum_i n_i(w)$ and $f_i = n_i(w)/n(w)$.

In a similar manner one can also define the mean radio emissivity along a given longitude as

$$\langle\varepsilon_\nu'(w)\rangle_l = 4\pi I_\nu(l)\,\Delta\theta \Big/ \Big\{\int_0^L F_{CR}(w)\,M_B^\alpha(w)\,ds \int_{-b}^{b}\eta_{CR}(z)\,\eta_{B_\perp}^\alpha(z)\,db\Big\} \quad ..\ (3)$$

where $M_B(w) = B_\perp(w)/B_\perp(\theta)$, the magnetic field strength at w with respect to that in the solar neighbourhood and $(\alpha - 1)$ is the radio spectral index. It can be noticed from equations (2) and (3) that the numerators are the observational data and the denominators can be calculated using various assumed forms of $F_{CR}(w)$ and $M_B(w)$. The main approach in this paper is to obtain simple forms for these so that the mean emissivities determined from equations (2) and (3) are either independent of 1 or distribute about a constant value.

## OBSERVATIONAL DATA

One of the important parameters of relevance to Gamma Ray production in the Galaxy is the distribution of the density of interstellar gas. We have made use of the distribution of gas in the galactic plane provided by Gordon and Burton [3] for both molecular and atomic hydrogen clouds for w $\gtrsim$ 2 kpc. An uniform density of .15 hydrogen atom. $cm^{-3}$ has been used for the intercloud medium upto 13 kpc [4]; beyond 13 kpc, the density is assumed to decrease similar to that of H I clouds. The distribution of gas perpendicular to the galactic plane is assumed to be Gaussian in shape. The values of $\sigma$ are taken to be 42 pc for molecular hydrogen and 111 pc for atomic hydrogen; these values are consistent with the measurements [5]. For intercloud medium the value of $\sigma$ is taken to be 233 pc [6]. The increase of $\sigma$ in the outer parts of the Galaxy for the intercloud medium and H I clouds is estimated with $d\sigma/dw$ = 23 pc/kpc [7].

The radio emissivity perpendicular to the galactic plane determined by Badhwar and Stephens [4] from their theoretical Model IV is found to be same as that deduced by Baldwin [8] using radio surveys. Therefore, we have made use of the distributions, $\rho_{CR}(z)$ and $M_B(z)$, provided by Badhwar and Stephens and assumed that they are independent of w. These can be approximated by exponential distributions with scale heights of about 2 kpc and 4 kpc respectively.

In this investigation, the longitudinal distribution of gamma rays from Cos-B data [9] has been used. The latitude extent of this line emission is $|b| \leq 6^\circ$. Since individual source contribution cannot be exactly removed from the observed line intensity, we have utilised the data from those regions, where there are no enhancement of the line intensity due to sources. In the case of radio distribution, 408 MHz surveys have been chosen [10-12] because at this frequency neither absorption nor thermal emission distorts the spectrum over the entire range of galactic longitudes. The value of the radio spectral index is found to be .65 at 408 MHz.

## RESULTS AND DISCUSSION

The longitudinal distribution of gamma rays from Cos-B data is shown in Figure 1, in which the identified sources [2] are indicated by arrows. The dotted curve A is the calculated line intensity of gamma rays for an uniform distribution of cosmic rays and for a gamma ray emissivity per hydrogen atom

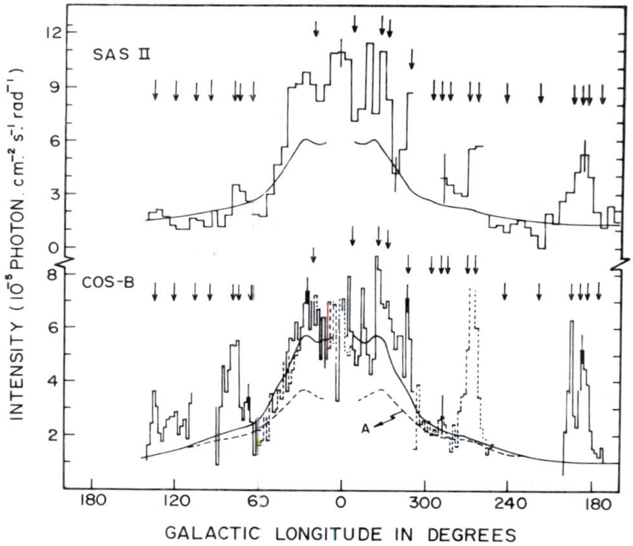


Fig. 1. Longitudinal distribution of gamma rays from SAS-II and Cos-B data; the arrows indicate the location of gamma ray sources. The dashed curve A is the calculated distribution of gamma rays using an uniform distribution of cosmic rays and the solid curves are by using a distribution derived in this investigation.

of $1.2 \times 10^{-25}$ photon/s. The curve falls short of the observed gamma ray distribution towards the inner parts of the Galaxy and one can infer that cosmic ray density should be larger towards the central region. The large scale variation of cosmic rays to account for this difference is deduced in the following manner.

A study [13] of the mean emissivity (eqn. 3) as a function of $<n>$ suggests that the radio emissivity decreases with w. It is found that, with $\{M_B(w).\ F_{CR}(w)\}$ $= \exp(1.82 - w/5.5)$ for $w \lesssim 10$ kpc and $= \exp(-12/w)$ for $w \gtrsim 10$ kpc, the calculated values of $<\varepsilon_\nu'>$ are distributed about a constant value. This constant emissivity is about $6.8 \times 10^{-41}$ watt/(m$^2$. hz) at 408 MHz and represents the local emissivity corresponding to a magnetic field strength of 4.8 $\mu$G. It may be mentioned here that similar decrease of radio emissivity with w is also noticed [14] in external spiral galaxies like M51 and NGC 6946.

In order to derive $F_{CR}(w)$ we make a simple assumption that $\rho_{CR} \propto B^2$ and we deduce $F_{CR}(w) = \exp(1-w/10)$ for $w \lesssim 10$ kpc. Substituting $F_{CR}(w)$ in eqn. 2, it is found [13] that $\langle \varepsilon_\gamma' \rangle_\ell$ distributes about a constant emissivity when $F_{CR}(w) \sim 1$ for $w \gtrsim 10$ kpc. The value of this constant emissivity per hydrogen atom for $E_\gamma \geq$ 100 MeV is $1.2 \times 10^{-25}$ photon/s, which is consistent with the value of $1.3 \times 10^{-25}$ photon/s through $\pi^o$ decay [15]. Using $F_{CR}(w)$, the variation of magnetic field is found to be $B_\perp(w) = 4.8 \exp(.5 - w/20)$ $\mu$G over the entire extent of the Galaxy.

We have calculated the line intensity of gamma ray emission as a function of $\ell$, using $\varepsilon_\gamma'$ and $F_{CR}(w)$ deduced in this analysis, and is shown by solid curve in Figure 1. The agreement between the calculated distribution and the Cos-B data is good. Most of the gamma rays are produced by cosmic ray nucleons of kinetic energy $\gtrsim$ 1 GeV and the energy of electrons contributing to the radio emission at 408 MHz are also $\gtrsim$ 1 GeV. Hence it can be inferred that the proton to electron ratio in cosmic rays of kinetic energy $\gtrsim$ 1 GeV remains the same over the entire extent of the Galaxy.

We have also shown in Figure 1, the SAS-II data [1] for comparison. The latitude extent of this data is $|b| \leq 10^o$. The solid line shown in this figure is the calculated distribution of gamma rays using the derived quantities from this analysis. One can notice that the calculated curve do not fit the observations, further, the observed flux values are systematically larger than those from Cos-B data towards the inner regions of the Galaxy. However, it was pointed out by Wills (during COSPAR) that such a large discrepancy may not exist if one uses the new Cos-B data.

## REFERENCES


1. C.E. Fichtel, Sp. Sci. Rev. 20, 191 (1977).
2. K. Pinkau, Nature, 277, 17 (1979).
3. M.A. Gordon and W.B. Burton, Ap.J. 208, 346 (1976).
4. G.D. Badhwar and S.A. Stephens, Ap.J. 212, 494 (1977).
5. W.B. Burton and M.A. Gordon, Ap.J. (letters), 207, L 189 (1976).
6. E. Falgarne and J. Legueux, Astron. & Ap. 25, 253 (1973).
7. P.L. Baker and W.B. Burton, Ap.J. 198, 281 (1975).
8. J.E. Baldwin, Proc. NASA Int. Gamma Ray Symp. (NASA X-662-76-154) p. 206 (1976).
9. K. Bennett et al., Proc. 12th ESLAB Symp. (ESA Sp. 124) p. 83 (1977).
10. A.J. Green, Astron. & Ap. Suppl. 18, 267 (1974).
11. C.G.T. Haslam, M.J.S., Quigley and C.J. Salter, M.N.R.A.S. 147, 405 (1970).
12. C.G.T. Haslam et al., Astron. & Ap. Suppl. 13, 359 (1974).
13. S.A. Stephens, Proc. 16th Int. Cosmic Ray Conf., Kyoto, 1979, OG-3.4.
14. P.C. van de Knit, R.J. Allen and A.H. Rots, Astron. & Ap., 55, 421 (1977); A. Segalovitz, Astron. & Ap. 61, 59 (1977).
15. G.D. Badhwar and S.A. Stephens, Proc. 15th Int. Cosmic Ray Conf., Plovdiv, 1, 199 (1977).

# "THE EUROPHYSICS STUDY CONFERENCE ON GAMMA-RAY ASTRONOMY AFTER COS-B" ERICE, ITALY, 17-23 MAY 1979

G. F. Bignami

*Laboratorio di Fisica Cosmica e Tecnologie Relative del CNR, 15 Via Bassini, 20133 Milan, Italy*

"The new lights that modern observations have thrown upon several interesting parts of astronomy begin to lead us now to a subject that cannot but claim the serious attention of every one who wishes to cultivate this noble science".

W. HERSCHEL, Esq. F.R.S.
(from a paper read on March 6, 1783)

## INTRODUCTION

The Study Conference was organized under the auspices of the EPS at the Centro di Cultura Scientifica "Ettore Majorana" in Erice, Italy and was held from May 17 to 23, 1979. It was attended to by over 65 scientists from Europe and the U.S. working in the field of high-energy astrophysics as well as in related fields like radioastronomy, cosmic rays, U.V. astronomy etc. The idea of a "Study Conference" is that of having as much discussion and exchange of ideas as possible, keeping to the bare necessary minimum the amount of formal data presentation (i.e. sessions that are just a series of papers with little time for discussion). This was found particularly useful and stimulating by the attendants, who all participated actively in the discussions (sometimes heated) and informal workshops triggered by the over 70 contributions (more than 1 per participant on average) presented. In what follows, a presentation will be given of the various scientific subjects that were discussed, the conclusions that have been reached and the ideas and recommendations for the future. No special order will be followed, nor will any attempt be made to give detailed credit for each contribution (other than those cases were the work presented is being published elsewhere), in line with the study-conference philosophy not to publish proceedings.
One of the highlights of the Conference, and a starting point for the works, has been the presentation of the latest set of data of the COS-B mission. The satellite was launched in August 1975 and it is still properly functioning, having, at the time of writing, practically doubled its expected lifetime and with its extension

approved at least throught 1979. The COS-B experiment and mission are described extensively in the literature [1,2], and the present day data coverage is complete as far as the galactic disc is concerned, with several extragalactic pointings also available (see the Caravane Collaboration contributions to these proceedings). The drammatic step forward of COS-B in gamma-ray astronomy is best depicted by figure 1, which shows, as a function of time, the actual number of high-energy photons ( $\geqslant$ 100 MeV) so far gather from the galactic disc by the main High-Energy Gamma-Ray Astronomy missions flown so far. (The implications and possibilities of HEGRA missions with total number of photons from the Galaxy in the range $10^5$ to $10^6$ ( $\geqslant 10^8$ eV) in the mid-late 80's will be discussed later on).

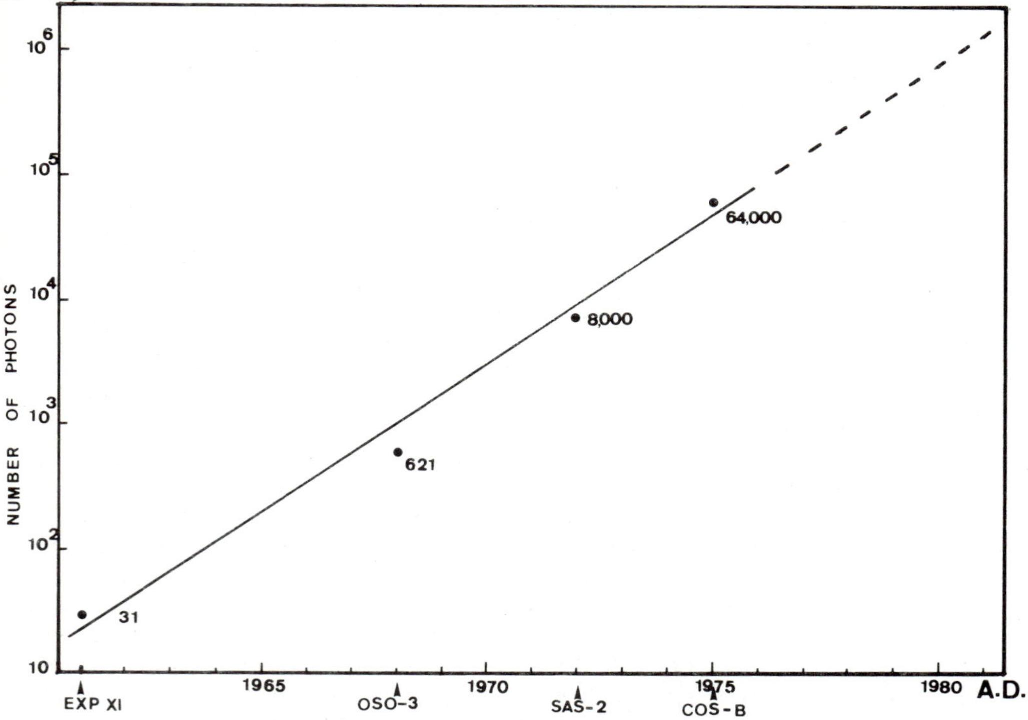


Fig. 1. Total number of photons ( $\geqslant$ 100 MeV) from the Galaxy as a function of the launch year for the major HEGRA missions flown so far [see ref. 3,4,5,6]

THE STUDY CONFERENCE

The Erice Study Conference has witnessed the first presentation of new COS-B results in three main fields: localized sources (including pulsars), spectral characteristics of the Galactic plane and source radiation, and quantitative assesment of the diffuse, mid-latitude radiation. Firstly, it is in order here to review what is to be considered a COS-B gamma-ray source: this is an excess of photon counts which is statistically significant with respect to the noise against which it is seen, and which is compatible with the instrument angular resolution or "point spread function".

Two comments should follow: the existence of a source is a signal-to-noise-ratio problem and thus contains an amount of arbitration in the setting of the confidence threshold level (e.g. 6 " $\sigma$ "), and secondly that these localized enhancements of radiation should not be referred at as "point" sources in the same way as, e.g., in X-ray astronomy, since their intrinsic angular extent can be of the order of 1° and still be unresolved by the instrument. With these provisos, a new list [7] of 29 COS-B gamma-ray sources was released at Erice which supercedes those of the published 1st COS-B catalogue [8] and of the 2nd COS-B list presented at the Munich Texas Symposium [6]. The Erice list is to be taken just as a progress report and does not yet have the characteristics of completeness and consistency of a catalogue. There has not been much time at the conference for more than a few immediate observations on this new list: the tendency of sources to be grouped around $b^{II}=0°$ is apparent (but 3 sources at high b are present, two of which have been identified, with the local dark cloud complex $\rho$ Oph and with the bright nearby QSO 3C273 [6,9] while the fluxes are still contained in about a decade, a consequence of the limited sensivity of the instrument and of the weak source fluxes. Great care should be exercised when trying to analyze the properties of these sources because of selection effects in the sample.

The properties of the 4 sources identified with radio pulsars have also been discussed, and are summarized in the recent work by Buccheri et al. [ref. 10].

The spectra of two of the COS-B sources (in the range 50 MeV - several GeV) have been given in detail for the Crab and Vela pulsars cases [see e.g. ref. 11]. Also for 3C273 the improved statistic has allowed the construction of a differential spectrum, best approximated by a power-law with a differential photon index of $\sim(-2.7 \pm .3)$. The effect of the improved statistic was also to reduce the error box down to $\sim$40' radius. For the other sources, weaker or in a heavily confused region of the plane, the spectral information is still rough, and can be based only on a ratio of the number of counts above and below 300 MeV. It is however clear that marked spectral difference are present in the galactic high--energy gamma-ray source population, one of the most spectacular examples being the source at $l^{II} \simeq 321$ which, albeit visible above 100 MeV, becomes much more significant $>$ 800 MeV or so. All the above source parameters suggest that possibly a variety of objects or processes are responsible for the UGO's (Unidentified Gamma Objects), as on the other hand is seen in the small fraction of sources that are identified: several pulsars, one (or possibly two) QSO's [see ref. 12], one dark cloud complex: quite an astronomical zoo, which is only likely to expand as more sources are (hopefully) identified. Several contributions were presented in Erice on the problem of the identification of the galactic UGO's both individually and as a population, without as yet much success. However, it has become apparent what UGO's are not, i.e. a known galactic population, like that of radio pulsars or SNR's. It was recognized that subsets of these (or other) galactic populations, seen with different selection effects could in principle be responsible for the emission. Individually, the identification work is difficult because of the size of the gamma-ray "error boxes". However, some of these, as in the case of Geminga (CG 195 + 4), are now down to an error radius of $\sim$ 20' of arc, and hopes have been expressed on the outcome of the X-ray astronomy esplorations now being performed with the Einstein Observatory. Quite some effort was also put into the evaluation of the contribution of "sources" to the overall galactic luminosity, reaching a consensus that they might be quite important, with percentages similar to those recently quoted in refs 13, 14 and 15.

Finally, the new result by COS-B presented at Erice also concerned the emission at mid-galactic latitude, $5° < |b| < 15°$ or so. It is important to note that only very few COS-B sources are seen at these latitudes, and yet the visibility for them is definitely better than near $b^{II} = 0°$. However, a very significant emission of gamma-rays is measured to come from such regions, and thus, since it does not show sources according to the previous definition, such an emission should be called "diffuse". Moreover, its $\sim$ cosec $b^{II}$ intensity law shows it to be of galactic origin. It is then natural to relate this emission to diffuse processes like interaction of cosmic-ray photons and electrons with IS gas, and correlations of the gamma-ray fluxes with tracers of such gas (i.e. $N_H$ column dendity and galaxy counts) have been presented. The results of this study are twofold: on one side it is possible to measure independently the local gamma-ray emissivity (per H atom per second) and at the same time to trace the amount of $H_2$ in the local ($5° < |b| < 15°$) region of the Galaxy, a very elusive gaseous component of the IS medium. [See e.g. refs 16 and 17]. The panorama of the data presentation for high-energy gamma-ray astronomy was completed by a summary of the final results of the NASA SAS-2 satellite, particularly important because of its capabily to disentangle the diffuse galactic and extragalactic components [ see e.g. ref 18]. Several (theoretical) sessions were devoted to models for gamma-ray sources, roughly from three points of view: gamma-ray emission from pulsars, galactic sources other than pulsars and extragalactic sources. It was clear that all of those models have to assume as a boundary condition that the $\sim$ keV luminosity of a galactic source must be at least one order of magnitude less than the $\sim$ 100 MeV luminosity (in the $10^{34}$ to $10^{36}$ erg $s^{-1}$ region, if the emission is isotropic), as well as analogous conditions on the emission at other wavelengths. Models for extragalactic sources must moreover take into account the recent low-energy gamma-ray observations of NGC 4151 [see refs. 19, 20 and 21] and hard X-rays from 3C273 [ref. 22], both presented at Erice in their most updated versions, and at the same time the limitations imposed by the gamma-gamma interaction cross sections, governing the relative sizes of the $\sim$ keV and $\sim$ 100 MeV sources. The consensus from the several contributions on pulsars was that they certainly encounter no difficulty in being gamma-ray sources of the required power ( $\sim 10^{33} - 10^{35}$ erg $s^{-1}$ when beaming is taken into account), but that most certainly several critical model parameters, such as the luminosity-period or the luminosity-frequency relations, have to be rediscussed in the light of the recent COS-B results. Detailed geometrical models [see ref. 23] presented were shown to be consistent with the observed gamma-ray light curves (remarkably similar) for the various pulsars, while at the same time not requiring any privileged orientation of the rotating neutron star with respect to the observer. Altogether, the gamma-ray emission is certainly a very important physical characteristic of at least some of the pulsars which populate our galaxy, as first suggested by the SAS-2 team and now extensively detailed by the COS-B results. Conversely, this subset of the galactic population of pulsars may, and in fact certainly does, contribute to the overall galactic gamma-ray emission for a fraction which is however still uncertain as other characteristic properties of those pulsars likely to be gamma-ray sources are still undefined.

In the session devoted to gamma-ray sources other than pulsars the possibility was explored that "small scale galactic system", generally of rather complex nature, could give size to gamma-ray emission through cosmic ray matter interactions, a feature common to this class of models. Associations between SNR's and OB associations [see ref. 24] where shown to have some degree of correlation with the COS-B sources and certainly to possess the energetic supply required of a galactic gamma-ray sources. The COS-B sources in the Cygnus region have been considered to be correlated with maxima of the recent CO data of ref. 25 folded with the COS-B point spread function [see ref. 26]. The special case of CG 353+16, tentatively identified by the COS-B collaboration with the Oph dark cloud complex, was discussed in the light of possible explanations of its apparently enhanged gamma-ray emission with respect to a "standard" [ref. 27] mechanism [see ref. 28]. A shock driven by the stellar wind of a young, blue star (HD147165) could represent the condition for a 1st order Fermi acceleration of the local cosmic rays [ref.29], or the same effects could be due to the shock which surrounds the cloud if it is moving supersonically with respect to the local intercloud medium [see ref. 30].

The extragalactic source models were characterized, for the case of NGC 4151, by a re-assesment of the multiple-stage Inverse Compton Scattering process of high-energy electrons in the photon field of this Seyferty nucleus. Such a process is seen to generate the required spectral shape, with its typical steepening in the MeV region [see ref.31]. More generally, active radio galaxies, and expecially double radio galaxies, were considered to be possible sites of very intense gamma-ray emission. Here the peculiar geometry of the radio lobes, which is considered a difficult phenomenon to explain in the framework of many current models, could very naturally stem from the energy being carried outwards by an intense beam of energetic gamma-rays, giving rise to gamma-gamma interactions with the 2.7°K blackbody radiation [see ref.32]. The extragalctic source models, together with the scantly available source data were also seen to be connected with the discussion on models for the diffuse background. The one truly diffuse mechanism proposed was that referring to the gamma-ray production in intergalactic space through Inverse Compton Scattering of universal microwave photons by relativistic electrons on the assumption that locally the relativistic electron spectrum remains unchanged over a long period of galactic development. The increase of the universal microwave radiation temperature with regression in time makes this contribution proportional (for q=0) to the square of the maximum redshift of integration. Then, in order to avoid an Olbers' Paradox, one has to consider various cut-off mechanism, i.e. change of electron spectrum is early times, on-set time of the galaxies [see ref.33]. The data on extragalactic gamma-ray sources, on the other hand, did not feature many novelties on the high-energy side (apart from the already quoted improved COS-B results from 3C273). A set of preliminary upper limits of emission from a number of interesting celestial objects was derived from the COS-B available data base. These upper limits are in the range from (several) x $10^{-7}$ photons ($\gtrsim$ 100 MeV) $cm^{-2}$ $s^{-1}$ to (very few) x $10^{-6}$ in the same units. They are entirely consistent (only at times marginally better) with those from the SAS-2 mission, obtained from the final analysis of the entire data base. In the case, several classes of objects can be identified, e.g. Seyfert galaxies, N and EL galaxies, BL Lacs, QSO's, for which the SAS-2 upper limits impose significant contraints in the shape of their spectra. Coupled with recent high-quality X-ray spectral data, this seems to have possibly a cosmological meaning, namely to provide a possible explanation of the diffuse extragalactic background in terms of a summation of sources [see e.g. ref. 34]. It is relevant to recall here the importance of the recent new data from NGC 4151 at low gamma-ray energies (providing now a fairly complete picture of the differential spectrum of this object up to the MeV region), of the results on Cen A, 3C273, NGC 4151 in very hard X-rays ( $\sim$ 100 keV), as well as of the direct measurements of the diffuse background at the same energies. All of these results, presented at Erice, will shortly be prepared for publication.

In the field of gamma-ray line astronomy, the object of a separate session at Erice, the contributions were both experimental and theoretical. Measurements reporting line emission at 511 keV from the central region of the Galaxy were presented [see e.g. ref. 35]. Such measurements were discussed in some length, also in the light of the recent history of celestial gamma-ray line detection. The 511 keV line was considered as a confirmed result, definitely credible at a confidence level of 5 to 6 " $\sigma$ ". Other lines had also been detected at a lower confidence level, but the credibility of this result is reduced greatly by the fact that the same data contain also "negative" lines, or intensity dips, of comparable significance.

The problem of the line emission from Cen A, reported in the literature a few years ago [ref. 36] was also discussed. Most recent data [see ref. 37] do not show convincingly the emission, at least at the intensity level previously reported, so that the well-known variability of NGC 5128 would have to be invoked to reconcile the different results.

The unusual phenomenon of a gamma-ray line flare coming from the general direction of the galactic anticentre was also reported [see also ref. 38]. The very interesting results was definitely shown to be not of instrumental origin, and remains a challenge for any straight-forward explanations, also because the arrival direction is poorly defined, with angular dimensions of the order of tens of degrees.

The experimental results presented above the gamma-ray line astronomy were reviewed in their theoretical context, with an aim to assess their astrophysical importance, mostly within the high-energy phenomena of our Galaxy [see ref. 39 and 40], such as, e.g. supernova dynamics and nucleosyntesis. Also, the presence of gamma-ray lines in a transient event may imply a substantial gravitational redshift as found in compact collapsed objects.

The study of gamma-ray bursts is one branch of gamma-ray astronomy that has recently witnessed very major steps forward, also reported at Erice. One could in fact say that it is really after such results [see also refs 41 and 42] that the study of gamma-bursts is showing a definite astronomical character. The collection of data from the "Signe" network, involving the Russian space probes (with French experiments aboard) Venera 11 and 12, Prognoz 6 etc., and such U.S. vehicles as ISEE, the Pioner-Venus Orbiter, HELIOS-B, etc., has greatly improved the sensibility and coverage for data collection, as well as arrival-direction reconstitution quality. As a result, the 15 events collected in about six months show a definite concentration towards the central regions of the plane of the Galaxy ! [ref. 42]. Tentative discussions on the Log N-Log S relationship were also presented, since the number of events begins now to render such an exercise meaningful. However, the difficulty here is also due to the fact that one may be in the presence of more than simply one population of sources within (or without) the galactic disc.

The truly exceptional event was that observed on March 5, 1979 by Venera 11 and 12, independently [see ref. 43]. Firstly, the arrival direction of the event could in this case be pinpointed accurately to be in the direction of Dorado, on the same line-of-sight as the Large Magellanic Cloud. In spite of this, the event is not necessarily to be considered extragalactic, but still galactic (distance $\lesssim$100 pc, luminosity $10^{37}$ erg $s^{-1}$ ) [again, see ref. 43]. The time structure of the event, clearly "periodical", i.e. featuring a train of pulses with period 8 s and total duration $\gtrsim$ 1 min, is also an absolute peculiarity, and the (very tentative) idea is put forward of a link between "gamma-ray bursts" and "X-ray bursters".

## WHAT TO DO NEXT ? (OR, CONCLUSION)

Such was the stimulating title of an Erice evening session, during which the participants tried to compress in few hours a discussion on their ideas (and hopes) for the future. Of course, the starting point was provided by the new experiments and missions in gamma-ray astronomy, now being planned or developed. In fact, a poster session had been especially devoted to this topic of experimental gamma-ray astronomy, with active participation of most of the groups involved. The experiments presented were both for future balloon and satellite projects, with emphasis on the payload of the NASA GRO mission. The participating scientists considered the merits of such a mission, with its complete payload covering homogeneously a vast energy range. The final unanimous recomendation from both the gamma-ray astronomers and the other scientists present was for a quick start of the GRO project, both because of its high scientific content and because of its right timing, with the gamma-ray astronomy community well trained and primed to go after the success of the SAS-2 and COS-B missions. However, as a complement to NASA's GRO, the participants considered with great interest the Franco-Russian mission GAMMA-1, presented at Erice by the French groups involved. This is a mission already under development with a launch scheduled in 1982 aboard a Russian spacecraft, with a high-energy instrument which may have a very significantly improved angular resolution; thanks to its novelty of design, and thus achieve important results on the localized sources of gamma radiation. The meeting considered suggestions for the viewing programme of such an experiment, and of course the various COS-B sources along the galactic disc plus strong extragalactic X-ray sources stood out as prime candidates for high-resolution studies with GAMMA-1.
After the presentation of the established (or last defined) missions for the future, a general discussion followed, yielding some guidelines for future work: experimentally, it was stressed that it is important and scientifically worthwhile to pursue the search for correlations between high-energy gamma-ray sources and sources at other wavelengths. These already include the HEAO-2 (Einstein Observatory) X-rays and the radio observations of various stations for pulsar searches, but it was recongnized that the recent CO data (in the millimetric range) and far infrared ( $\sim$100 $\mu$m) surveys also deserved attention as some degree of correlation is present with the gamma-ray data.

The results of gamma-ray line astronony, as well as of low-energy gamma-ray astronomy, point to the need for bigger and better instruments, before detailed astronomical correlations can be tried. On the other hand, the recent results obtained on gamma-ray bursts have produced a dramatic quality jump, with the beginning of detailed astronomy becoming apparent, as mentioned above. Here, for the first time, correlations with high-precision X-ray measurements are being tried as in the case of the Dorado event. It is obvious that in the case of gamma-ray bursts, as opposed to the case of high-energy gamma-ray astronomy where long integration times are needed, istantaneous correlated measurements at other wavelengths should be pursued, including radio observations.

From the point of view of the theoretical investigation, it has become apparent that gamma-ray astronomy, in its various aspects, has now advanced experimentally so rapidly that satisfactory explanations of most of the data are still missing or incomplete. The emission of high-energy gamma-radiation from the Galaxy is now characterized by localized (not necessarily point) sources: in most cases these are UGO's which are here to stay but on which we know very little. Of course some general processes and models (see above) are available and may actually be at work; however, the case of pulsars is indicative: while general models of emission are available for a rotating neutron star, the recent COS-B results have shown that plenty of basic work has to be redone, e.g. in the direction of the period-luminosity relationship, and so far only a geometrical explanation of the shape of the light curve has been succesfully achieved.
The theoretical investigation on the other aspects of gamma-ray astronomy is also at present meeting with difficulties, with limited success being claimed in areas like extragalactic sources and background, where one should honestly recognize that the experimental constraints are still quite loose.

It would certainly be very difficult to draw general conclusions from the Erice '79 Study Conference. The participants at the "What to do next?" session had the same feeling, in the end. What certainly has to be considered valuable are the guidelines for the work, experimental and theoretical, that emerged from the common discussion. Also it was found very positive that gamma-ray astronomy is now established within the astronomical community as are channels relating unique information, the importance of which is much improved by correlation with sister branches, once, with some effort, the barrier posed by different "languages" is overcome.

REFERENCES

1. G.F. Bignami, G. Boella, J.J. Burger, P. Keirle, H.A. Mayer-Hasselwander, J.A. Paul, E. Pfeffermann, L. Scarsi, B.N. Swanenburg, B.G. Taylor, W. Voges, R.D. Wills, Space Science Instrumentation 1, 245 (1975)

2. L. Scarsi, K. Bennett, G.F. Bignami, G. Boella, R. Buccheri, W. Hermsen, L. Koch, H.A. Mayer-Hasselwander, J.A. Paul, E. Pfeffermann, R. Stiglitz, B.N. Swanenburg, B.G. Taylor, R.D. Wills, Proc. of the 12th ESLAB Symposium ESA SP-124, 3 (1977)

3. W.L. Kraushaar, G.W. Clark, G. Garmire, H. Helmken, P. Higbie, M. Agogino, Astrophys. J. 141, 845 (1965)

4. W.L. Kraushaar, G.W. Clark, G.P. Garmire, R. Borken, P. Higbie, C. Leong, T. Thorson, Astrophys. J. 177, 341 (1972)

5. C.E. Fichtel, R.C. Hartman, D.A. Kniffen, D.J. Thompson, G.F. Bignami, H.B. Ogelman, M.E. Ozel, T. Tumer, Astrophys. J. 198, 163 (1975)

6. H.A. Mayer-Hasselwander, K. Bennett, G.F. Bignami, R. Buccheri, N. D'Amico, W. Hermsen, G. Kanbach, F. Lebrun, G.G. Lichti, J.L. Masnou, J.A. Paul, K. Pinkau, L. Scarsi, B.N. Swanenburg, R.D. Wills "Proc. IX Texas Symposium on Relativistic Astrophysics, Munich", in press (1979)

7. R.D. Wills, K. Bennett, R. Buccheri, P. Caraveo, N. D'Amico, W. Hermsen, G. Kanbach, G.G. Lichti, J.L. Masnou, H.A Mayer-Hasselwander, J.A. Paul, B. Sacco, B.N. Swanenburg - These Proceedings -

8. W. Hermsen, B.N. Swanenburg, G.F. Bignami, G. Boella, R. Buccheri, L. Scarsi, G. Kanbach, H.A. Mayer-Hasselwander, J.L. Masnou, J.A. Paul, K. Bennett, J.C. Higdon, G.G. Lichti, B.G. Taylor, R.D. Wills, Nature 269, 494 (1977)

9. B.N. Swanenburg, K. Bennett, G.F. Bignami, P. Caraveo, W. Hermsen, G. Kanbach, J.L. Masnou, H.A. Mayer-Hasselwander, J.A. Paul, B. Sacco, L. Scarsi, R.D. Wills, Nature 275, 298 (1978)

10. R. Buccheri, K. Bennett, G.F. Bignami, N. D'Amico, W. Hermsen, D.J. H. Huizing G. Kanbach, G.G. Lichti, J.L. Masnou, H.A. Mayer-Hasselwander, J.A. Paul, B. Sacco, B.N. Swanenburg, R.D. Wills, Astron. and Astrophys., in press (1979)

11. G.G. Lichti, G.F. Bignami, R. Buccheri, P. Caraveo, N. D'Amico, W. Hermsen, G. Kanbach, J.L. Masnou, H.A. Mayer-Hasselwander, J.A. Paul, B. Sacco, B.N. Swanenburg, R.D. WIlls, This volume

12. K.M.V. Apparao, G.F. Bignami, L. Maraschi, H. Helmken, B. Margon, R. Hyeling, H.V. Bradt, R.G. Dower, Nature 273, 450 (1978)

13. G.F. Bignami, P. Caraveo, L. Maraschi, Astron. and Astrophys. 67, 149 (1978)

14. R.J. Protheroe, A.W. Strong, A.W. Wolfendale, P. Kiraly Nature 277, 542 (1979)

15. R. Rothenflug, P. Caraveo, Astron. and Astrophys., in press (1979)

16. F. Lebrun, J.A. Paul, Astron. and Astrophys.65, 187 (1978)

17. F. Lebrun, Astron. and Astrophys., in press (1979)

18. C.E. Fichtel, G.A. Simpson, D.J. Thompson, Astrophys. J. 222, 833 (1978)

19. G. Di Cocco, G. Boella, F. Perotti, R. Stiglitz, G. Villa, R. Baker, R.C. Butler, A.J. Dean, S.J. Martin, D. Ramsden, Nature 270, 319 (1977)

20. V. Schönfelder, Nature 274, 344 (1978)

21. F. Perotti, A. Della Ventura, G. Sechi, G. Villa, G. Di Cocco, R.E. Baker, R.C. Butler, A.J. Dean, S.J. Martin, D. Ramsden, submitted to Nature, (1979)

22. F.A. Primini, B.A. Cooke, C.A. Dobson, S.K. Howe, A. Scheepmaker, W.A. Wheaton, W.H.G. Lewin, W.A. Baity, D.E. Gruber, J.L. Matteson, L.E. Peterson, Nature 278, 234 (1979)

23. E. Massaro, M. Salvati, R. Buccheri, submitted to M.N.R.A.S. (1979)

24. T. Montmerle, Astrophys. J., in press (1979)

25. H. Cong, NASA Technical Memorandum 79590 (1978)

26. R.J. Protheroe, A.W. Strong, A.W. Wolfendole, M.N.R.A.S. in press (1979)

27. H.J. Black, G.G. Fazio, Astrophys. J. 195, L23 (1973)

28. G.F. Bignami, G.E. Morfill, submitted to Astron. and Astrophys. (1979)

29. M. Cassé, J.A. Paul, Proc. of the 16th Cosmic Ray Conference Kyoto (1979)

30. M.A. Forman, G.F. Bignami, G.E. Morfill, H.J. Völk Proc. of the 16th Cosmic Ray Conference Kyoto (1979)

31. K. Pinkau, Proc. of the 16th Cosmic Ray Conference Kyoto (1979)

32. B.N. Swanenburg, Astron. and Astrophys. 70, L71 (1978)

33. R. Schlickeiser, K.O. Tielheim, Nature, in press (1979)

34. G.F. Bignami, C.E. Fichtel, R.C. Hartman, D.JH. Thompson, Astrophys. J. in press (1979)

35. M. Leventhal, C.J. Mac Callum, P.D. Stang, Astrophys. J. 225, L11 (1978)

36. R.D. Hall, C.A. Meegan, G.D. Walraven, F.T. Djuth, R.C. Haymes, Astrophys.J. 210, 631 (1976)

37. M. Leventhal, private communication

38. A.S. Jacobson, J.C. Ling, W.A. Mahoney, J.B. Willett, NASA TM 79619 eds. T.L. Cline and R. Ramaty (1978)

39. R. Ramaty, R.E. Lingenfelter, Nature 278, 127 (1979)

40. R. Ramaty, NASA TM 99619, eds. T.L. Cline and R. Ramaty (1978)

41. T.L. Cline, U.D. Desai, G. Pizzichini, A. Spizzichino, J.H. Trainor, R.W. Klebesadel and H. Helmken NASA preprint (1979)

42. E.P. Mazets, S.V. Golenetzkii, V.N. Il'inskii, V.N. Panov, R.L. Aptekar, Yu.A. Gur'jan, I.A. Sokolov Z. Ya. Sokolova, T.V. Kharitanova Lebedev Institute Preprint N° 610 (1979)

43. E.P. Mazets, S.V. Golenetzkii, V.N. Il'inskii, V.N. Panov, R.L. Aptekar, Yu.A. Gur'jan, I.A. Sokolov, Z. Ya. Sokolova, T.V. Kharitanova Lebedev Institute Preprint N° 617 (1979)

**SESSION 6**

# Techniques and Prospects

*Chairman:* **M. ODA**

# THE IHXR80 HARD X-RAY EXPERIMENT

P. Ubertini, L. Boccaccini, C. La Padula, M. Mastropietro, G. Medici, R. Patriarca and F. V. Polcaro

*Laboratorio di Astrofisica Spaziale, CNR, CP 67 - 00044, Frascati, Italy*

## ABSTRACT

Large area Multiwire Proportional Chambers for hard X-ray Astronomy have been succesfully employed in the HXR76 and HXR78 balloon borne experiments flown from the CNR Milo Base (Sicily) respectively on-board transatlantic and transmediterranean flights.
The detectors have shown competitive performance with respect to more conventional arrays of NaI (CsI) scintillator crystals in terms of effective area and background discrimination at lower cost and weight.
Lower efficiency at higher energies but better full-range spectral resolution have been obtained.

## INTRODUCTION

The employment of stratospheric balloons to lift scientific payloads to flotation altitude in the higher atmosphere (0.8 - 4 mbar), is a good tool to perform competitive experiments in the hard X-ray range [1]. In fact the possibility to carry heavy experimental arrays for long periods [2,3] allows the use of complexe payloads including stabilized platforms to point detectors with large sensitive areas. In spite of these advantages, severe problems have arisen. In particular the sensitivity of the experiments is drastically reduced by the high X- and $\gamma$-ray atmospheric background induced on the detectors and by the only partial transparency of the residual atmosphere, that absorbs low energy X-ray photons (up to 30-100 keV).

Starting from these considerations our group has developed a large-area high-pressure xenon-filled multiwire proportional chamber (MWPC) to be employed in long duration balloon flights. The scientific aim is the study of the hard X-ray emission from active nuclei of galaxies (Seyfert, BL-Lac, Radiogalaxies), QSO, galactic collapsed stars in binary systems, transient events, $\gamma$ bursts, etc.

In this paper we summarize the characteristics of the payload that will be employed in a hard X-ray experiment to be flown during summer 1980 from the base of Hyderabad (India), in the framework of an Italy- India collaboration.

## PAYLOAD DETAILS

The payload configuration is basically the same as that succesfully employed in a transmediterranean flight during summer 1978 from the Milo Base (Italy). It consists of two large area MWPCs, mounted with the same axis, on a two-stage pointing system driven by a magnetic attitude sensor and on-board prestored scanning program. The experimental arrangement is shown in Figure 1 during test and calibration.

Fig. 1 The payload arrangement. The steering system and the two detectors are clearly shown.

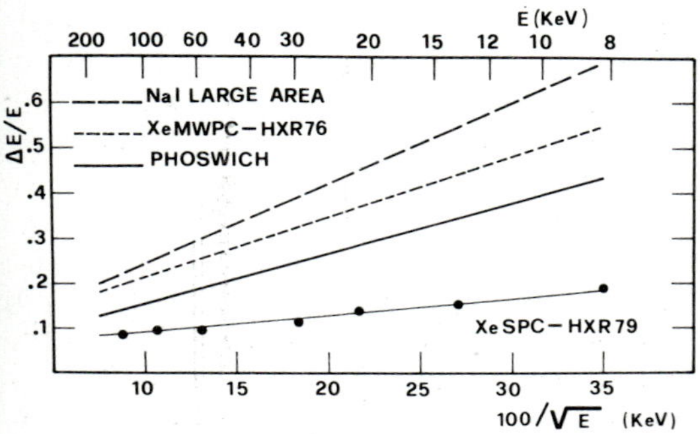


Fig. 2 Spectral-resolution performance (FWHM) of the HXR79 - SPC compared with NaI scintillator, MWPC flown during 1976 and Phowsich detector.

### Detectors

The detectors to be employed in the IHXR80 experiment are two large-area spectroscopic proportional chambers (SPC). They are the last generation of MWPC developed in our Laboratory, obtained by improvement of the detector succesfully employed in the HXR76 [4] and HXR78 [5] experiments extensively described elsewhere [6] . Our main effort was finalized to obtain a very high-purity filling mixture in order to improve the overall performance of the detector. In particular the use of an "on-line" purification system maintains the poisonous oxygen impurity level inside the counter, at less than 0.1 ppm. So far we have obtained SPCs showing an increase of spectral resolution (at 30 keV) by a factor of three in respect to the MWPCs of the HXR76 payload. In particular the performance characteristics of the detectors to be flown on-board the HXR79 experiment, from the Milo Base next August, are shown in Figure 2. The data are compared with results from NaI scintillators [6] , and Phoswich detector (Scheepmaker, Private communication). Each chamber has a geometrical area of 30 x 30 $cm^2$, reduced to 800 $cm^2$ by the collimator, with sensitive volume of about $10^4$ $cm^3$. The internal configuration of the chamber is optimized to have different in-flight performances. In fact one of the chambers is divided into 15 independent cells with 35 x 60 $mm^2$ cross section, and the anode signals are independently preamplified, amplified and conditioned by the validation-logic electronics. In this way is performed a very high spectral resolution (up to 10% at 30 keV and up to 8% at 100 keV), and a low induced background counting rate. Unfortunately cells of this width give a poor geometrical efficiency (about 55%) as shown in Figure 3 and 4. The efficiency is basically independent of the anode high tension and of the pressure. The anodes are golded molybdenum wires 40 μm in diameter. In Figure 5 a,b is shown, for the same detector, the voltage output (obtained at chamber gain about 2000) normalized to fit the telemetry requirements, at 3 and 5 atmospheres. The gas mixture is xenon - isobutane (95%-5%). The typical figure in energy resolution versus anode high tension is shown in Figure 6 (E = 30 keV, P = 5 atm). In Figure 7 is shown the detector response to an $^{241}$Am calibration source. The 30 keV fluorescence line and the partially resolved 25 keV line are indicated. The second detector is finalized to obtain a higher efficiency by dividing the internal volume in 65 independent cells of 21 x 21 $mm^2$ cross section, electronically wired in groups of 4 anodes. In this way the geometrical efficiency increases to about 94%. The energy resolution achievable in this configuration is about 20% less than the one showed in Figure 2 due to the different anode gain. In both detectors the anticoincidence is obtained with a three-walls active shield around the sensitive volume, and with internal anticoincidence rejection. A rise-time discrimination system is under development and extrapolating results from earlier experiments [4,6] the expected residual background should be about 3 x $10^{-4}$ ct $cm^{-2}$

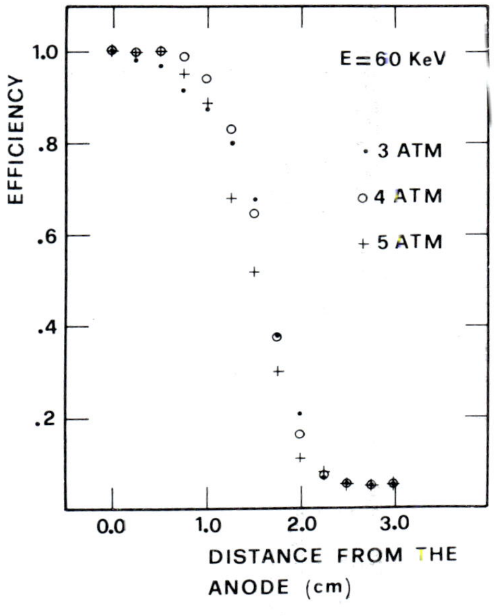


Fig. 3 Experimental detector efficiency at different pressure versus the distance normal to the wire. Cell width 35 mm x 60 mm.

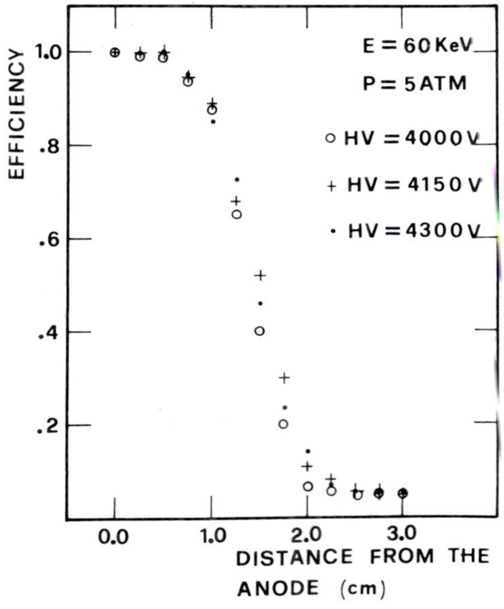


Fig. 4 Experimental detection efficiency at different anode high tension versus the distance normal to the wire. Cell width 35mm x 60mm.

$s^{-1}$ $keV^{-1}$, at 4 mb and Milo Base coordinates. The field of view of the detectors is limited by multilayer passive collimators (Sandwich walls: 0.1 mm Cu, 0.3 mm Pb; 0.1 mm Cu) with acceptance angle of 5° x 5° and 15° x 15° FWHM. The counter array is mounted on the central cage of the steering system on the gondola.

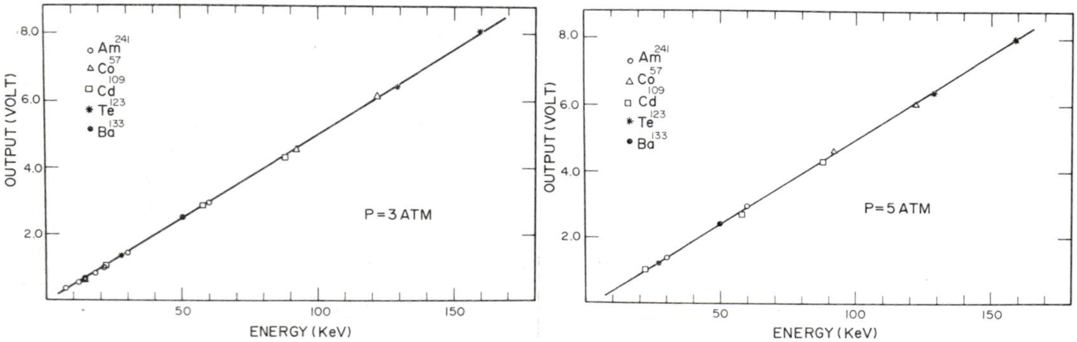


Fig. 5a,b Voltage output (normalized to fit telemetry requirements) obtained from the SPC 1 at different photon energies, and pressure.
a:P = 3 atm; b:P = 5atm.

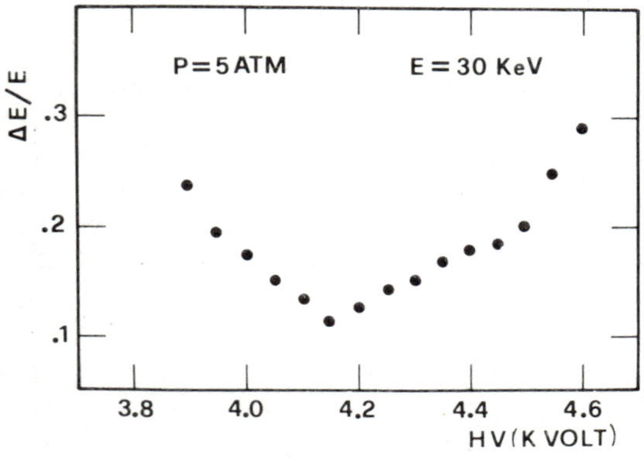


Fig. 6 Energy resolution versus anode high tension.

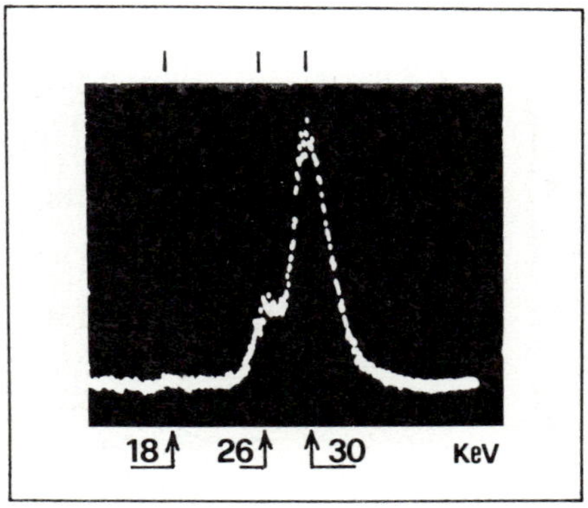


Fig. 7 SPC 1 spectral response at various photon energies. HV = 3450 V.

Pointing System

A two-stage gondola is employed to obtain source pointing and scanning, background measurements, and in-flight detector calibration with two $^{241}$Am sources. In particular a first stage is used to decouple the gondola from the rotation of the balloon. The payload is coupled with a cardan joint and an active bearing system controlled by a magnetometer and a torque motor that allow the complete gondola to point to magnetic north with a precision of about 0.5 deg. The total weight to be azimuthally stabilized including the loaded ballast-control system, power supply, telemetry etc, is about 700 kg. On the azimuth-controlled gondola is mounted a ground pre-programmed steering system to control the elevation and azimuth axes. On the central frame of this device are mounted the two detectors. The pointing error of the steering system was checked during a previous flight and is of the order of 6 arcmin rms for both platform axes.

## CONCLUSION

The use of SPCs with large sensitive area, high background discrimination, and high spectral resolution, at low cost and weight is a good chance to perform competitive experiments in the hard X-ray range. In particular we remark the high sensitivity to line features, recently observed in binary systems and transient sources [7,8].

## References

1. L. E. Peterson, R. M. Pelling and J. L. Matteson, Space Sci. Rev. 13, 320 (1972).
2. J. T. Delury, G. W. Lascombe, D. Ramsden, R. Baker, ESA SP-115, 265 (1976).
3. M. Malavasi, J. Medina, J. Sequeiros and A. Soubrier, ESA SP-135, 383 (1978).
4. G. Auriemma, L. Angeloni, B. M. Belli, A. Bernardi, D. Cardini, E. Costa, A. Emanuele, F. Giovannelli, and P. Ubertini, Ap. J. L7, 221 (1978).
5. C. La Padula, G. Auriemma, L. Boccaccini, C. Marchetti, M. Mastropietro, G. Medici, R. Patriarca, F. V. Polcaro and P. Ubertini, VII IFAC SYMPOSIUM, Oxford, 2-7 July, 1979, In Press.
6. E. Costa, G. Auriemma, L. Boccaccini, A. Emanuele, R. K. Manchanda, M. Mastropietro, G. Medici and P. Ubertini, Nucl. Inst. and Meth. 156, 57 (1978).
7. J. Trumper, W. Petsch, L. Reppin, W. Voges: Ap. J. 219, L 105 (1978).
8. S. H. Prando, E. A. Boldt, S. S. Holt, R. E. Rotschild and P. S. Seremitsos: Ap. J. 225, L 53 (1978).

# THE NASA GAMMA RAY OBSERVATORY (G.R.O.)

Ph. Durouchoux

*Section d'Astrophysique, Centre d'Etudes Nucléaires de Saclay, France*

## ABSTRACT

Many satellites or specific experiments onboard satellites have been devoted to gamma-ray astronomy. Interesting questions are to know what is scheduled for the future by NASA in gamma-ray astronomy, what are the main goals which are expected to be reached, and which progress have been accomplished in detector technologies.

The gamma-ray observatory (G.R.O.) is a project to carry gamma-ray astronomy forward in the 1980's. The G.R.O. is planned as a free-flying satellite to be launched from the space shuttle, carrying 5 large gamma-ray experiments aimed at providing data for a major advance in gamma-ray astronomy. These experiments, with their scientific objectives and sensitivities, are described in this paper.

## THE G.R.O. OBJECTIVES

The main objectives of the experiments onboard the satellite are the following :
. a study of the dynamic evolutionary forces in compact objects such as neutron stars and black holes as well as gamma-ray emitting objects whose nature is yet to be understood.
. a search for evidence of nucleosynthesis -the fundamental building process in nature- particularly in the environment of supernovae.
. the study of intense gamma-ray bursts of many types, whose origins remain a mystery.
. the exploration of our galaxy in the gamma-ray range, particularly with regard to regions difficult to observe at other wavelengths, the question of the origin and dynamic pressure effects of the cosmic rays, and structural features particularly related to high-energy particles.
. the study of the nature of other galaxies in the high-energy realm and especially the extraordinary ones such as radiogalaxies, Seyfert galaxies, BL Lacertae objects and quasars.
. the study of cosmological effects through the detailed examination of the diffuse radiation and the search for primordial black-hole emission.

## SPACECRAFT DESCRIPTION

The G.R.O. will be a large spacecraft weighing over 8 000 kg. It will be designed to accomodate 5 large instruments with gamma-ray detectors that require sustained pointing toward gamma-ray sources in space.

The planned operating life in orbit is 2 years. G.R.O. will be a 3-axis stabilized spacecraft with a circular orbit, at an altitude of about 400 km (orbital period is 96 min), and at an inclination of 28.5 degrees to the Equator.

## EXPERIMENTS SELECTED FOR G.R.O.

In the table 1 are summarized the different experiments selected by NASA for the observatory.

TABLE 1 Summary Of The Experiments In G.R.O.

| NAME | INSTITUTIONS (Co-I) | ENERGY | PURPOSE |
|---|---|---|---|
| GAMMA RAY EXPERIMENT<br>PI : J. Kurfess (N.R.L.) | . N.R.L.<br>. Northwest University | 0.1 - 10 Mev | . Detection of nuclear line radiation<br>. Observation of continuum emission |
| IMAGING COMPTON TELESCOPE "COMPTEL"<br>PI : V. Schönfelder (M.P.I.) | . M.P.I<br>. Univ.of Leiden<br>. Univ.of New Hampshire<br>. SSD,ESA | 1 - 30 Mev | Study of :<br>. Point sources<br>. Diffuse emission from the galaxy<br>. Cosmic diffuse flux<br>. Broadened Line Emission |
| GAMMA-RAY SPECTROSCOPY (GRSE)<br>PI : L. Peterson (U.C.S.D.) | . U.C.S.D.<br>. Saclay/Toulouse<br>. G.S.F.C.<br>. J.P.L | 0.03 - 10 Mev | Measurements of $\gamma$-Ray lines<br>. Intensities<br>. Time evolution<br>+ Continuum emission<br>. Discrete sources<br>. Galactic plane |
| GAMMA RAY TELESCOPE<br>PI : C. Fichtel (G.S.F.C.) | . G.S.F.C.<br>. Stanford Univ.<br>. M.P.I.<br>. Grumman Aerosp. corp. | 20 Mev-2 Gev | Study of high energy $\gamma$- ray sources<br>. Intensity<br>. Energy spectrum<br>. Position<br>. Time variation |
| TRANSIENT EVENT MONITOR<br>PI : G. Fishman (M.S.F.C.) | . M.S.F.C. | 60 - 600 Kev | . Detection of Transient events and Bursts<br>. Time measurements<br>. Location |

## THE GAMMA-RAY EXPERIMENT

NRL (J. Kurfess)
Northwestern University
Royal Naval College

### 1. Scientific objectives

The main objective of this experiment is to undertake sensitive, detailed observations of low-energy gamma-ray sources. Of particular interest is the unique information that can be derived from the detection of the nuclear line radiation which is expected from many

objects. Observation of this radiation will, among other things :
. provide observational evidence that nucleosynthesis occurs in supernovae
. determine whether novae are powered by thermonuclear explosions
. provide data on the internal structure of neutron stars and on the accretion process onto black holes and neutron stars
. help determine the origin of gamma-ray bursts
. infer the intensity of low-energy cosmic-rays and the density of matter throughout the Galaxy
. elucidate the nature of the solar-flare mechanism through observations of nuclear reactions in the flare region

In like manner, observations of continuum emissions, in addition to providing information on the above, will :

. help determine the origin of the recently discovered high-energy gamma-ray sources
. elucidate the emission mechanism in pulsars and X-ray bursters
. help determine the energy source in Seyfert galaxies and quasars

## 2. Detectors

The approach is illustrated in fig.1. Four identical high-sensitivity scintillation detectors are independently mounted on one-axis orientation systems. The detectors are optimized for line and continuum sensitivity in the 0.1 - 10 MeV region, but also provide a significant capability for continuum observations > 10 MeV. The offset pointing capability is used to provide the background observations that are essential to the success of any low-energy gamma-ray experiment. The gamma-ray detectors use NaI(Tl)-CsI(Na) phoswiches in an actively shielded configuration to maximize the sensitivity for the weight and volume available. An $\sim 9°$ field-of-view is defined by an active, drilled-hole CsI(Na) collimator. The pertinent details regarding the detectors are summarized below :

. type of detectors : 4 identical NaI-CsI phoswiches, actively shielded
. aperture area (total) : 1752 $cm^2$
. effective area : 1507 $cm^2$ (0.51 MeV) ; 566 $cm^2$ ( 4.43 MeV)
. field-of-view : 9° FWHM ( 0.51 MeV)
. energy range : 0.1 - 10 MeV gamma rays (primary objective) ; 10 - 50 MeV (secondary objective) ; > 10 MeV solar neutrons (secondary objective)
. energy resolution : 8.0% (0.661 MeV) ; 3.2% (4.43 MeV)
. time resolution : 8 s in normal mode ; 4 ms in pulsar or burst mode

## 3. Experiment sensitivities

The order of magnitude improvement in sensitivity that is achieved relative to current experiments is due in part to the large detector area (the effective area >1 MeV is at least a factor of 15 greater than the gamma-ray detector on HEAO-1) and the long observation times available (1-2 weeks). However, much of the improvement is due to the observational technique. For most observations, two detectors will be pointed at the source while the remaining two detectors will be offset from the source by -15° to acquire simultaneous background measurements. The source/background roles of the detectors are alternated on a time scale of $\sim 1$ minute using the orientation systems. With this technique the systematic problems associated with the time-dependent backgrounds produced by changes in the geomagnetic latitude and by radioactivity induced during South Atlantic Anomaly traversals are minimized. It is important to realize that the limiting sensitivities of the HEAO gamma-ray spectrometers, both of which incorporate wide field-of-view experiments on slowly-scanning satellites, are seriously compromised by the background uncertainties due to the inability to obtain source/background comparisons on a short time scale.

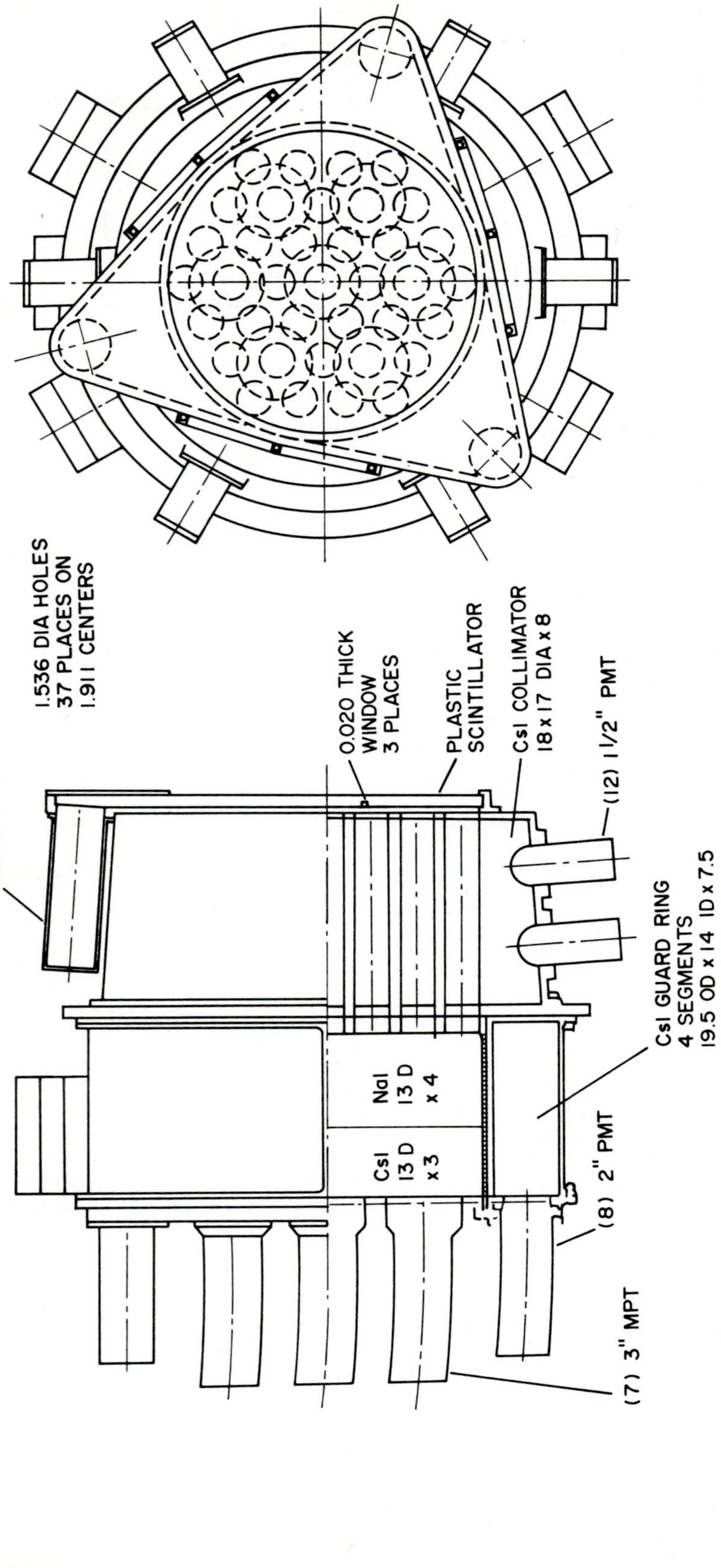


Fig.1 Schematic Of An Individual Gamma-Ray Detector

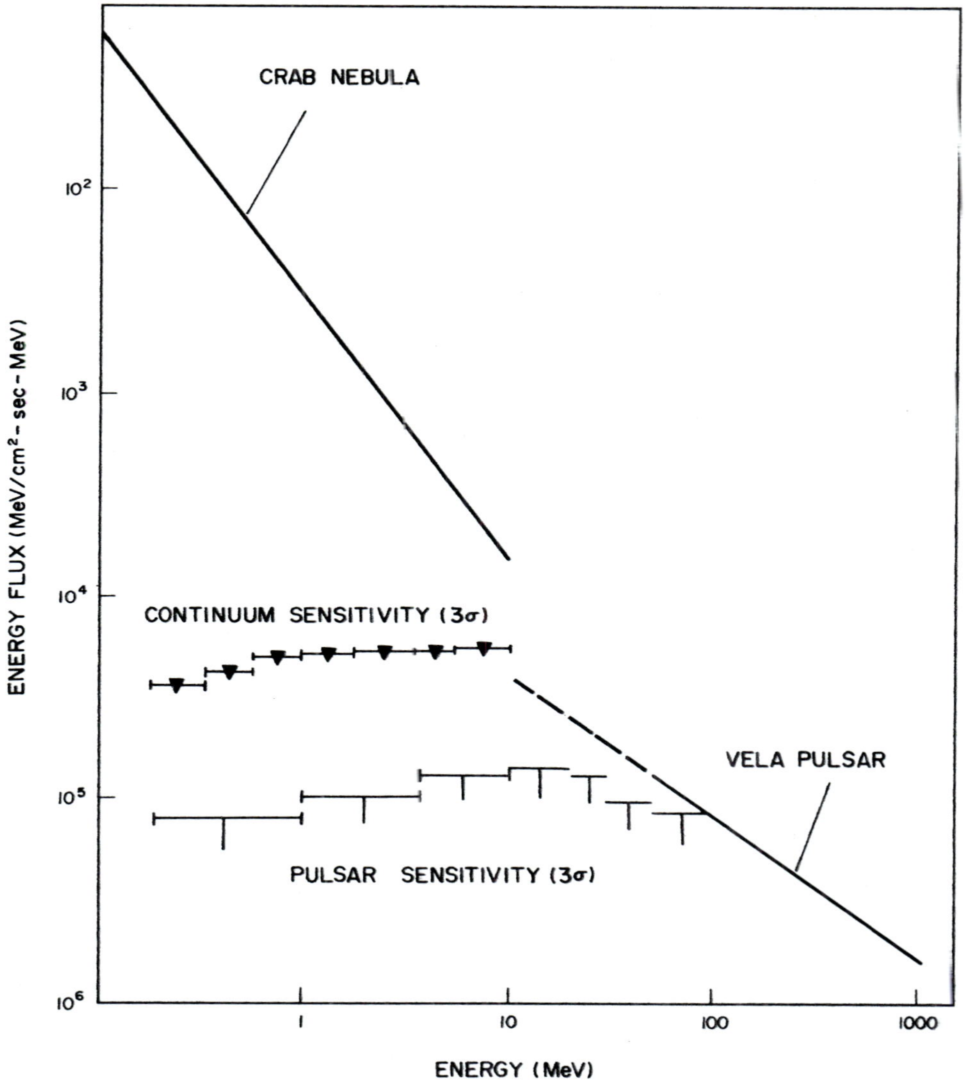


Fig. 2 Estimated Continuum Sensitivities Of The Tamma-Ray Experiment For $10^6$ s.

For sources which exhibit intrinsic variability on time scales of 1 minute or less (e.g., pulsars, burst sources and solar flares) the instrument sensitivity can be increased by pointing all four detectors at the source. This technique will provide maximum sensitivity to time variable phenomena.

Details regarding experiment sensitivities lines and continuum are summarized below and continuum and pulsar sensitivities versus energy are plotted in fig.2.

Sensitivities :
. 0.1-10 MeV line gamma rays ($10^6$ s) : $1.5 \times 10^{-5}$ gamma/$cm^2$-s
. 0.1-1 MeV continuum gamma rays : $5.10^{-3}$ Crab
1-10 MeV continuum gamma rays : $2.10^{-2}$ Crab
. gamma-ray burst : $5 \times 10^{-8}$ erg/$cm^2$
. solar flare line gamma rays ($10^3$ s flare) : $5 \times 10^{-4}$ gamma/$cm^2$-s
. solar flare neutrons ( $\geq$10 MeV) : $5 \times 10^{-3}$ n/$cm^2$-s

## IMAGING COMPTON TELESCOPE ("COMPTEL")

M.P.I. (V. Schönfelder)
University of Leiden
University of New Hampshire
SSD, ESA

Experience at all wavelengths in astronomy has shown that with a carefully designed low-background instrument, the scientific return is always high when exploring a new energy domain and there are many surprises in the data acquired. Recent examples of this are the SAS-2 and COS-B high-energy gamma-ray instruments which have discovered gamma-ray pulsars as well as a new class of gamma-ray objects.

Previous investigations in the 1 - 30 MeV gamma-ray region have been motivated by the knowledge that unique astrophysical phenomena may become observable in this range between X-ray energies and high-energy gamma rays. Many attempts have been made to measure the flux and spatial distribution of these intermediate-energy photons using balloon- and satellite-borne instrumentation. However, the results obtained to date have largely been conflicting and unconvincing compared with the high quality measurements made at energies above and below this range.

The basic difficulty is one of instrumental background which is present in all shielded and unshielded single-element detectors and even in earlier versions of the Compton telescope. Suppression of this instrumental background to an acceptable level is therefore the key to successful measurements in this energy range.

This proposal demonstrates how the concept of an imaging Compton telescope, brought to the state of the art level, can overcome these background difficulties and provide unprecedented sensitivity and spatial resolution in this energy range.

### 1. Scientific objectives

The energy domain from 1-30 MeV is one of the least well understood parts of the electromagnetic spectrum. Definitive observations in this energy range will help to interpret a range of astrophysical problems from those of compact stellar objects to the universe as a whole. In this energy range the gamma-ray production mechanisms undergo a transition from low-energy processes such as thermal emission to a variety of high-energy processes like inverse Compton radiation, $\pi^\circ$ radiation, nuclear radiation, etc. Studies in this transition energy range will therefore provide sensitive tests and constraints on the various production mechanisms.

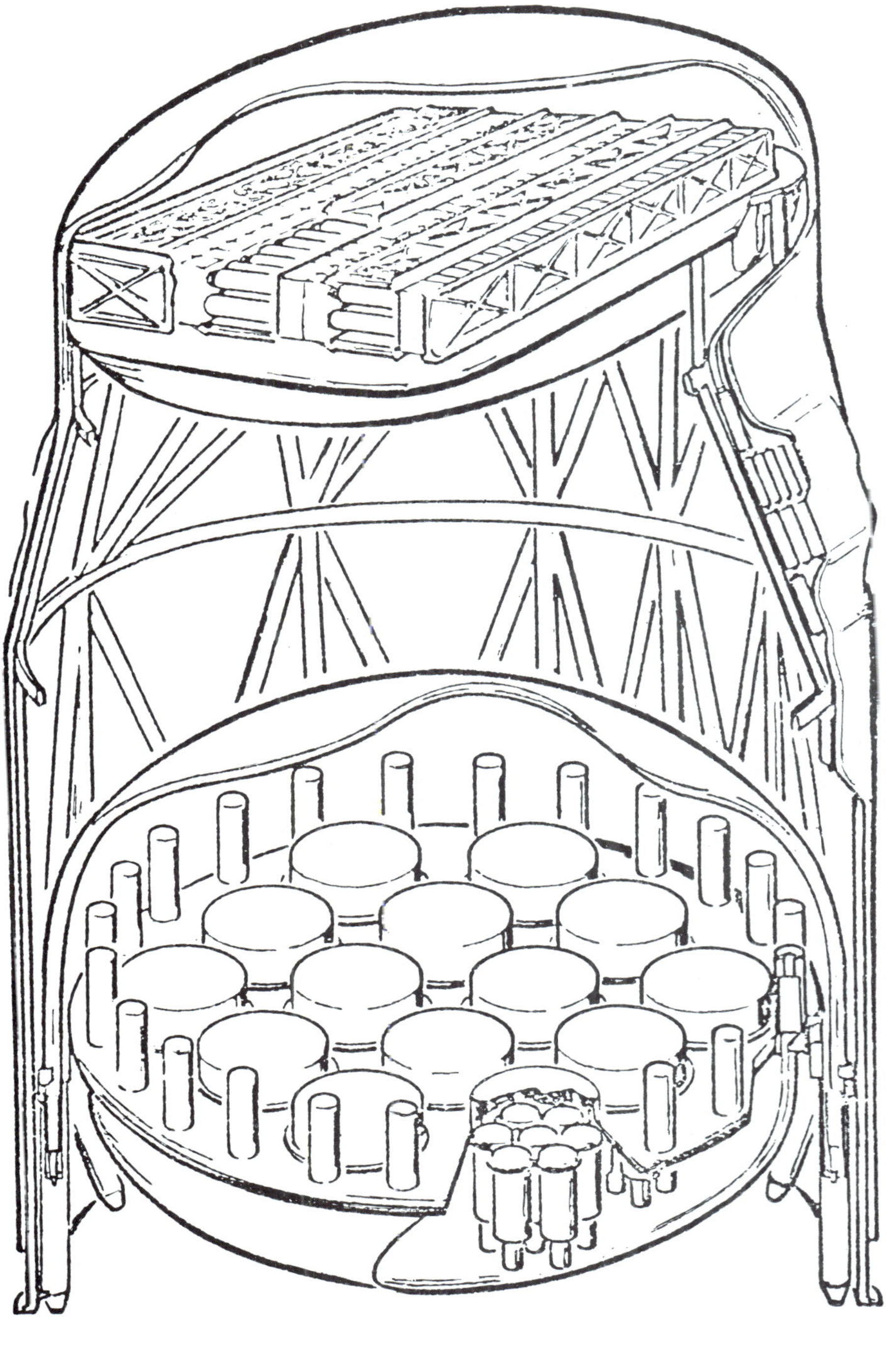

Fig. 3 A Schematic Drawing Of The COMPTEL

The imaging character, the broad effective aperture and the low background characteristics of the imaging Compton telescope are matched to the principal scientific objectives. These can be broadly grouped under the following headings :

. study of point sources
. study of diffuse emission from the Galaxy itself
. study of the cosmic diffuse flux
. study of broadened line emission

## 2. The telescope

A schematic drawing of the telescope is shown in fig.3. The telescope consists of two large-area scintillation detectors, an upper one in which the infalling gamma ray is Compton scattered, separated by 1.5 m from a lower one, in which the scattered gamma ray is absorbed. Both detectors are nearly entirely surrounded by a thin anticoincidence shield of plastic scintillator, which rejects charged particles. Between both detectors are four small plastic scintillation detectors containing weak $Co^{60}$ sources ; these are used as gamma-ray calibration sources which can be electronically gated.

Upper detector : this consists of 120 individual aluminum tanks of liquid scintillator NE 213 Each tank is of 6 x 6 $cm^2$ area and 15 cm depth and viewed by one 5 cm photomultiplier. The total geometrical area of the upper detector is 4200 $cm^2$.

Lower detector : this consists of 14 cylindrical blocks of polycrystaline NaI(TI) (Poliscin 1) of 7.5 cm thickness and 28 cm diameter, which are mounted on a groundplate within a diameter of about 1.50 m. Each NaI-block is seen from below by seven 7.5 cm multipliers. The total geometrical area of the lower detector is 8620 $cm^2$.

Anticoincidence shields : each shield consists of two 1.5 cm-thick domes of plastic scintillator NE 110. Each dome is viewed by 24 photomultipliers. A summary of the telescope parameters is given below :

| | |
|---|---|
| Energy range | 1-30 MeV |
| Sensitive area | 20-50 $cm^2$ |
| Angular resolution | 2.0°-6.0° (FWHM) |
| Energy resolution | ~10% FWHM |
| Geometrical factor | 5-30 $cm^2$ sr |
| Field of view | ~1 sr |

## 3. Experiment sensitivities

Using the derived telescope properties, the sensitivity for the detection of point sources and for extended features like the galactic plane, can be calculated in a straightforward manner. In the calculation it was estimated that for a fixed pointing of the spacecraft only 50% observation time remains as a result of earth occultation and passage through the radiation belts. Results are summarized in the table 2.

## GAMMA RAY SPECTROSCOPY EXPERIMENT (G.R.S.)

UCSD (L. Peterson)
CEA-SACLAY/CESR-TOULOUSE
GSFC
JPL

The objective of this gamma-ray astronomy experiment is to obtain a better understanding of nucleoynthesis, accretion onto dense objects, strong cosmic magnetic fields, the interaction of cosmic rays with astronomical material, and other high-energy astrophysical phenomena. The energy range of 0.03-10 MeV is covered with an instrument which considerably surpasses previous instruments in sensitivity, energy resolution, and angular resolution. The instrument is capable of detecting both line and continuum radiation from

TABLE 2 COMPTEL Telescope Sensitivity (5 $\sigma$ Confidence Level)

| Type of Observation | Point Sources $T=1.3\ 10^6$ s | Point Sources $T=10^3$ s | Galactic Survey $T=1.3\ 10^6$ s | Galactic Survey $T=1.3\ 10^6$ s |
|---|---|---|---|---|
| Objectives | Deep survey Spectra locations variability | Flare Phenomena | Integrated over $\Delta l^{II}=20°$ Detailed Spectra | Integrated over $\Delta l^{II}=5°$ Detailed Structure |
| Energy range (Mev) | $(cm^{-2}s^{-1}Mev^{-1})$ | $(cm^{-2}s^{-1}Mev^{-1})$ | $(cm^{-2}s^{-1}rad^{-1}Mev^{-1})$ | $(cm^{-2}s^{-1}rad^{-1}Mev^{-1})$ |
| CONTINUUM | | | | |
| 1-2 | $4.6\times10^{-5}$ | $2.1\times10^{-3}$ | $1.7\times10^{-4}$ | $6.1\times10^{-4}$ |
| 2-5 | $1.0\times10^{-5}$ | $4.5\times10^{-4}$ | $3.6\times10^{-5}$ | $1.3\times10^{-4}$ |
| 5-12 | $1.2\times10^{-6}$ | $9.6\times10^{-5}$ | $5.1\times10^{-6}$ | $1.8\times10^{-5}$ |
| 12-30 | $1.8\times10^{-7}$ | $5.4\times10^{-5}$ | $8.8\times10^{-7}$ | $3.1\times10^{-6}$ |
| LINES | $(cm^{-2}s^{-1})$ | $(cm^{-2}s^{-1})$ | $(cm^{-2}s^{-1}rad^{-1})$ | $(cm^{-2}s^{-1}rad^{-1})$ |
| 0.8 | $3.2\times10^{-5}$ | $3.3\times10^{-3}$ | $1.1\times10^{-4}$ | $4.1\times10^{-4}$ |
| 2.2 | $1.8\times10^{-5}$ | $1.0\times10^{-3}$ | $6.6\times10^{-4}$ | $2.4\times10^{-4}$ |
| 4.4 | $7.2\times10^{-6}$ | $6.5\times10^{-4}$ | $2.9\times10^{-5}$ | $1.0\times10^{-4}$ |
| 6.1 | $2.8\times10^{-6}$ | $5.0\times10^{-4}$ | $1.2\times10^{-5}$ | $4.0\times10^{-5}$ |

point, extended, and diffuse sources, whether their temporal behavior is steady, periodic, or transient. Energy-versus-time curves can be obtained for pulsars and gamma-ray bursts. The instrument also measures polarization of gamma rays.

## 1. Scientific objectives

Many scientific objectives may be obtained by observing cosmic gamma-ray lines and continua over the 0.03-10 MeV range. In a developing, but sensitivity-limited field such as MeV-range gamma-ray astronomy, attainable objectives and instrument concepts are intertwined. The observational objectives for the proposed GRO Gamma-Ray Spectroscopy Experiment are :

. measure the intensities and time evolution of gamma-ray lines from discrete galactic and extragalactic sources with a sensitivity of $\sim 10^{-5}$ photon $(cm^2\text{-}s)^{-1}$ and with an energy resolution $\Delta E/E \sim 10^{-3}$.

. measure the spectra and time variations of continuum emission from these and other discrete sources with a sensitivity ( $\Delta E/E \sim 1 \times 10^{-3}$ photon $(cm^2\text{-s-MeV})^{-1}$ at 0.1 MeV, 6 x $10^{-4}$ at 1 MeV, and 5 x $10^{-5}$ at 10 MeV.
. measure transient ( $\Delta t \leq 1$ s) line and continuum emission due to time-varying phenomena from objects such as X-ray and gamma-ray burst sources, pulsars, and other compact sources.
. determine the extent and luminosity of line and continuum emission from the galactic plane with a sensitivity of $\sim 2 \times 10^{-5}$ photon $(cm^2\text{-s-rad})^{-1}$ for lines and $\sim 1.2 \times 10^{-3}$ photon $(cm^2\text{-s-MeV-rad})^{-1}$ for continuum at 1 MeV with an angular resolution of ~3°.
. measure the line and continuum spectrum of the diffuse cosmic background.

These observations directly attack some of the outstanding problems in astrophysics today, such as acceleration of nucleons and electrons in cosmic plasmas, nucleosynthesis and supernova explosions, origin and propagation of cosmic rays in the galaxy, and the nature of compact stellar objects and galactic nuclei. The combination of line and continuum measurements can reveal such unique characteristics as the ratio of accelerated protons and electrons in non-thermal sources, or the proton and electron temperatures in compact thermal objects.

## 2. Instrument concept

The instrument is shown in figure 4. At the center of the instrument is an array of 18 germanium detectors which are cooled by a solid-cryogen refrigerator. The detector assembly is surrounded on all sides by an active shield. This sectioned 10 cm-thick CsI(Na) scintillator, viewed by photomultiplier tubes, provides anticoincidence signals for charged particles and gamma rays entering from all directions, except for the coarse-aperture openings imediately above each detector. The aperture shield, consisting of a thin plastic scintillator and photomultiplier tubes, completes the $4\pi$ sr charged-particle anticoincidence. Six of the coarse-aperture openings contain narrow-aperture passive low- energy collimators. Gamma rays enter through the rotating modulator which consists of 7 tiltable CsI(Na) slats in a rotating assembly. The rotational motion produces at each detector flux modulations which contain gamma-ray intensity and direction information. The modulations are deconvolved in ground data analysis and produce gamma-ray intensity maps while suppressing detector background effects.

A summary of the experiment parameters is given below :

| | |
|---|---|
| Energy range | 0.03 to 10 MeV |
| Energy resolution | 1.4 keV at 0.1 MeV ; 2.2 keV at 1 MeV |
| Total detector area | 600 $cm^2$ for18 coaxial detectors |
| Detection efficiency | 83% at 0.1 MeV ; 25% at 1 MeV |
| Active detector volume | 2800 $cm^3$ |
| Angular resolution | 3° |
| Position determination | 3° (threshold source); $<1°$ (for $>10\ \sigma$ detection) |
| Field of view | 30° full width at half maximum |
| Time resolution | 1 ms |

## 3. Experiment sensitivities

The instrument proposed for the Gamma-ray observatory is designed to have a threshold sensitivity a factor of 10 better than similar instruments on HEAO-1 and HEAO-C. This is obtained approximately by an increase in active volume by a factor 10, reduction in background by a factor of 2, and an increase in observing time by a factor of 5. The volume/area factor is obtained by a larger number (18) of cooled germanium detectors of larger volume (156 $cm^3$ each) ; and the lower background by a thicker anticoincidence shield (approximately 10 cm average of CsI(Na)) and a more favorable orbit than HEAO-C. The source observing time is increased by the combination of stationary pointing foreseen for the G.R.O. and the large field of view (approximately 30°) with a rotating modulator. Compared to a simple collimated aperture with 3° FWHM, the rotating modulator provides a

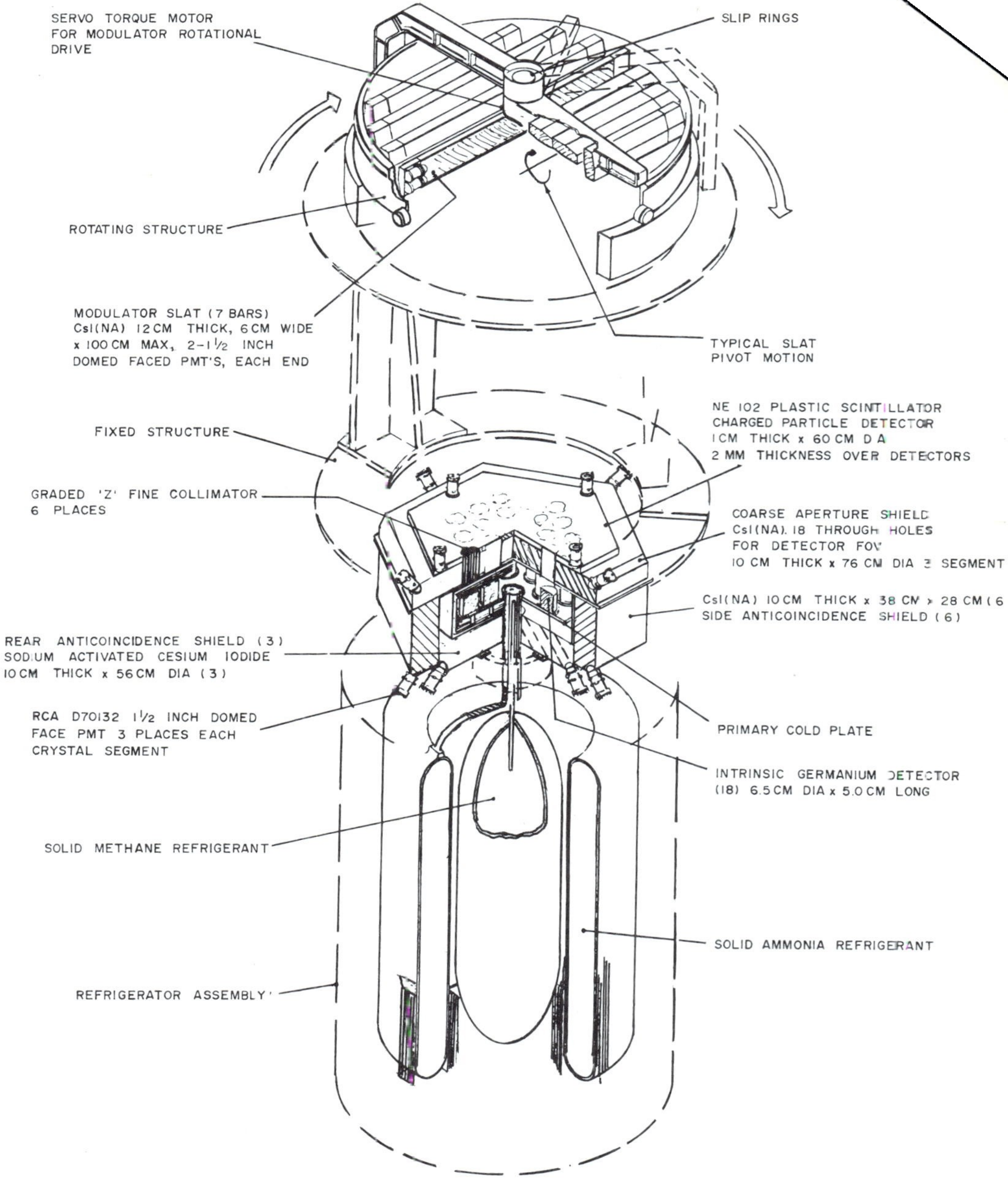


Fig. 4 Gamma Ray Spectroscopy Experiment

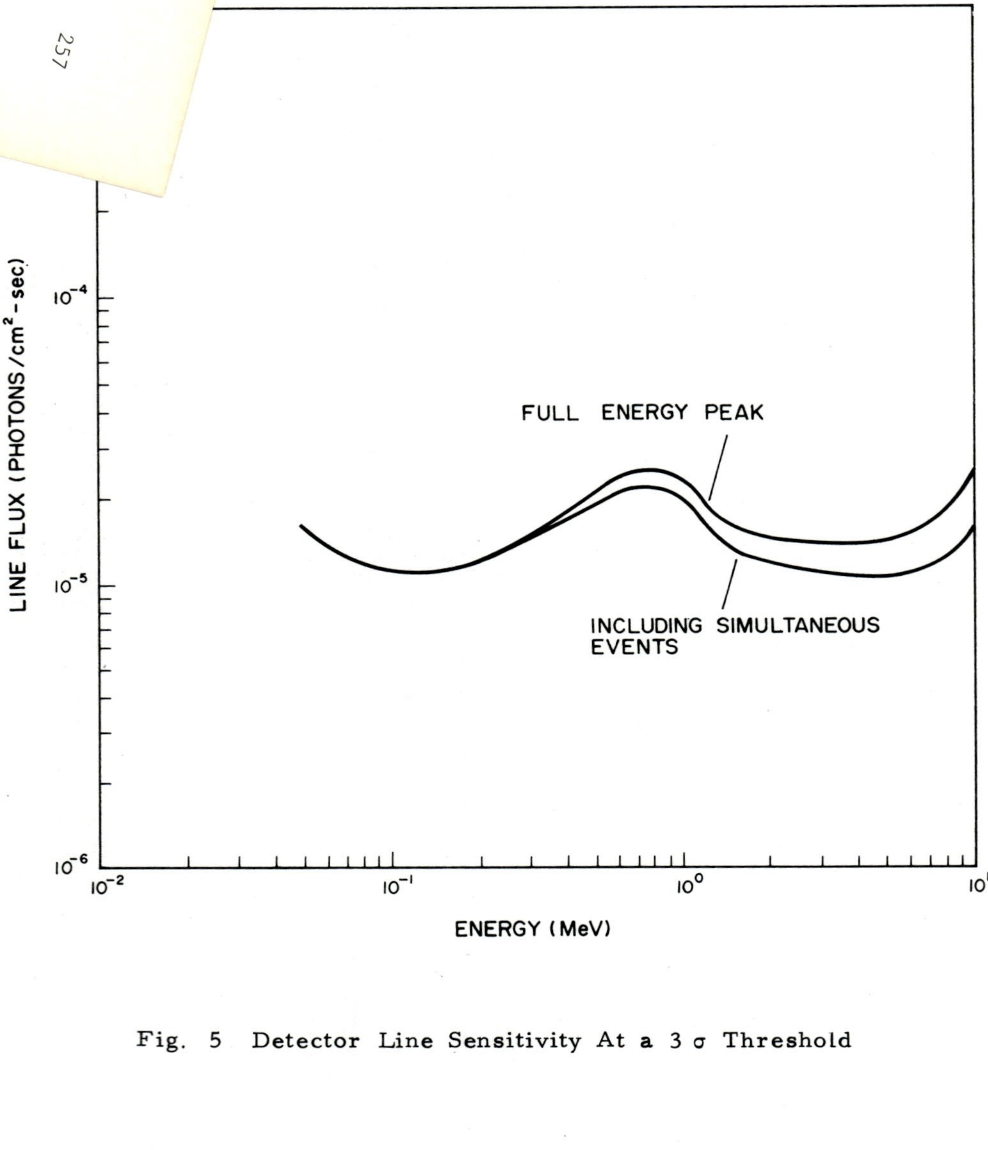


Fig. 5 Detector Line Sensitivity At a 3 $\sigma$ Threshold

"multiplex advantage" while maintaining the effective angular resolution of 3° inherent in the proposed instrument. For additional background reduction, and to eliminate source confusion at low energies ($E \leq 200$ keV), 6 of the detector apertures are additionally collimated to 6° FWHM with passive graded-Z collimators inserted into their shield aperture openings. The detector line sensitivity at a $3\sigma$ threshold for $10^6$ s of net observation is shown in fig. 5, whereas the threshold continuum sensitivity is $10^{-3}$ photon/cm$^2$ -s-MeV at 0.1 MeV and $6 \times 10^{-4}$ at 1 MeV.

## GAMMA RAY TELESCOPE

GSFC (C. Fichtel)
Stanford University
MPI
Grumman Aerospace Corporation

The broad objective of this experiment is to make a major advancement in high-energy (>20 MeV) gamma-ray astrophysics with the aid of a gamma-ray telescope with much larger sensitivity and much better angular and energy resolution than instruments previously flown. The study of the gamma-ray sky reveals the sites of the most energetic interactions occuring in astrophysics. These include the interactions of high-energy electrons with matter, photons, and magnetic fields ; high-energy nuclear interactions, which reveal concentrations of matter and cosmic rays ; matter-antimatter annihilation and nuclear and fundamental-particle interactions in the early universe. Because these interactions are generally associated with the dynamic, non-thermal processes in nature, gamma-ray astrophysics provides an excellent opportunity to learn about the forces of change occuring throughout the universe. In addition, since gamma-rays have a very low interaction cross section, they have a very high penetrating power and can reach the Earth from essentially any part of the Galaxy or Universe.

### 1. Scientific objectives

Gamma-ray astronomy can provide new information on a wide variety of subjects, including the distribution of cosmic rays in the Galaxy, galactic structure, supernova explosions, neutron stars, black holes, other galaxies, and the nature of the Universe.
Beginning with compact objects and proceeding to the Galaxy, other galaxies, and the Universe the specific objectives of this experiment are :

. to search for localized gamma-ray sources in the energy range from 20 MeV to 30 GeV and to measure the intensity, energy spectrum, position, and possible time variations of each.

. to improve the knowledge of the locations of known high energy gamma-ray sources not identified with objects seen at other frequencies.

. to examine supernova remnants and search for evidence of cosmic-ray particle acceleration revealed by high-energy gamma-rays, and, assuming there are cosmic rays, to study their expansion.

. to search for high-energy gamma-ray bursts, to study the spectra of low-energy gamma-ray bursts with high statistical accuracy, and to observe gamma-ray lines from solar flares.

. to obtain a detailed picture of the diffuse high-energy gamma-ray emission from the Galaxy which in turn will provide information of fundamental importance in the study of the dynamics of the Galaxy. This study should provide a high contrast picture of galactic structure, including spiral arm segments, clouds, the galactic center, and the extended disk (or "halo").

. to determine the relative importance of cosmic-ray electrons and cosmic-ray nucleons throughout the Galaxy by studying the gamma-ray spectrum above 20 MeV and, by combining the results with the continuum radio data, to obtain a better estimate of the galactic magnetic fields.

. to detect and examine the high-energy gamma-ray emission from other galaxies, including both normal galaxies and galaxies which are exceptionally luminous at other wavelengths.

. to study the diffuse celestial radiation in the energy range above 20 MeV including its energy spectrum and the degree of uniformity, both on a small scale and a broad scale. These measurements are of particular importance in relation to several cosmological models.

## 2. Instrument

The proposed telescope is shown in fig. 6. It represents an evolution of a technique that has been pursued over approximately the last fifteen years.The basic concept is that a gamma ray entering the telescope within the selected angle has a reasonable probability of converting into an electron pair in one of the thin plates between the spark chambers in the upper portion of the telescope. At least one of the electrons is detected by the directional time-of-flight coincidence system and triggers the track-imaging system which provides a picture of the gamma-ray event, and initiates the analysis of the energy signal from the NaI(TI) crystal. An event is recorded when a signal from the multi-element directional time-of-flight coincidence system indicates that a downward moving charged particle is detected in anticoincidence with a signal from the large scintillator surrounding the upper portion of the telescope system. In general, incident charged particles are rejected by the anticoincidence dome and low-energy backward-moving charged particles which do not reach the anticoincidence dome are rejected by the time-of-flight measurement. Events other than the desired gamma rays, such as those gamma rays interacting in the thin pressure vessel inside the anticoincidence scintillator, are rejected in the subsequent data analysis.

The directional telescope consists of two levels of a four-by-four scintillator array with selected elements of each array in a time-of-flight coincidence. The elements of the two scintillator planes to be placed in coincidence are selected to avoid wide angles that might have undesirable aspects, such as having other parts of the spacecraft in the field of view. The directional time-of-flight coincidence system determines that there are forward-moving charged particles in the telescope system within the desired solid angle. The anticoincidence dome extends over the top and sides of the gamma-ray conversion region and the upper scintillator array of the time-of-flight system, but not into the bottom region where secondaries from gamma rays might cause rejection of these desirable events. Experience with similar instruments as well as calculations have been used to determine the optimum position for the bottom of the anticoincidence dome.

As seen in fig. 6, the upper spark-chamber assembly consists of 28 spark-chamber modules interleaved with 27 0.02 radiation-length plates in which the gamma ray may convert into an electron pair and in which the initial direction of the electrons may be determined. The lower spark-chamber assembly, between the two time-of-flight scintillator planes, allows the electron trajectories to be followed, provides further information on the division of energy between the electrons, and permits seeing the separation of the two electrons for very high-energy gamma rays.

The energy of the gamma ray will usually be determined from measurements made in an eight radiation-length thick, 76 cm x 76 cm square NaI(TI) scintillator crystal below the lower time-of-flight scintillator plane. The energy resolution of the proposed experiment is about 15% (FWHM), which is sufficient to meet the objectives of the proposed investigation. In the lower energy range the energy information from the crystal may be supplemented by multiple-scattering observations in the spark chamber assembly. These measurements also furnish information on the division of energy between the secondary electrons, which is valuable in reconstructing the arrival direction of the primary gamma ray from the trajectories of the electrons. Moreover, for events where one secondary particle enters through the side of the telescope and does not enter the NaI(TI) crystal (evident from the spark chamber pictures), scattering measurements give an estimate of the energy not included in the crystal measurement.

## 3. Experiment sensitivities, field of view and angular resolution

For a discrete source near the pointing direction of the detector, the effective area of the detector is approximately the active area ($6.4 \times 10^3$ cm$^2$) multiplied by the detection efficiency for gamma rays of a given energy. The detection efficiency in turn is the product of the efficiency for pair production, the efficiency with which at least one of the particles can trigger the coincidence telescope, and the efficiency with which the gamma-ray interaction may be recognized from the resulting picture.

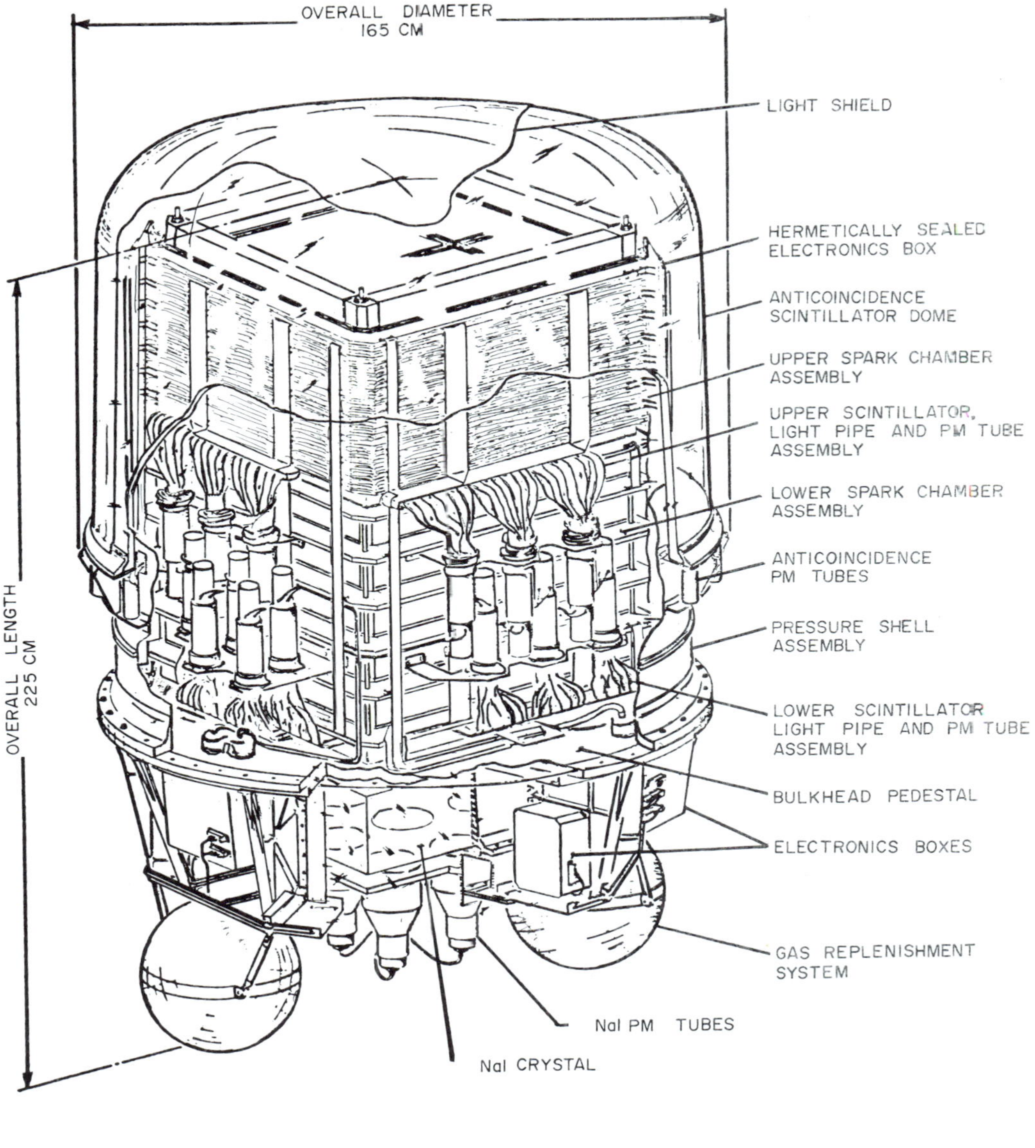


Fig. 6 View Of The High Energy Gamma Ray Telescope

In the normal wide-angle mode, the geometric solid angle of the experiment is 0.48 sr., and this solid angle is very nearly the effective solid angle at high energies.

The arrival direction of individual gamma rays is estimated from the information in the spark chamber picture. The single gamma-ray angular resolution is defined as the r.m.s. error projected on a plane (the average angle is 20% smaller and the three-dimensional r.m.s. angle is 40% larger). Calculations give values of the angular resolution of 1.6° at 100 MeV, 0.6° at 500 MeV, and 0.2° at 2000 MeV. It should be noted again that the pair production kinematics alone give angular errors of typically about 40% of these values. Angular resolution will be measured in detail as part of the calibration program. The ability of the instrument to determine the location of a point source or the shape of an extended source depends not only on the individual gamma-ray accuracy, but also on the source strength and energy spectrum, the instrument sensitivity, and the diffuse gamma-ray background against which the source appears. Assuming an accuracy of 2 arcmin in the spacecraft aspect, the positions of point sources will be measured to within 10 arcmin for most parts of the galactic plane, assuming source spectra similar to known sources. The positions of strong sources such as CG 195+4, or those well off the galactic plane, will be determined to better than 5 arcmin.

## TRANSIENT-EVENT MONITOR

MSFC (G. Fishman)

Among the recent dramatic surprises in high-energy astonomy have been the temporal variations on almost all time scales in X-ray and gamma-ray emission. Bursts, flares, and perhaps other as yet undiscovered transient phenomena could not be properly studied without a wide-field monitor to signal their presence. The discovery of gamma-ray bursts by small omnidirectional detectors, after numerous more sophisticated instruments were flown on spacecraft, testifies to the power of a wide-field detector for studying intense but infrequent events.

It is appropriate that a spacecraft named the Gamma-Ray Observatory should have as one of its major objectives a comprehensive study of gamma-ray bursts and other transient phenomena. Since the major fraction of the energy of gamma-ray bursts is near 100 keV, they are best detected and studied in that region, although other wavelength regions are expected to contribute to an understanding of them if identification of the sources is made. It is proposed with this experiment to monitor gamma-ray bursts and other transient phenomena with unprecedented sensitivity in the energy range of 60 keV to 600 keV

### 1. Scientific objectives

In summary, these are aimed at providing the Gamma Ray Observatory with a monitor of the entire unocculted sky for transient events and bursts. Specific objectives are the following :

. measure the frequency-versus-size distribution of gamma-ray bursts over a burst energy range of $6 \times 10^{-8}$ to $6 \times 10^{-4}$ erg/cm$^2$
. provide a single-station determination of burst directions within several degrees
. measure time variations of the intensity and spectra of strong bursts and other transient events on time scales down to $10^4$ s
. provide a near-Earth station for an interplanetary burst-timing network for strong bursts to derive arc-second directions
. continuously monitor stronger X-ray and gamma-ray sources such as Cyg X-1 for transient phenomena

### 2. Detectors

It is proposed to mount 12 independent detector modules on the faces of a dodecahedron, a 12-sided figure whose faces are pentagons. The dodecahedron, one of the five Platonic solids, has properties that are advantageous to the design of the Transient Event Monitor.

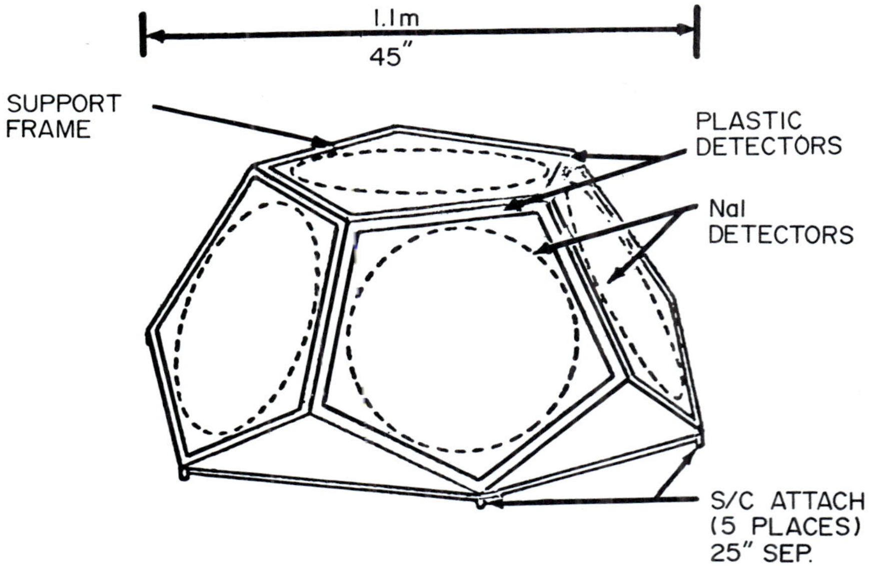


SIX-DETECTOR OPTION

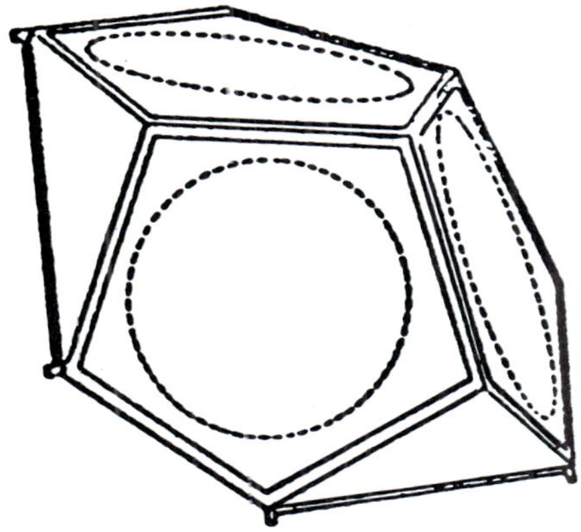

THREE-DETECTOR OPTION

Fig. 7 Three And Six Detector Options

For this figure, all edges and all vertices are identical, equalizing loads and perhaps simplifying construction. The faces of all adjacent detectors are separated by 63°26', providing uniform sensitivity and directional accuracies for deriving single-station burst directions. Two configurations are considered and the choice will be determined during the study phase :

Two six-detector arrays

If two half-dodecahedron arrays of detectors were located on opposite sides of the G.R.O. structure, a complete view of the unocculted sky would be possible regardless of spacecraft orientation. Each burst or transient would be viewed by at least six detectors. A drawing of the six detector array is shown in figure 7.

Four three detector arrays

If the mounting of two six detector arrays in satisfactory locations cannot be accomodated, an arrangement of four three detector arrays would satisfy the objectives. Figure 6 also shows such an arrangement, which is derived by removing three detectors from the six detector array.

The twelve 48 cm-diameter detectors will be fabricated from sodium iodide, NaI(Tl), which has a high scintillation efficiency, a large photoelectric cross section in the energy range of operation, and an acceptable scintillation decay time. The optimum thickness for the detector modules is a compromise between gamma-ray efficiency and background effects. A thickness of 1.27 cm (0.5 in., 4.7 g/cm$^2$) was chosen, corresponding to the 1/e range of 250 keV photons. The photopeak efficiency at 600 keV is approximately 10 percent. Since the nuclear interaction length for fast protons and neutrons is $\geq 40$ g/cm$^2$, the internal background due to radioactive spallation products and neutron effects is nearly linearly proportional to the thickness. For a thicker detector, the increased efficiency at higher energies would be offset by larger internal background rates and more frequent false triggers. Also, phosphorescence background is expected to increase with crystal thickness. Therefore, crystals much thicker than that proposed will have greatly reduced sensitivity to weak bursts.

The projected sensitive area in any direction is 5500 cm$^2$.

3. Sensitivity

The sensitivity to weak bursts will be limited by the background rate. The major sources of background include :

. earth albedo (atmospheric) gamma rays and neutrons
. spacecraft-generated gamma rays and neutrons from charged-particle interactions (cosmic rays and trapped protons)
. activitation of detector crystals
. diffuse background and discrete sources
. charged particle induced phosphorescence

The sensitivity for a 10 s gamma ray burst is $6 \times 10^{-8}$ erg cm$^{-2}$ whereas the sensitivity to longer lived transients is 1/3 Crab in 10 s assuming the same spectrum as the Crab. The high sensitivity of the proposed detector array will allow the measurement of the size distribution (log N - log S curve) over four decades in energy ($6 \times 10^{-8}$ to $6 \times 10^{-4}$ erg cm$^{-2}$).

Most of the matter of this paper comes from the proposals submitted to NASA by the different institutions (NRL, MPI, UCSD, GSFC and MSFC).

# THE INDIA/SOVIET GAMMA RAY ASTRONOMY PROGRAMME

S. V. Damle

*Tata Institute of Fundamental Research, Homi Bhabha Road, Bombay 400 005, India*

ABSTRACT

The Indo-Soviet Collaborative Programme in Gamma Ray Astronomy (GRISP) was launched in 1976 under an agreement between Indian Space Research Organisation (ISRO, Govt. of India) and INTERCOSMOS, Academy of Sciences of USSR. The programme envisaged in the first instance investigation from balloon platforms, with the possibility of continuing collaborative work from satellite-borne gamma ray telescopes.

A series of balloon flights for the Anna-6 and Natalya-I oriented gamma ray telescopes were conducted in India by the National Balloon Facility of the Tata Institute of Fundamental Research, from Hyderabad during 1977 autumn and 1979 spring seasons. This paper reviews the status of the GRISP programme.

## INTRODUCTION

Gamma Ray studies yield unique new information directly connected to the nuclear component of cosmic rays far away from the earth. Such information unobtainable by optical, radio or x-ray astronomy studies is important for high energy astrophysics including the problem of the origin of cosmic rays. Because of low fluxes and high background, gamma ray investigations are however difficult to make. However, thanks to SAS-II and COS-B investigations, (Fichtel et al [1], Bennet et al [2] and Pinkau [3] and references therein), Gamma Ray Astronomy has been now established as an observational science. Although the best platform for studies in Gamma Ray Astronomy, is clearly the satellite platform, balloon-borne investigation still has some role to play especially in studies of specific point sources and for proving the future satellite-borne instruments. With these aspects in mind, a joint Indo-Soviet Gamma Ray Astronomy Programme (GRISP), first using balloon platforms and later satellite platforms, was initiated in 1976 under agreement between Indian Space Research Organisation (ISRO, Govt. of India) and INTERCOSMOS, USSR Academy of Sciences. This programme is part of a general collaboration in space sciences between India and USSR.

The balloon flights were planned to be launched from Hyderabad, India ($17^{o}$N) to take specific advantage of the location near geomagnetic equator (vertical cut-off rigidity $\simeq 16.8$ GV) where the atmospheric gamma ray background is much lower than that experienced at higher latitudes.

Three research laboratories - one from India and two from Soviet Union - are participating in this programme. They are : Tata Institute of Fundamental Research (TIFR), Bombay, India; Moscow Physical Engineering Institute, Moscow, USSR; and P.N. Lebedev Institute of Physics, Moscow, USSR. The Soviet side was responsible for constructing two Gamma Ray Telescopes - Anna-6 using optical spark chambers, and Natalya-I using digital wire spark chambers; the Indian side was responsible for programmable azimuth orientation platforms, PCM/FM telemetry for data and had the full responsibility for providing balloons and conducting the balloon flights.

The first phase of this programme, namely, the balloon-borne investigation for point sources and atmospheric background in 10-300 MeV range has been now partially completed; two balloon flights each for the Anna-6 and Natalya-I gamma ray telescopes mounted on azimuth orientation platforms have been conducted during 1977 autumn and 1979 spring, from Hyderabad. The additional flights for study of splash, reentrant and 'horizontal' charge particle albedo have also been made under this programme. The results from the first flight of the albedo experiment have already been reported at the XXIst COSPAR meeting [4]. We shall not deal with results of albedo experiment here. The GRISP collaboration is expected to continue in future with satellite-borne Gamma Ray Telescopes to be launched on USSR satellites sometime in early eighties.

This paper gives a review of the status of GRISP.

## MISSION OBJECTIVES AND TELESCOPE SYSTEMS

The principle scientific objectives of GRISP are to study galactic and extragalactic gamma ray point sources including pulsars (e.g. CRAB, CP1133, Gamma 195 + 5, CYG X-3, NGC4151, 3C273) and to study atmospheric and cosmic diffuse background in the energy range 10 MeV to 300 MeV. To cover this wide energy range,two different telescope systems were used in overlapping energy regions. Anna-6 gamma ray telescope comprising optical spark chambers covers the energy range 40 MeV to 300 MeV while the low mass Natalya-I telescope using digital wire spark chambers and wire proportional counters for triggering the telescope,covers the energy range 4 MeV to 100 MeV. Technical details of Anna-6 and Natalya-I gamma ray telescopes have been described earlier by Galper et al [5, 6].

Figure 1 shows schematically the Anna-6 telescope. The principal elements of this telescope are - (i) anti-coincidence plastic scintillation counters $A_1$ and $A_2$, (ii) coincidence plastic scintillation counters $C_1$ and $C_2$, (iii) plexiglass Cerenkov counter C, and (iv) spark chambers - SpCh1, SpCh2 and SpCh3. Counters $C_1$, $C_2$ and C act as triggering telescope for the system. The electronic scheme selects $\overline{A}_1\overline{A}_2C_1C_2C$ events occuring due to gamma ray conversion in steel electrodes of SpCh1 and in lead conversion plates $K_1$ and $K_2$; the total conversion thickness is 0.9 radiation length. The direction of the gamma ray is measured

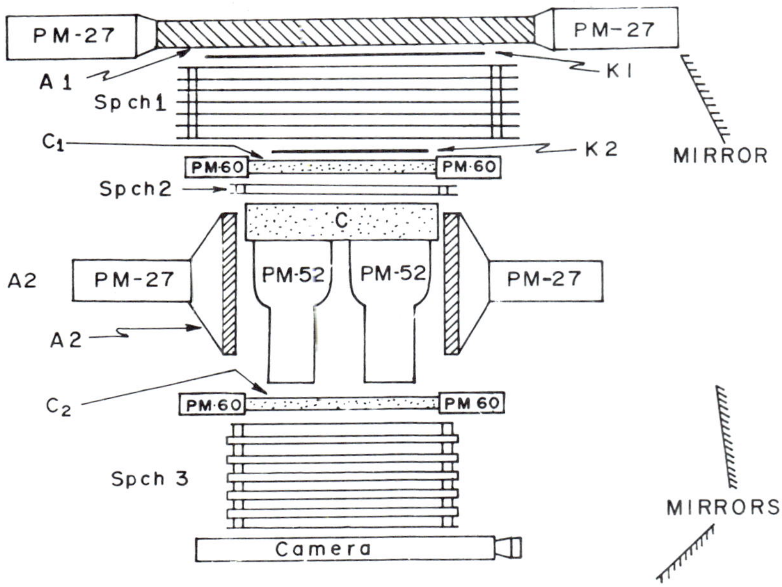


**Fig. 1. Schematic diagram of ANNA-6 telescope.**

using the electron/positron tracks in a six gap spark chamber SpCh1 and in the single gap chamber SpCh2. The spark chamber SpCh3 which has a total of 3.6 radiation lengths comprising of lead electrodes of the spark chamber serves for detecting cascade showers. The camera system records the stereo pictures of the spark chambers along with the charged particle rates of counter $A_1$ and coincident rate $C_1C_2C$ and housekeeping information on photographic film. The response of the telescope for gamma rays was calculated by a Monte Carlo method. The telescope has also been calibrated using a tagged gamma ray beam from an accelerator. The sensitive area of the telescope is 250 $cm^2$; the geomatric factor for isotopic flux is 115 $cm^2$ sterad. The energy range covered is 40 to 300 MeV. The angular resolution is about 3 degrees for energy (E) greater than 200 MeV. At balloon altitudes over Hyderabad, the minimum detectable flux at 5$\sigma$ level for energy greater than 40 MeV is $2 \times 10^{-5} cm^{-2} sec^{-1}$ for four hour exposure.

**Figure 2 shows a schematic diagram of the Natalya-I gamma ray telescope. The principle element of the telescope is a multiwire spark chamber with ferrite core readout memory. The main parts of the telescope are:**
**(i) 18 spark gaps (SG1 to SG18) sandwiched with lead converters; 18 spark gaps are grouped into 3 spark chambers (SC1, SC2 and SC3) each comprising of six spark gaps; the thickness of converters within spark chambers being 0.01,**

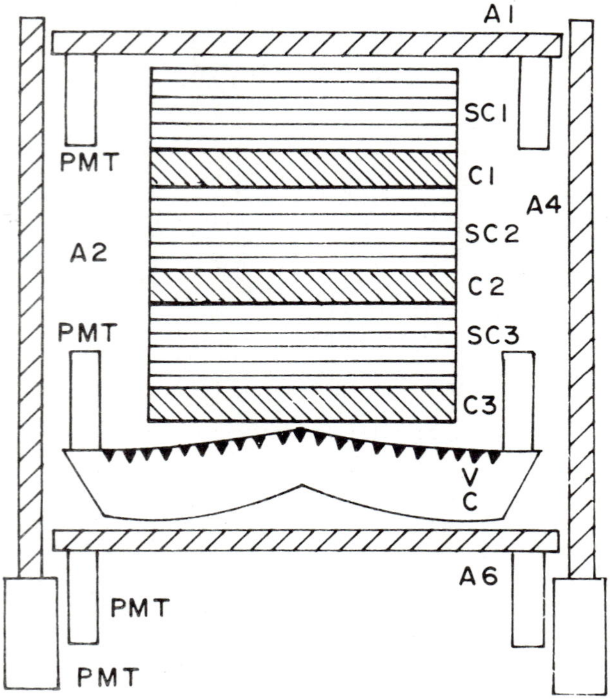


**Fig. 2. Schematic diagram of NATALYA-I telescope.**

0.01 and 0.02 radiation lengths respectively; (ii) the multiwire proportional counters C1, C2 and C3 which together act as the basic triggering telescope of the system; (iii) the plexiglass directional Cerenkov counter C which blocks particles entering the telescope from backward direction. The telescope is surrounded on all sides by plastic scintillation anticoincidence counters $A_1$ to $A_6$. Basic trigger of the telescope is $C_1 C_2 C_3$ with no pulse in anticoincidence counters $A_1$ to $A_6$. An important new feature of the telescope is the low mass triggering telescope consisting of wire proportional counters $C_1 C_2$ & $C_3$. Because of this feature the threshold to gamma radiation detection is considerably reduced. The minimum detectable gamma ray energy is about 5 MeV. The sensitive area of the telescope is 450 $cm^2$; the geometrical factor for isotropic flux is 250 $cm^2$ ster. The energy range is 4 to 100 MeV and the angular resolution is about 8 degrees at 10 MeV and about 2 degrees for energy greater than 25 MeV. At balloon altitudes over Hyderabad, the minimum detectable flux at 5$\sigma$ level for energy greater than 10 MeV is $10^{-5}cm^{-2}sec^{-1}$ for a four hour exposure.

The telescopes Anna-6 and Natalya-I are mounted on a platform whose azimuth orientation is programmable to track sources during the flight. The azimuth orientation system is a null seeking servo employing reaction wheel control with flux gate magnetometer as the basic element for sensing the position. The holding accuracy of the azimuth orientation platform is $\pm$ 0.2 degrees. The telescope aspect information, the gamma ray event timing information and other housekeeping information are telemetered to the groundstation on PCM/FM telemetry link. Two separate orientation platforms with telemetry subsystems were fabricated - one each for Anna-6 and Natalya-I gamma ray telescopes.

## BALLOON FLIGHTS

The first series of three balloon flights for GRISP were conducted during October/November 1977 from TIFR's National Balloon Facility, Hyderabad. The second series of three balloon flights for the two telescopes - Anna-6 and Natalya-I - were conducted during March/April 1979, from the same location.

During the 1977 balloon flight programme the Anna-6 flight was successful and we obtained results on the atmospheric background at balloon altitudes over Hyderabad, and were able to place an upper limit of $1 \times 10^{-5}$ $cm^{-2} sec^{-1}$ (for E greater than 100 MeV), for gamma ray emission from galactic pole regions. Preliminary results of this flight were reported to the XXIst COSPAR meeting [7]. The atmospheric background for E greater than 40 MeV was found to be about $1.8 \times 10^{-3}/cm^2$ ster sec. (This observation is in agreement with calculations by Daniel and Stephens [8]). The observed background rate is about 10 times lower than the one observed over northern latitudes; this confirms our earlier expectations that the gamma ray astronomy investigations from the equatorial latitude of Hyderabad would be very fruitful in improving the minimum detectable flux.

During the 1978 balloon flight programme, the principle objective was to study the reported 4.8 periodicity for the Cygnus X-3 source using Anna-6 telescope and to study the extragalactic source NGC4151 using the low mass Natalya-I telescope. We have about 6-1/2 hour exposure for Cyg X-3. The analysis of data is in progress.

## FUTURE PLANS

The first phase of GRISP, namely, the investigations using balloon flights from Hyderabad, is now more or less complete and has given useful data for finalising the design of the satellite-borne telescope systems. It has also provided interesting data for Cyg X-3. The second phase of the GRISP collaboration, namely, the satellite-borne experiment is expected to be taken up in the near future. Under this second phase we hope to plan a Natalya-I type low mass telescope to be launched on a Soviet satellite. An important feature as noted before of the Natalya-I telescope is its ability to extend the lower energy limit of the gamma ray telescope to about 4 MeV. We believe that such a system will give interesting results in the hitherto unexplored energy range of **4-40 MeV.**

ACKNOWLEDGEMENT

A collaborative programme like GRISP naturally involves cooperation from a number of organisations. We acknowledge the assistance rendered by various government and non-government agencies. Without the interest shown and cooperation rendered by ISRO and INTERCOSMOS this programme could not h been carried out. We gratefully acknowledge their assistance. We thank the TIFR Balloon Flight Group under Shri R.T. Redkar for conducting balloon flights.

REFERENCES

1. Fichtel C.E., Hartman R.C., Kniffen D.A., Thomson D.J., Bignami G.F., Ögelman H., Özel M.E. and Tumer J., Ap.J., 198, 163 (1975).

2. Bennet K., Bignami G.F., Buccheri R., Hermsen W., Kanbach G., Lebrun F., Mayer-Hasselwander H.A., Paul J.A., Piccinotti G., Scarsi L., Soroka F., Swanenburg B.N., and Wills R.D., Proc. of XIIth ESLAB Symp. ed. R.D. Wills and B. Battrick, ESA SP-124, 83 (1977).

3. Pinkau K., Nature, 277, 17 (1979).

4. Gorshkov V.S., Gusev V.N., Kruchkova T.N., Kurnosova L.V., Logachev V.I., Platanov G.B., Rosonorov L.A., Samoilenko V.T., Topochiev N.P., and Fradkin M.I. (USSR) and Damle S.V., Gokhale G.S., Kunte P.K. and Sreekantan B.V. (India), Paper E.4.17 Proc. XXIst COSPAR meeting, Innsbruck (1978) (published in Advances in Space Exploration Vol. 5, Pergamon Press London, 1979).

5. Galper A.M., Kirillov-Ugryumov V.G., Kurochkin A.V., Luchkov B.I. and Yurkin Yu. T., Proc. XIIIth Int. Conf. on Cosmic Rays, Denver, 1, 427 (1973).

6. Galper A.M., Grigoryev V.A., Kirillov-Ugryumov V.G., Kotov Yu.D., Kulakov Ye. V., Smirnov Yu.V. and Yurov V.N., Proc. XIVth Int. Conf. on Cosmic Rays, Munich, 9, 3143 (1975).

7. Galper A.M., Kirillov-Ugryumov V.G., Kotov Yu.D., Kurnosova L.V., Kurochkin A.V., Leikov N.G., Fradkin M.I. and Yurkin Yu.T. (USSR) and Damle S.V., Gokhale G.S., Kunte P.K. and Sreekantan B.V. (India), Paper E.4.16, Proc. XXIst COSPAR meeting, Innsbruck, (1978) (published in Advances in Space Exploration Vol. 5, Pergamon Press London, 1979).

8. Daniel R.R. and Stephens S.A., Rev. of Geophysics and Sp. Physics, 12, 233 (1974).

# BACKGROUND EFFECTS BETWEEN 250 km AND 1100 km

D. R. Parsignault

*Physics Research Division, Emmanuel College, Lexington, Massachusetts, U.S.A.*

## INTRODUCTION

As with all detectors used in remote sensing experiments, the instrument sensitivity in an X-ray or $\gamma$-ray observation is limited by the background noise. The main source of noise is background events induced by cosmic rays (mostly protons) which are incident on the top of the atmosphere at a rate of 0.2-0.5 $cm^{-2}$ $sec^{-1}$, depending on the geomagnetic latitude. The cosmic rays interact directly with the detector. They also interact with the satellite itself (structure, electronics) and with the earth atmosphere, producing $\gamma$ radiations. These $\gamma$ radiations in turn interact with the detector.

Another source of background for an instrument in orbit is the low-energy electrons (few tenths of eV to few MeV). These electrons, at least those with E $\gtrsim$ 50 keV, are present mostly in well-defined regions above the Earth, i.e., in the South Atlantic anomaly (SAA) and the auroral zones. Leakage of the trapped particals outside these regions occurs sporadically.

We present data on the cosmic-ray-induced and electron-induced background obtained with a pair of large-area, argon-filled, proportional counters (Referred to as LAD1 and LAD2), part of the Hard X-ray Experiment (HXX), flown on board the Astronomical Netherlands Satellite (ANS). These data can be used to estimate the expected total background counting rates (i.e. before rejection) in the energy range of 1.4-7.2 keV of any proposed proportional counter experiment in Earth orbit. To apply these results to other detector geometries, one simply multiplies the observed ANS background rates by the ratio of the total surface area of the proposed proportional counter to that of the ANS proportional counter.

## THE INSTRUMENTATION

ANS was launched on August 30, 1974, into a polar orbit. Due to a malfunction in the first stage of the Scout rocket, the satellite achieved an elliptical orbit, instead of the anticipated 550 km circular orbit, with a perigee $\sim$250 km and an apogee $\sim$1100 km. The point of perigee precessed at a rate of about 3° per day. During the year and one-half of operation, the radiation environment above the Earth between these two altitude limits was sampled for all longitudes and latitudes.

A description of HXX, its calibration and operation, has been previously given

[1]. The relevant parameters of each of the proportional counters are listed in Table 1.

TABLE 1. Parameters of the Proportional Counters (LAD 1 or LAD 2)

| | | | |
|---|---|---|---|
| Length: | 29.0 cm | Depth: | 3.7 cm |
| Width: | 3.6 cm | Total surface area: | 450 $cm^2$ |
| Window: | Be, 50 μm | Gas fill: | Ar (90%), $CO_2$ (10%), 1.4 atm |

The LAD was collimated with a coarse and fine collimator system which restricted the field of view to 10 arcim × 3° FWHM. The geometrical factor for each detector was $\Omega A = 0.03$ sr $cm^2$.

Figure 1 represents a pictorial view of HXX. LAD 1 is situated adjacent to the honeycomb skin and the thermal blanket of the satellite, in the -Y direction. LAD 2 is shielded in the +Y direction by the u.v. telescope and associated electronics.

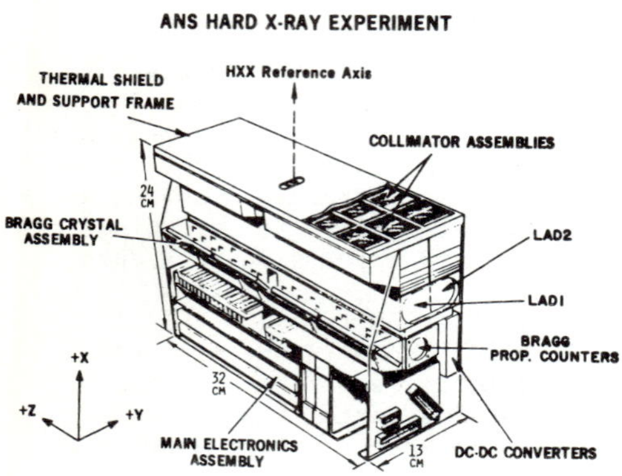


Fig. 1

## THE MAPS

The maps shown here represent the non-X-ray background detected by LAD 2 (Figures 2 and 3) as a function of 2 positions above the Earth. The data were grouped in bins extending 20° in longitude and 4° in latitude. The number given in each bin is the count rate (counts per second as would be measured before most of the non-X-ray events are rejected by the pulse shape discrimination).

The r.m.s. fluctuations of the count rates in each bin for $-32° \lesssim$ lat. $\lesssim +32°$ are 10-15%. For $32° \lesssim |\text{lat.}| \lesssim 60°$, the r.m.s. fluctuations are $\sim$ 15-25%. In the high latitude region, the r.m.s. fluctuations of the non-X-ray background increase with increasing altitude and latitude.

High-energy electrons will penetrate the detectors not only through the front windows but also through the exposed side of LAD 1 and LAD 2. Calculations show that the cut off energy, $E_c$ for LAD 1 in that direction is equal to $\sim$ 1 MeV. $E_c$ for LAD 2 in that same direction is $\sim$ 3 MeV. Figure 4 shows the count rates detected by LAD 1 minus the count rates detected by LAD 2 at 900-1000 km. These net count rates represent the number of electrons in the 1-3 MeV range which penetrate the thermal shield of the satellite and the side wall of LAD 1 and interact with the gas of LAD 1 ($A\Omega \simeq 150$ sr $cm^2$). These count rates are highly variable and the r.m.s. fluctuations are as high as 100%.

As seen from Figures 2 and 3, no data exist over the region of the SAA. The HXX instrumentation was turned off in the SAA and during its passage through the auroral zones. The determination of the extents of these high radiation environment regions were obtained from World Maps of Constant B,L, and Flux Contours, by E.G. Stassinopoulos [2], and later through the experience obtained with ANS itself.

## CONCLUSIONS

Some pertinent observations can be made from the ANS data:

(1) Few electrons with $E_e \gtrsim 1$ MeV are detected in the band $-32° \lesssim$ lat. $\lesssim +32°$, except in the SAA. The background counts could be produced by electrons with 80 keV $\lesssim E_e \lesssim 1$ MeV penetrating the front window, but are most probably due to the cosmic ray-induced interactions.

(2) At |lat.| $\gtrsim 32°$, high energy electrons (E > 1 MeV) penetrating the skin of the satellite are being detected.

(3) The large r.m.s. fluctuations in the count rates at |lat.| $\gtrsim 32°$ are most probably due to time fluctuations of the electron fluxes.

(4) Between 250 km and 800 km, and in the $-32° \lesssim$ lat. $\lesssim +32°$ band, the altitude dependence of the radiation environment is small. However, the extent of the SAA increases with altitude.

## REFERENCES

1. H. Gursky, H. Schnopper and D. Parsignault, Astrophys. J. (Letters) 201, L127 (1975).
2. E.G. Stassinopoulos, NASA SP-3054 (1970).

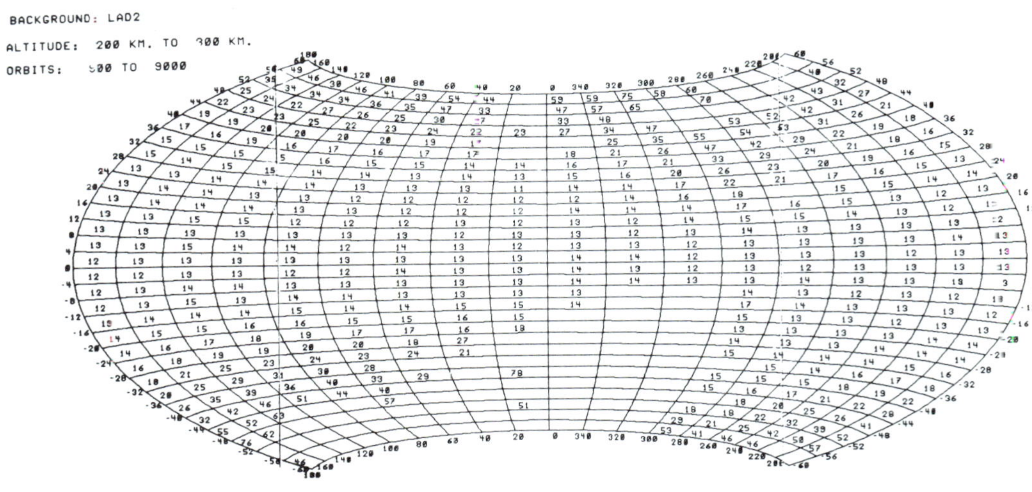


Fig. 2

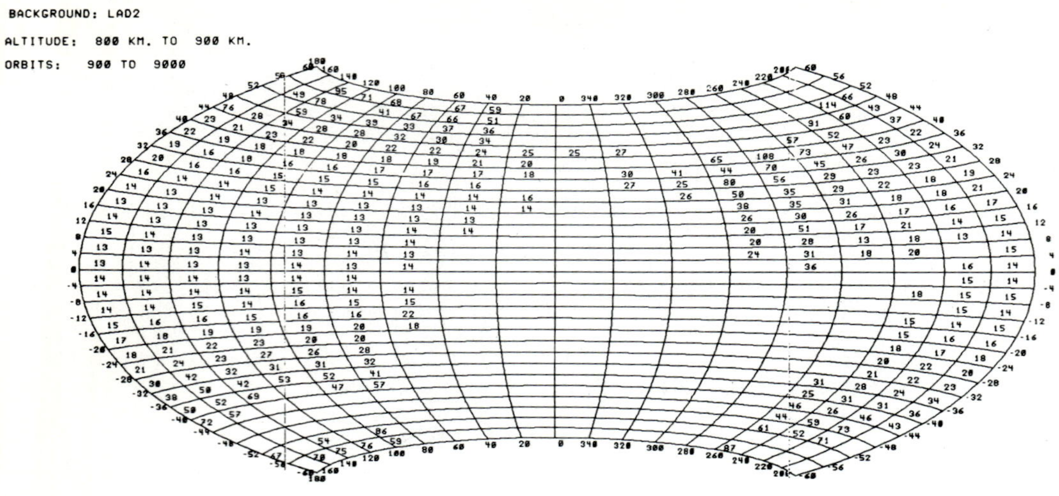
BACKGROUND: LAD2
ALTITUDE: 800 KM. TO 900 KM.
ORBITS: 900 TO 9000

Fig. 3

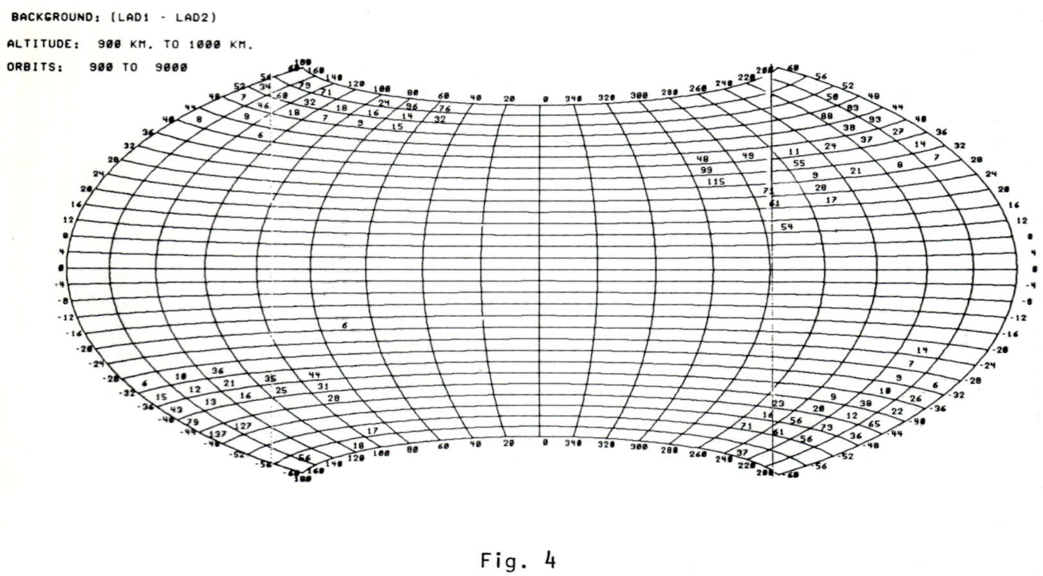
BACKGROUND: (LAD1 - LAD2)
ALTITUDE: 900 KM. TO 1000 KM.
ORBITS: 900 TO 9000

Fig. 4

# AUTHOR INDEX